**Nuclear Proliferation
and the Near-Nuclear
Countries**

Nuclear Proliferation and the Near-Nuclear Countries

Edited

by

Onkar Marwah and Ann Schulz

Ballinger Publishing Company • Cambridge, Mass.
A Subsidiary of J. B. Lippincott Company

This book is printed on recycled paper.

Copyright © 1975 by Ballinger Publishing Company. All rights reserved. No part of this publication may be reproduced, stored in a retrieval system, or transmitted in any form or by any means, electronic mechanical photocopy, recording or otherwise, without the prior written consent of the publisher.

International Standard Book Number: 0–88410–605–5

Library of Congress Catalog Card Number: 75-31882

Printed in the United States of America

Library of Congress Cataloging in Publication Data

Main entry under title:

Nuclear proliferation and the near-nuclear countries.

    Papers presented at a conference held at Clark University Mar. 13–14, 1975.
    Bibliography: p.
    1. Atomic weapons and disarmament—Congresses. 2. Atomic energy—Congresses. I. Marwah, Onkar S. II. Schulz, Ann.
JX1974.7.N818   327'.174   75-31882
ISBN 0–88410–605–5

Contents

Preface	xi
List of Tables	xiii
List of Figures	xv

Introduction—Nuclear Proliferation: To Bell the Cats or Catch the Mice?	1
The NPT: Objectives of the Co-Sponsors	3
The NPT: Objectives of the Non-Nuclear Weapons States	5
The NPT: Tactics and Roles of Sponsors and Nonsignatories	6
The NPT: Paradigm for Arms Control Agreements, 1961–1975	9
The NPT: Nuclear Weapon and Nuclear Weapon Threshold States in the Future	12
The Clark University Conference	13

PART ONE

Chapter One
The Emergence of a New Second Order of Powers in the International System—*Saul B. Cohen*

	19
The Emergence of Hierarchical Orders	20
Geopolitical Regions and Second Order Powers	23
Criteria for Second Order Power Status	28
Some General Principles and Conclusions	29

Chapter Two
The United States—Soviet Arms Race:
SALT and Nuclear Proliferation
—Stefan H. Leader and Barry R. Schneider   35

Introduction   35
Vertical Proliferation—The US—Soviet Strategic Weapons Race   36
The Superpowers and Nuclear Spread   52

Chapter Three
The Nonproliferation Treaty and the Nuclear Aspirants:
The Strategic Context of the Indian Ocean
—Robert M. Lawrence   59

Introduction   59
What Factors Caused Three Indian Ocean States to Reject
  the NPT?   59
How Might an Extrapolation of Current Superpower Activities
  in the Indian Ocean Influence the Position Taken Regarding
  the NPT by Indian Ocean Aspirants?   73
How Might Possible Future Activities of the Superpowers in the
  Indian Ocean Influence the Position of the NPT Adherents
  and Nonadherents?   75
Conclusions   76

Discussion Essay: Proliferation and the International
Strategic System   79

The Race Between the Superpowers   79
Regional Issues   81

PART TWO

Chapter Four
Risks of the Nuclear Fuel Cycle and the Developing Countries
—Christoph Hohenemser   85

Introduction   85
The Nuclear Fuel Cycle   89
Risks of the Nuclear Fuel Cycle   95
Research on Catastrophic Nuclear Risks   103
Conclusions: Implications for the Developing Countries   107
Acknowledgments   110

Chapter Five
Commercial Nuclear Technology and Nuclear Weapon Proliferation—*Theodore B. Taylor* — 111

Nuclear Explosives Technology — 116
Profile of Worldwide Nuclear Technology, 1982 — 118
Safeguards Against Diversion or Theft of Nuclear Materials Provided to Other Countries by the United States — 120

Chapter Six
India's Nuclear Policy—*K. Subrahmanyam* — 125

Third World Perceptions of Nonproliferation and the NPT — 126
International Perceptions of Indian Nuclear Capability — 138

Discussion Essay: Nuclear Safety, Weapons and War — 149

Risk Assessment — 149
Safety — 150
Weapons — 150
Safeguards — 151

PART THREE

Chapter Seven
Nuclearization and Stability in the Middle East —*Steven J. Rosen* — 157

Loss of Faith in the Conventional Balance of Power — 158
The Nuclear Deterrence Approach — 161
Is a Nuclear Middle East Inevitable? — 176

Chapter Eight
Determinants of the Nuclear Option: The Case of Iran—*Anne Hessing Cahn* — 185

General Background — 186
Present and Future Capabilities — 187
Military-Strategic Considerations — 194
Domestic Considerations — 198
International Considerations — 200

Chapter Nine
South Africa's Foreign Policy Alternatives and Deterrence Needs—*Edouard Bustin* — 205

Foreign Policy Alternatives — 206
Advantages and Disadvantages of "Going Nuclear" — 219

Chapter Ten
Japan's Response to Nuclear Developments: Beyond "Nuclear Allergy"—*Yoshiyasu Sato* — 227

Introduction — 227
Evolution of Nuclear Policy — 228
Peaceful Uses of Nuclear Energy — 232
Elements of Constraints — 239
The Nonproliferation Treaty and Japan — 246
Conclusion — 253

Chapter Eleven
Brazil's Nuclear Aspirations—*H. Jon Rosenbaum* — 255

The Accord — 255
Safeguards — 257
Motives — 261
Capabilities — 262
Intentions — 266
Conclusions — 270

Chapter Twelve
Incentives for Nuclear Proliferation: The Case of Argentina—*C. H. Waisman* — 279

Introduction — 279
Argentine Nuclear Capabilities — 280
Argentine Nuclear Policy — 282
Policy Options for Near-Nuclear Countries — 284
The Argentine–Brazilian Interaction — 286
Internal Incentives — 289

Discussion Essay: The Typical Nuclear Proliferator — 295

Technology and Manpower — 296

Strategic Needs and Proliferation 297
Internal Politics 299

Epilogue: The NPT Review Conference, Geneva, 1975 301

Appendixes

A — The Non-Proliferation Treaty 317
B — Final Declaration of the Review Conference of the Parties to the Treaty on the Non-Proliferation of the Nuclear Weapons 323
C — Status of the Nonproliferation Treaty 335

Bibliography 337

Notes on Contributors 345

Conference Participants 349

Preface

The chapters in this volume were prepared as contributions to a conference on nuclear proliferation held in March, 1975, at Clark University, Worcester, Massachusetts. The Clark conference was one of several gatherings prior to the nuclear Non-Proliferation Treaty (NPT) review conference of May of that year.

The NPT addresses itself to several dimensions of nuclear proliferation—the security needs of non-nuclear states, the vertical proliferation of nuclear weapons by the superpowers, and the application of safeguards procedures. These issues underlie the structure of the present volume, with the addition of a chapter on nuclear safety problems. Each of these issues also was discussed at the Geneva review conference. A brief epilogue takes a preliminary look at the conference proceedings.

The editors were concerned first with looking at proliferation from the viewpoint of the nearly nuclear states. To the extent that nuclear energy is the stimulus to proliferate plants and technology, much still remains to be learned. More discussion is also needed of the military, strategic and tactical implications of various degrees of proliferation. Nonetheless, we do think that the chapters included add to the increasing literature which addresses international political questions from a perspective beyond that of the superpowers. The chapter authors have undertaken that task in various, unique ways.

The conference participants [see appended list] also contributed to lively discussions of each paper. Brief summaries of the discussions, which unfortunately cannot reproduce intonation, follow each of the three sections. Throughout each stage of the conference and the process of putting the volume together, both authors and discussants were consistently helpful. Responsibility for the discussion essays, communications with the contributors, and copy-reading were the responsibility of editor Schulz; the responsibility was a pleasant one to assume. The introduction and epilogue were written by editor Marwah.

The conference was in many respects a group project. The editors wish to express sincere appreciation to Clark University and particularly to President Mortimer H. Appley and Deans Leonard Berry and Frank Puffer for their support of the conference. Similarly, we want to thank our colleagues Bob Kates, Roger Kasperson, Dick Ford and Christoph Hohenemser for their financial assistance.

Once the financial viability of the conference was assured, the next task was to organize and run the conference. Fortunately, our students at Clark have many talents, which they willingly shared. For their help and good cheer, we want to particularly mention: Massood Abolfazli, Ira Bloom, Susan Burke, Russell Cooper, Sandy Friese, Nancy Goody, Percy Hintzen, James Kelman, Leonard Kreppel, David Lake, Bernard Ntegye, Neal Shact, Neil Simon, and Peter Truwit.

List of Tables

2–1	US–Soviet Nuclear Weapons Race–Strategic Nuclear Weapons	36
2–2	Strategic Weapon Launchers	38
2–3	Force Loadings–Nuclear Weapons	39
2–4	Accuracy	40
2–5	United States Megatonnage	43
2–6	Soviet Strategic Megatonnage	44
2–7	US Nuclear-Capable Guided Missiles, Aircraft and Ships Given and Sold to Foreign Governments, 1950–1973	53
4–1	Capital and Annual Costs of a Nominal Nuclear Weapons Program as Estimated in 1960	87
4–2	Plutonium Production from Civilian Nuclear Power–Projection versus Reality	88
4–3	Installed Nuclear Capacity in 1974 and 1980	90
4–4	Typology of Catastrophic Nuclear Risks–Light Water Reactor Fuel Cycle	99
4–5	Catastrophic Nuclear Risk Described in Terms of Classes of Initiating Events	100
4–6	Typology of Catastrophic Nuclear Risk–Canadian Heavy Water Fuel Cycle	102
4–7	Revised Consequences for LWR Accident Having a Probability of 5×10^{-6} per Reactor Year	106
7–1	Combat Aircraft	159
7–2	Tanks	159
7–3	Survivability of Missile Silos	167
8–1	Iran's Military Capabilities, 1966–1975	196
9–1	South African Defense Expenditures	207
9–2	South African Military Personnel	213
9–3	South African Armaments	214

12-1	Argentine Production Capacity of Plutonium, According to Different Sources	281
12-2	Indicators of Modernization in Argentina and Brazil, Circa 1965	287
12-3	Annual Growth Rates of GDP and Manufacturing Sector 1965-1973	288

List of Figures

1–1	Map	22
2–1	Comparative Effectiveness in Terms of Destructive Area	45
2–2	SALT I at a Glance	48
4–1	Pressurized Water Reactor Fuel Cycle	92
4–2	Boiling Water Reactor Fuel Cycle	92
4–3	Light Water Reactor Fuel Cycle	93
4–4	Three Stages in the Development of a Nuclear Disaster and the Linkages Between Them	97

Introduction

Nuclear Proliferation: To Bell the Cats or Catch the Mice?

Though extensive and by now a decade old, much of the nuclear proliferation debate has been conducted on a general level. Specifically, it has not been grounded in the concerns and the anxieties of those states possessing both aspirations to power and the necessary technology. To these states, limitations on military capabilities formulated by the great powers have not appeared meaningful. The debate entails a perceptible division of the world. It has directed one set of questions at potential nuclear weapons powers and another at the existing nuclear states. Forms of abstinence which apparently do not apply to the nuclear "haves" are urged upon potential nuclear weapons powers.

The preceding character of the nuclear weapons debate is more natural than unusual, reflecting as it does the power hierarchies growing out of the Second World War. The organizational structure of the UN with its five veto-carrying members of the Security Council and an advisory General Assembly of the world's nations provides a formal legitimacy to the invocations and premises of the nuclear debate. Some states, it can be argued, *are* more equal than others and have been so recognized by the other states. Indeed, one could make a persuasive case to the effect that both the nuclear debate and actions flowing from its deliberations have not remained frozen reflections of 1945 but continue to mirror the dynamism of international power realities. Britain, a co-sponsor of the Nuclear Nonproliferation Treaty, 1968 (hereafter NPT), is not today a party to the series of SALT negotiations between the US and the USSR. Nor, presumably, would France or China be eligible participants in SALT had they co-sponsored the NPT or engaged in other worldwide disarmament negotiations. Clearly, there exists an hierarchy of power among the current nuclear weapons states as substantive as that sought to be impressed between the latter as a group and the rest of the world.

It is easier to understand changes in military power certainties than to evaluate changes in political power uncertainties. Unlike the assumptions made in textbooks, military and political power do not necessarily go together, except in some ultimate and abstract sense. In the postwar period, a steady growth in the weapons' inventories of the great powers has been accompanied by a steady erosion, relatively, of their ability to influence or control worldwide political events. The immense military power of the US could have obliterated the Indochinese, but the latter could not be prevented from attaining their political objectives. In two weeks of the closing month of 1970 India, aiding in the creation of Bangladesh, wrecked a 25 year US input of policy and resources for stabilizing a balance in the South Asian subcontinent. President Sadat ejected 20,000 Soviet advisors from Egypt in 1971 without significant retaliation by the USSR. And in 1973, the OPEC group administered the heaviest of shocks to an international monetary and economic system dominated by the rich and militarily powerful states.

While too much need not be read into instances of the weak successfully using their political or economic power against the strong, neither should these occurrences be ignored as exceptions to the rule. Fitfully, sometimes by chance but at others by cautious design, a range of second order states are modifying the political operations of an international system which in strategic terms continues to be bipolar or perhaps bi-multipolar [1]. By a combination of the limited use of force, the fudging of alliance structures and the claim to pre-eminent regional interests, or by articulating the collective needs of smaller entities, these states have brought confusion to the stakes or roles of various great powers in the international strategic system. By so doing, they have blurred the picture of an international system managed in terms of calculable military equations, and invoked instead a situation where the high managers of the system must bid competitively and politically for support despite retaining the sanction of overwhelming force.

The negotiations which led to the drafting of the NPT and a roll call of its subsequent acceptance or rejection by different states reinforces the image of an international system organized hierarchically in terms of power but given over in some strategic measure to the political demands of a particular category of its constituents. The reality for the co-sponsors of the NPT is that in spite of their military power many (but not all) states have been persuaded to sign the treaty. The reality for the international system is that the humanitarian intentions behind the NPT—shared, presumably, by sponsors, signatories and nonsignatories alike—have to contend with what the nonsignatories label as (for them) the treaty's negative political, strategic and technological implications. The reality of the NPT as it stands today is that those states most able in varying degrees to embark upon national nuclear weapons programs either have not signed, have signed but not ratified (which is the same as nonsignature) or may be suspected of "unsigning" the treaty. Among these states are Argentina, Brazil, Chile, Egypt, India, Indonesia, Iran, Israel, Japan, Pakistan, South Africa, South Korea, Spain and Taiwan.

Whether the number of nuclear weapons powers is to remain frozen at five (six if India is included) or increase to 20 or more becomes, therefore, a matter for bargaining and perhaps compromise between the existing and the nascent nuclear weapons states.

THE NPT: OBJECTIVES OF THE CO-SPONSORS

What was the extent of the compromises offered by the sponsors of the NPT in return for the adherence of all states, especially the nuclear threshold states, to the treaty? The NPT was unusual inasmuch as it laid down two sets of rights and obligations for two categories of states. In lieu of abjuring from the production or acquisition of nuclear weapons and accepting IAEA-supervised verification of their nuclear programs, the rest of the world's states were assured of the following responses, by the co-sponsors of the NPT (the US, the USSR and Britain) on behalf of all nuclear weapon states (". . . one which has manufactured and exploded a nuclear weapon or other nuclear device prior to January 1, 1967"):

1. Seek the discontinuance of all (underground) nuclear weapon tests as a corollary to the Partial Test Ban Treaty of 1963 which had banned such tests under water, above ground or in outer space (preamble to the NPT).
2. (All) states would refrain from the threat or use of force in accordance with the UN Charter (preamble to the NPT).
3. Make available to others on a nondiscriminatory basis the potential benefits from any peaceful applications of nuclear explosions. An appropriate international body would be set up to allow for such transfers (Article V).
4. Pursue negotiations "in good faith" to cease the nuclear arms race at the earliest date and move toward nuclear disarmament (preamble and Article VI of the NPT).

Being the culmination of a ten year effort, the provisions and purport of the NPT have been surrounded by a vast amount of analytical literature bearing on the subject of nuclear arms control for others [2]. Beyond the clauses of the NPT but within the political arena, nonsignatories have been asked to respond to a series of interlocking questions.* First there are the primary questions, such as:

1. Does the country really need nuclear weapons for its security?
2. Can the country afford the costs of the weapons and delivery systems?
3. Will the weapons system be technically and industrially feasible within the means of the country?
4. Can the country make use of peaceful nuclear explosives?
5. Is the country seeking nuclear weapons for prestige purposes?

*The sets of questions have been assembled from a reading of the general literature on nuclear proliferation (see Bibliography).

The secondary questions are leveled as point-counterpoint or as general evaluations of a state's nuclear aspirations. They arise in case one or more of the primary questions are rationalized in the affirmative for a particular country. The genre of these secondary questions may be stated as follows:

1. Not gainsaying a requirement for nuclear defenses, is it justifiable for nuclear threshold states to further destabilize the international system by statistically increasing the chances of an accidental or irresponsible nuclear war through the fact and example of nuclear proliferation?
2. Not denying that the costs could be born, is it reasonable for the near-nuclear state to invest the vast resources eventually needed for a credible nuclear posture—and in weapons that may never be used?
3. Not underrating the technical and industrial competence of the nuclear weapons aspirant, is it worthwhile, in alternative use analogies, for it to employ (scarce) skilled manpower and industrial capacity for destructive rather than constructive purposes?

If the secondary questions are also reasoned in support of a country's claim to possess nuclear weapons, then a third category of questions can be posed and made operative in respect of that country. Inspired by common-sense and ethical perceptions about the future of international society, these questions suggest that a near-nuclear state's desire for "going nuclear" be evaluated in the context of a non-nationalist pragmatism or on the basis of universalist criteria. They expound that:

1. The current nuclear weapons powers, particularly the US and the USSR, are doing all that is circumstantially possible to control the (strategic) arms race, and that these efforts should not be further complicated by a horizontal increase of nuclear weapons states.
2. Internationalist concerns demand of all states, but particularly the nuclear threshold states, a rejection of all blandishments or provocations for the development of nuclear arsenals amongst themselves. The existing nuclear powers are an unfortunate reality, and it is necessary for the rest of the world to understand the complex of tensions which disallows them from eliminating their nuclear arsenals.
3. It would seem impossible in any case for latecomers in the nuclear game to match the resources or nuclear effort of the US and the USSR. Hence, a minor accretion of prestige or a minimal level of nuisance value at best could attend the acquisition of nuclear status by other states. Common sense demands that newcomers weigh their nuclear options accordingly.

The preceding sets of questions are not rank-ordered or raised with the same intensity for all nuclear weapons aspirants. They provide, instead, the bases for raising a selective combination of concerns subject to time and the particular nuclear aspirant being reviewed.

The formal objectives of the NPT's co-sponsors can be viewed as the acceptance by all other states of an international regime that restricted at five the number of nuclear weapon powers. With the adherence of 95 states to the treaty it might be said that the documents' incidence of success, if modest, remains positive, and further, that the treaty—not an end in itself but a means to an end—has played an educative role and helped to define problems in an area of complex and competing national calculations. However, the tacit objectives of the treaty must be seen as the provision of an agreeable compact between the first and second order states. Here, the treaty is largely irrelevant. Indeed, as an arms control effort the NPT may have provided incentives to states in terms of the very consequences it sought to prevent. Its draft spelled out in fairly unambiguous language the low level of obligations which the co-sponsors were willing to assume. As such, the NPT may have been a learning process in the opposite direction for those states most crucial to its success.

THE NPT: OBJECTIONS OF THE NON-NUCLEAR WEAPONS STATES

There were three bases for the refusal by non-nuclear weapons states to formally sign away their right to national nuclear weapons programs. First, there were disagreements within the provisions of the NPT as drafted. Second, there were reservations beyond the NPT regarding the implications of what had been left out of its provisions. Third, there existed internecine disputes among near-nuclear countries which made the assent of some states to the NPT conditional on that of perceived opponents. In other words, the NPT in itself was one and not the only premise by which non-nuclear states evaluated their future options.

The NPT was presented for signature by its co-sponsors to the rest of the world's states in August 1968. Within a few days of the event, states objecting to its terms met in a Conference of Non-Nuclear Weapons States (August-September 1968). Registering their general complaint as an imbalance of mandatory obligations between nuclear weapon and non-nuclear weapon states, the nonsignatories called for more specific undertakings in nuclear arms control by the NPT sponsors as a measure of their real intent. The treaty's sponsors were asked to provide more visible evidence of:

1. plans to accord "high priority" to a comprehensive ban on the testing of nuclear weapons;
2. efforts to halt the development of nuclear weapons and delivery systems;
3. moves to reverse the production of fissile materials for weapons purposes; and
4. steps to reduce and subsequently eliminate nuclear weapons stockpiles.

The NPT sponsors were not willing to commit themselves to nuclear weapons control measures more precise or obligatory than those indicated in the treaty. These measures, as outlined earlier, were in the nature of exhortations that urged nuclear weapons states to "seek" a comprehensive nuclear test ban, and "pursue" the goal of nuclear disarmament. Timetables and, failing that, concrete if initial steps toward achieving the preceding objectives were excluded from the compact. As a result, non-nuclear weapons states considered mandatory abjurations on themselves as too heavy a trade-off against discretionary undertakings by the NPT sponsors.

The implications of the NPT beyond its clauses were more perplexing. What the treaty's sponsors were unable to enjoin upon themselves was interpreted by the non-nuclear states as a clue to the sponsors' appreciation of the international system. It was thus apparent to the non-nuclear states that five years after the Partial Test Ban Treaty of 1963, the nuclear weapons powers viewed their own security needs as requiring the continued testing of nuclear weapons, and conceived as non-negotiable on an international obligatory basis steps toward their own nuclear disarmament. Further, that the problem of nuclear weapons proliferation would be defined in terms of the number of nuclear powers but not the number or increasing sophistication of such weapons systems in the possession of these powers. In return, the non-nuclear weapon states were being required to desist unconditionally from the manufacture of nuclear weapons, and also from such research and experimentation as would provide them with a weapons option. As a logical corollary, all the nuclear research establishments of the non-nuclear states would be open to international inspection. The same conditions would not apply to the weapons research activities of the treaty's sponsors [3].

Some states raised special contention within and beyond the treaty's provision relating to the threat or use of force. Within the NPT, these states asked for a specification that *nuclear* threats or use—as against the vaguely worded abstention from *general* threats or use of force—would not be inflicted by states possessing the weapons on those who gave up the right to their manufacture. Outside the treaty, one particular state, India, asked for a joint and public guarantee of nuclear support from the US and the USSR, were China to resort to nuclear threats against it. In the complexity of their global tensions the sponsors of the NPT were hesitant to commit themselves beyond an undertaking of informal assurances [4].

Finally, and quite apart from the provisions of the treaty, states such as Pakistan and Egypt withheld their signatures or ratification until such time as India and Israel became full parties to the NPT.

THE NPT: TACTICS AND ROLES OF SPONSORS AND NONSIGNATORIES

The roles played, bargains struck, threats made or resisted, and ploys used for assent or disagreement as occurred in corridors and chancelleries during the NPT negotiations provide an interesting example of relationships between first and second order states.

Among the co-sponsors of the NPT the role of the US was crucial in maintaining a steady momentum for signatures to the treaty. Britain, though an equal co-sponsor of the NPT, worked in a less obtrusive and at times a surrogate capacity to the US, e.g., in offering to stand in with a nuclear guarantee for India in place of the latter's demand for a US—Soviet guarantee. The Soviet Union adopted a basically silent position in public, but covertly leaned as strongly as the US on recalcitrant states with which it had influence. Meanwhile, the media and academic institutions in the two former co-sponsoring states created a heavy volume of static in support of the NPT. The propaganda effort by the Soviet Union was more circumspect, but in the few foreign language Soviet journals available beyond its borders, a "commentator"—usually a euphemism for a high Soviet official—made much the same points in support of the treaty as his British and American counterparts [5].

France and China were the two nuclear weapons powers who refused to, or could not, be included as co-sponsors of the NPT. Both of them in varied language denounced the treaty as a hegemonic superpower plot to control the world. While their rejection of the treaty reflected discontent with the workings of the international system and the roles assigned to them in it, France and China articulated, albeit unconsciously, many of the resentments felt by lesser states but left unsaid because of the latter's dependent relationships with the treaty's co-sponsors. As such, and from the beginning, there existed a split in the ranks of the nuclear weapons powers that reduced the legitimacy of the treaty, provided allies to the nuclear threshold nonsignatories and abetted their resolve to hold back signatures.

It is unlikely that the Soviet Union would have been a party to the NPT had it not received some tacit understanding that West Germany would sign. The case of Japan, with its sophisticated technological potential, was somewhat similar to West Germany inasmuch as, like the latter, its assumption of nuclear weapons status would quickly and substantially destabilize the existing international system. Therefore, a sustained effort at coaxing Japanese signature was mounted by the US on the strength of its close working and security arrangements with Japan. Eventually, both West Germany and Japan signed the NPT. Indeed, they could hardly not have signed without immediately raising alarm about their intentions, carrying the onus as they did from World War II—and in Japan's case with the negative experience of nuclear destruction visited upon two of its cities. However, of the two states, Japan has not ratified the NPT, as a result of which its signature remains unattested [6]. By adopting such procedure, a sophisticated nuclear threshold state has escaped the opprobrium of nonsignature while continuing to maintain a legal option to change its nuclear status in the future—with the slight inconvenience, no doubt, of having to publicly withdraw signature to the NPT, or more likely, of construing it as nonapplicable at will.

It is interesting to note that all non-nuclear members of the European Economic Community—Belgium, West Germany, Italy, Luxembourg and the Netherlands—signed the treaty in 1968 but none ratified it until 1975. The inference is persuasive that all the EEC non-nuclear states adopted a commonly agreed policy. While bending formally to the needs of a junior position in the Western alliance and sensitive to geopolitical tensions on the European continent, they adopted a wait and see attitude and maintained their legal right to change nuclear policy until recently. Finally, among all the noncommunist industrial states only the lesser ones—Australia, Canada, Denmark, New Zealand, Norway and Sweden—signed and ratified the NPT without hesitation. At the time, non-European nuclear threshold states watched the negotiating platforms of Europe's ex-imperial countries with a special wariness.

The earlier response of non-nuclear states in Eastern Europe contrasted with that of the Western industrial states. Bulgaria, Czechoslovakia, East Germany, Hungary, Poland, Rumania (and Yugoslavia) all signed and ratified the NPT when presented in 1968—reflecting perhaps their inferior bargaining positions vis-à-vis the Soviet Union.

From among the non-European nuclear threshold states, one or two were more vociferous in their criticism of the treaty. They were singled out for special attention by the treaty's co-sponsors. India is the most obvious instance in this category. Its path from nuclear innocence to the nuclear explosion of May 18, 1974 has been investigated by scholars whose studies analyse the ambivalence of existing arms control agreements and show how other states may achieve nuclear status similar to India's without breaking either the letter or the spirit of any international legal obligations [7]. While the actual decision in favor of the Indian nuclear explosion appears to have been taken during the Bangladesh War, an Indian nuclear explosion probably would have occurred in any case [8].

The response of some other nuclear threshold states to strategic environments perceived as uncertain may be similar to India's and fall partly beyond the scope of the NPT. Israel, already reputed to have assembled nuclear weapons, could face an Egyptian—or probably a joint Arab—riposte by way of nuclear capabilities in the future. India has to assume that Pakistan will work towards nuclear capability, and President Park Chung Hee has hinted at nuclear weapons for his country were the US to waver in support of South Korea in its confrontation with North Korea [9]. Brazilian and Iranian worldwide purchases of nuclear technology, particularly efforts at securing uranium enrichment facilities, are viewed as commensurate to acquisition of weapons options for the future.

In systemic terms, the Indian nuclear example may be more symptom than cause, highlighting NPT insufficiencies with respect to many states rather than affecting them. The NPT, in fact, was a modest document invested with powers of success beyond its terms, and certainly beyond the rhetoric accompanying it. As one important omission, it failed to provide any effective basis for a bargain to be struck between the two most crucial levels of the international hierarchy of states. Instead, it sought to freeze the hierarchy by the most difficult means—voluntarism—on those states most naturally spoilers in a centrally managed system.

Another way in which the NPT may be viewed, as proponents then and now suggest, is by evaluating it as a first step toward diffusing international tensions rather than as a panacea for all the world's cross-cutting hostilities. In a realistic construct, it is argued, the NPT is a measure of earnest intent on the part of the co-sponsors and not the declaration of a moratorium on great power rivalries. Quite so, and thus it could not be expected that lower level disputations, many complementary and some connived at by the treaty's co-sponsors, could be swept away under the NPT.

Great power efforts at nuclear détente suggest that the process of nuclear arms negotiation is going to be slow and, in the initial stages, peripheral to substantive control of nuclear weapons or their delivery systems. In the intervening period, control may require increase in nuclear weapons systems, reductions may lead to qualitative improvements in weapons technology and curbing sales of nuclear techniques or plant may narrow their availability at lucrative rates to special national bidders. It is possible, however, that to some extent, the major nuclear powers are themselves victims rather than perpetrators of the preceding state of affairs. In either case, the experience that they have provided the nuclear threshold states in efforts at national or international arms control since the NPT came into force in 1970 is probably a negative one.

THE NPT: PARADIGM FOR ARMS CONTROL AGREEMENTS, 1961–1975

With the exception of the Treaty for the Prohibition of Nuclear Weapons in Latin America, 1968 (Tlatelolco Treaty), and the NPT, all the other significant arms control agreements since 1961 are designed to seek prohibitions amongst the nuclear weapons powers. As a matter of practical reality, they involve the intentions mainly of the United States and the Soviet Union. The question arises, therefore, as to the extent and nature of the arms restraints accepted by the two global powers since 1961. The major arms control agreements since 1961—in fact, since the Second World War—are:

1. *The Antarctic Treaty,* 1961, leading to the demilitarization of the Antarctic continent.
2. *The "Hot Line" Agreement* between the US and the USSR, 1963, to reduce the possibilities of a miscalculated nuclear attack.
3. *The Partial Test Ban Treaty,* 1963, prohibiting nuclear tests in outer space, above ground and under water.
4. *The Outer Space Treaty,* 1967, banning emplacement of nuclear weapons in outer space or on other celestial bodies.

5. *Treaty For The Prohibition of Nuclear Weapons In Latin America*, 1968, as implied by the title.
6. *Treaty on the Nonproliferation of Nuclear Weapons (NPT)* 1968, whereby all except five states would surrender the option to make nuclear weapons.
7. *Seabed Arms Control Treaty*, 1972, prohibiting the enplacement of nuclear weapons in the sea depths.
8. *Convention on Biological Weapons*, 1972, prohibiting the development, production and stockpiling of bacteriological or toxin weapons.
9. *Strategic Arms Limitation Talks* (SALT), I and II, 1969, and other agreements under its aegis, such as the one at Vladivostok in 1974, seeking to place limitations on the strategic weapons systems of the US and the USSR.
10. *Treaty on the Limitation of Anti-Ballistic Missile Systems* (ABM Treaty), 1972, restricting the aforesaid weapons to two sites each in the US and USSR (subsequently reduced to one site each for the two countries).
11. *Nuclear Threshold Test Ban Treaty*, 1974, whereby the US and the USSR agreed to limit the destructive yield of underground nuclear tests at 150 kilotons after 1976.

Items 1, 4 and 7 are in the nature of "nonarmament" agreements, concern primarily the capabilities of the US and the Soviet Union, and while important in themselves, impose a spatial abstention rather than call for any reduction, control or inhibition of existing nuclear weapons systems. Item 2 relates to a procedural issue in an "enemy partners" scenario, reducing the probability of an accidental nuclear exchange between the US and the Soviet Union but hardly their complement of weapons or conditions of hostility. Item 3, like Item 6, calls for voluntary surrenders by the world's states—ostensibly of nuclear tests but with the hope that it would presage abstention from national nuclear weapons programs. In return, the nuclear weapons powers, France excluded (China was not yet a nuclear power), agreed to drive their nuclear tests underground (France and China did not). That nothing more was given up is evidenced by the fact that the incidence of nuclear tests increased after 1963, pointing to the reality of an intensified arms race among the nuclear powers. Item 5 is a restraining act by states in a geographical zone, and while welcomed by all, does not abate or otherwise affect the central arms race of the nuclear weapons powers [10]. Item 8 has led to the elimination of a particular type of weapon by the major powers, and an abstention by the others. However, by their nature of production, biological compounds can be assembled swiftly for use as weapons, and the processes, once learnt, cannot be unlearnt. Their nonstockpiling by all states including the major powers represents—as with poison gas in World War I—an unavoidable acceptance of the unknown and the uncontrollable [11]. Items 9, 10 and 11 are exclusive pacts between the two global powers and are generally interpreted as procedures to streamline rather than decelerate their arms race. Item 9 envisages, through the Vladivostok protocol, a "capping" of the strategic weapons arms race by doubling the American and quadrupling the Soviet current complements of such weapons systems [12]. Item 10 was an agreement resulting from a joint American–Soviet cost benefit analysis: country-wide ABM systems were judged as beyond the financial means of even the global powers. And Item 11 has, in effect, sanctioned the continued underground testing of nuclear weapons without any time limit [13].

The role of the two major nuclear powers in such arms control measures as exist is not only pre-eminent—which would be inevitable—it is almost exclusive. In any effective scale of actions possible in the present, the rest of the world's states, including the non–nuclear states, can be no more than advisory, expectant or hopeful actors. It might seem logical to assume, in consequence, that some modest but visible attempts at decelerating their own arms race would provide more than compensatory strategic dividends to the major nuclear powers without reducing their own perception of strategic parity. However, a level of rapport that would allow considerations of the preceding type does not as yet exist between the two global powers. As a result, other states have probably calculated that, for the moment, the global powers will accept only such weapons control measures as are (1) unfeasible within their economic means; (2) applied to others, or other geographical zones; (3) impractical in the context of their mutual deterrence needs as redefined from year to year; (4) pragmatically beyond the reach of any calculus of deterrence, e.g., biological weapons. There are, in consequence, strong incentives and scope for the secondary states to achieve benefits in the area carved out by global-power or all-nuclear power rivalries—and the former are unlikely to give up their advantages without valid quid pro quos from the nuclear weapons powers [14].

The benefits already secured by the nuclear threshold states flow from complex and at times competitive calculations. Security arrangements between the US and West Germany have not prevented the two from being commercial nuclear rivals. Therefore, what Brazil could not receive from the former in uranium enrichment technologies it has persuaded the latter to supply. India has been plugged into the international nuclear technology circuit for years, supplementing its research base from wherever possible. Iran is today following a similar procedure, with the added lure to suppliers of immediate cash payments instead of credit arrangements. Some states have noted that countries such as Japan, Italy and West Germany, though nonratifiers of the NPT (the latter two until recently), have received easier access to guarded nuclear technologies from allied NPT sponsors than have those who signed and ratified the treaty. Others suspect surreptitious transfers of nuclear technology or materials to particular states outside of the NPT, e.g., Israel. Again, there are anxieties that conflicting strategic objectives in West Asia, coupled with remunerative commercial and military sales in the region, may eventually lead to the extraction, by important local states, of side payments from the global powers in the form of sensitive nuclear technologies [15]. Equally, some nuclear threshold states may resort to an inverted form of nuclear blackmail by threatening "intermediate nuclear technology" sales in the area, were barriers to be raised in their own purchases of technology from more technologically sophisticated states. Many NATO member states have accepted enriched uranium supply contracts from the Soviet Union as a future hedge against feared US cost increases, stricter terms of supply or inability to provide the same in the future [16].

The co-sponsors of the NPT, and other countries with a sophisticated base in nuclear technology (France and West Germany), have already met in London to coordinate their future nuclear export policies. It is possible that, in agreement, their nuclear commercial rivalries may be muted through a formula for sharing the financial benefits and invoking more stringent safeguard and safety procedures for worldwide nuclear exports. However, a number of nuclear threshold states will view the nuclear export restrictions as inspired equally by oligopolistic profit motives and a discriminatory technology curb against them. One result could be that countries such as Argentina, Brazil and India would redouble their efforts to acquire the technologies before further restrictions are imposed, coordinate the mutual exchange of such nuclear techniques and materials as they possess, and jointly resort to informal mechanisms for their transfer from the nuclear sophisticates. Some delay, not prevention, appears possible.

THE NPT: NUCLEAR WEAPON AND NUCLEAR WEAPON THRESHOLD STATES IN THE FUTURE

The evaluation is that solutions for many of the issues raised in the preceding analysis lie outside the scope of the NPT regime, or indeed, beyond the experience of the workings of other postwar arms control agreements. For a variety of reasons, serious nuclear weapons aspirants probably cannot be stopped from embarking on national nuclear weapons programs.

If the chances of halting the super arms race of the global powers are remote, it becomes necessary to move to the next stage of the nuclear proliferation debate. The future world system must be conceived as one consisting of a number of nuclear weapons states, arranged perhaps in step-level functional terms, possibly in serial order according to nuclear capabilities, probably in their regional influence characteristics. It is, moreover, insufficient and somewhat unreal to open and close the proliferation issue with a priori assumptions about the destabilization effects of "a world of nuclear powers." From the perceptions of the unequal states which cannot be ignored and may not be prevented from "going nuclear," the world could not be more unstable than it already is.

A restructuring of the nuclear proliferation debate, as suggested, may provide certain benefits. Initially, it may lead to the construction of premises that, in arms control matters, make sense for both the satisfied and the spoiler states of the international system. The chance might then exist for isolating issues of relevance to both orders of states. Bargains can be considered more easily in the interface of the needs of both nuclear weapon and nuclear weapon threshold states. While nuclear weapons are different from stockpiles of dynamite or TNT, it is possible to consider the risks of their spread among states in specific rather than general terms.

In the parallel area of nuclear safety and risk assessment it might be worthwhile to suggest that a new treaty regime, disassociated from the NPT, could more easily invite the participation of all states. Dangers from inadequate safety procedures as a consequence of the proliferating availability of fissile materials and nuclear power plant technology involve every state possessing a nuclear reactor. The anxieties in this respect should be common. They could be formally separated from the political or strategic disputes occasioned by the other clauses of the NPT. As it is, the issue of nuclear safety is tied too closely to nuclear safeguard concerns, which through IAEA inspection procedures tend to emphasize preventions against suspect weapons proliferators rather than to encourage safety measures for all nuclear-related facilities in the world.

The process of rethinking might begin with a restatement of the view that there are few political innocents in the international power game and that over time their numbers appear to decrease. Also, in the area of nuclear technological competency, the number of states, limited for the moment, is likely to increase. Finally, the responsibility for ameliorating the dangers in the strategic and the technological environments are shared equally by the global and the challenger states.

THE CLARK UNIVERSITY CONFERENCE

The conference was convened to evaluate the impasse that presently exists in the nuclear arms control field. It was felt that the dialogue between nuclear weapons states and potential proliferators needed to be raised on a common set of premises. The cross-purpose objectives of various states were viewed as significant in themselves so long as the international system sanctioned hostility amidst cooperative measures. Further, the hierarchy of current power capabilities or relationships did not provide either the nuclear weapons powers or the threshold states with pre-eminent claims to moral intent—particularly in the control of mass destruction weapons. Both categories of states had intervening preferences in their arms control stances. All states could also be prisoners of autonomous processes beyond their control as regards problems of nuclear safety and safeguards. These issues required, at a minimum, to be understood across the needs of individual states. If the subsequent opinion was that the conflict of interests and processes could not be tamed, then arms control negotiators as well as nuclear technologists would have to be urged to seek solutions for the future within the preceding realities. The starting point for their labors (and not the end) would then be a world composed of many nuclear powers and many forms of nuclear technology-related hazards.

The papers assembled in this volume explore these themes by stages. Part One considers the political-strategic framework in which regional and global powers interact in the present. Saul B. Cohen analyzes the geopolitical context in which the second order states are mounting their challenge to the current international system. The results, he suggests, may be more positive than are usually imagined. Stefan H. Leader develops the issue by stating that second order states, far from being uninterested, are keen observers of and draw important lessons from the superpowers' unbridled arms race. Robert M. Lawrence evaluates the great power–second order power tensions in a particular region. With the mapping of the Indian Ocean region as another arena for their confrontation, the superpowers, he feels, are inevitably going to hasten nuclear weapon procurement programs in the important littoral states.

Part Two focuses upon the problems arising from the spread of nuclear weapon technology. Christoph Hohenemser approves of an interim solution to safety problems, adopted by Canada, which would allow for exports to developing countries of nuclear power plants with truncated fuel cycles. Theodore Taylor considers the multiplicity of routes by which states or groups of individuals could acquire nuclear explosive materials and the expertise to handle them. Tightening international nuclear safeguards and bilateral assistance terms without making them discriminatory are urgent problems for exporting countries. K. Subrahmanyam expresses the disenchantment with great power arms control strategems felt by the leaders and elites in one state already suspected of having taken the nuclear weapon decision. He argues that arms control as presently construed has not meant international stability.

Part Three is devoted to the strategic anxieties felt by a number of nuclear aspirant states and the chances and consequences of their acquiring nuclear weapons. Steven J. Rosen argues that a regional nuclear deterrence situation between Israel and the Arab states could have positive results. According to Edouard Bustin, South Africa may acquire nuclear weapons for reasons that have little to do with internal or external threats to the regime. The incentive, he suggests, lies in the means to pressure the Western countries especially the United States, into irrevocable support for the minority regime's continued existence. Anne Cahn considers the chances of a nuclear weapons lobby developing in Iran, the country's technical and strategic options over time, and the possibilities of its "unsigning" the NPT. Yoshiyasu Sato details the reasons which now seem to preclude a Japanese venture into nuclear weapons. However, he feels that Japanese leaders might rethink their options if changes were to take place in their country's relationship with the United States, or if geopolitical events were to redesign the Korean peninsula.

H. Jon Rosenbaum evaluates Brazil's plans for nuclear cooperation with West Germany and the impact of that arrangement upon proliferation and relations with the United States. Carlos Waisman discusses the potential for nuclear rivalry between Latin America's two biggest states, and the possibility that Argentina will embark on a peaceful nuclear explosion program similar to India's, particularly in light of Argentinian political instability.

Part Four returns to international systemic concerns and examines nuclear arms control trends in the post–NPT period. Enid C. B. Schoettle traces US attitudes toward nuclear proliferation. Onkar Marwah assesses the NPT Review Conference held in Geneva in May 1975.

The participants in the Clark Conference were keenly aware that their limited effort only begins the search for a redefinition of nuclear arms control discussions. Their purpose would be served if the focus in their concerns suggests to others the need for construing disarmament procedures in frameworks of equal validity to great and second order states alike.

NOTES TO INTRODUCTION

1. See Richard Rosecrance, "Bi-polarity, Multi-polarity, and the Future," *Journal of Conflict Resolution* X (3), September 1966: 390–406.

2. See Bibliography.

3. In the parry and thrust of arms control negotiations, one of the NPT sponsors offered to "permit the International Atomic Energy Agency to apply its safeguards to all nuclear activities... excluding only those with direct national security significance." *Documents on Disarmament,* 1967, U.S. Arms Control and Disarmament Agency pub. 46 (Washington D.C.: U.S. Arms Control and Disarmament Agency, 1968), pp. 613, 615. The offer was rejected as being insufficient by the non-nuclear weapon states.

4. The preamble to the NPT contains a general statement urging all states against resort to the threat or use of force "in accordance with the U.N. Charter." Given the veto powers of the UN Security Council's permanent members—who also were the nuclear powers within the treaty's provisions—the injunction in the preamble was possibly read in reverse by some nuclear threshold states: that is, the right to issue (nuclear) threats or use of force was to be *maintained* and legitimized by the NPT.

5. See "Commentator's" views, "An Important Aspect of Disarmament," *International Affairs* (Moscow), January 1967, pp. 61–63; and "Questions of Nuclear Non-Proliferation," *International Affairs* (Moscow), October 1967, pp. 72–74.

6. The Federal Republic of Germany recently ratified the NPT along with other members of Euratom.

7. See, for sensitive analyses in this respect, William Epstein, "The Proliferation of Nuclear Weapons" *Scientific American* 232 (4), April 1975: 18–33; and Ian Smart, "Non-Proliferation Treaty: Status and Prospects," in *NPT: Paradoxes and Problems,* Anne W. Marks ed. (Washington, D.C.: Arms Control Association and Carnegie Endowment for International Peace, 1975), pp. 19–30.

8. According to unconfirmed news reports, the positioning of a US naval task force in the Bay of Bengal led by the nuclear-weapons-armed *U.S.S. Enterprise* played an important role in the Indian decision to conduct a nuclear

explosion. While American reasons for the expressive show of force are complex, the psychological humiliation of the affair must have been immense on the Indian side. No event since independence in 1947 had as overtly and dramatically presented to India a view of its powerlessness.

9. "Korea Denies Nuclear Weapons Research Plan," *The Washington Observer,* July 15, 1975.

10. Further, Brazil and Argentina have both refrained from full accession to the Tlatelolco Treaty. Both states also interpret the treaty as permitting peaceful nuclear explosions.

11. The ambiguity of the prohibition on biological weapons is better assessed by contrasting it with the hesitancies expressed by the major powers in prohibiting chemical warfare weaponry. Unlike the former, chemical weaponry may be relatively "more controllable," and safer to handle and stock as "binary" compounds.

12. See, for a detailed analysis, Milton Leitenberg, "The Race to Oblivion," *The Bulletin of the Atomic Scientists* XXX (7) September 1974: 8–20.

13. Peaceful nuclear explosions are not forbidden by the Nuclear Threshold Test Ban Treaty, 1974. It also appears that the global powers are in the process of reducing the size of nuclear warheads for the next generation of MIRV missiles. Hence, the 150 kiloton barrier on weapons testing is in keeping with future strategic weapon systems needs.

14. See George H. Quester, *The Politics of Nuclear Proliferation* (Baltimore: The Johns Hopkins University Press, 1973), for a country-by-country review of utilities desired by nuclear threshold states.

15. Many nuclear threshold states do not "distinguish" prohibitions sought on nuclear weapons and technology from the major powers' profligacy in the worldwide sales of conventional weapons. In their calculations, sales of conventional weapons by the nuclear powers are a part of the "destructive intentions" spectrum in respect of themselves.

16. Apart from their use in weapons, nuclear fissile materials are becoming highly profitable commercial items. The sales value to the US of a variety of reactor fuels and attendent technologies figured at $421 million in 1974. They are projected at $5 billion in the 1980s. See Paul Lewis, "A Nuclear Peace–Profit Motive," *The New York Times,* July 20, 1975, sec. F, p. 6.

Part One

Chapter One

The Emergence of a New Second Order of Powers in the International System

Saul B. Cohen

In the past two years, a number of national claimants to second order power status have emerged. This is the most significant development in international relations since the restructurization of the great power world following World War II. Establishment of a discreet group of second order powers affects the status and capacities for action of the first order nations that have long dominated these "new boys on the block." Moreover, other states that were recently on a parity plane are now subordinate to the newcomers. We are witness to the emergence of a new level of the international political hierarchy—the second order.

In contrast to the three decades that were needed for the fashioning of the multipolar great power level, only a few years will be required for full development of the second tier. The rapid pace and unexpectedness of the emergence have exaggerated our perceptions of the process as disorderly and chaotic. If we place this development in perspective, however—i.e., as a stage of dynamic equilibrium—we may find the process comprehensible and even reassuring. I do not subscribe to such fears as expressed by Henry Kissinger, who has warned that the world faced a choice between *order* and *anarchy,* depending upon how national states chose to behave in this interdependent world [1]. This American secretary of state spoke for those whose views of the world are narrowly shaped by a "rage for order" and fear for disorder, by the security of the known and the anxiety of the unknown [2]. To pose perfect balance or chaos as the sole alternative outcomes of change forces, however, is to ignore the nature of the process, which is intrinsically plastic. In times of reordering, our hopes for absolute order are as illusory as our fears of anarchy. If we accept dynamic equilibrium as a goal for the international system, then by this very definition there will be periods when short term instability is necessary if dynamism is to be maintained within the system.

THE EMERGENCE OF HIERARCHICAL ORDERS

The concept of geopolitical equilibrium is, after all, a concept that is not imbedded in nature. It is the product of man's efforts to abstract observations about nature and apply them to the men-in-environment system. In viewing the geopolitical world as a system which, as with any set of life organisms, becomes increasingly complex, differentiated, specialized, integrated and hierarchical in a developmental sequence [3], we must be prepared to put up with periods of disorder to satisfy our desires for change. The main theme of my past work in geopolitical theory has been a model of a world divided and striving for balance that is complex in terms of its multiplicity and diversity of power nodes [4]. Such a world is hierarchical, in terms of the spatial nesting of geopolitical regional functions within broader geostrategic regions. It is integrated, in terms of the interdependence of the various axes of cooperation and competition.

For nearly three decades, the complexity that has evolved has been mainly at the geostrategic level, the first order of the hierarchy. Events within the geopolitical regions were dependent upon geostrategically related decisions. Multiple great power nodes replaced the binodal US—USSR competition, until today four first order powers and an emergent fifth, with global links and global aspirations, are reckoned within any policy that is geostrategic in scope. These nodes are the United States, the USSR, Common Market Europe, the Chinese People's Republic and, emergently, Japan. For ideological reasons China rejects being cast as a superpower, but it surely ranks as such. (The Chinese characterize themselves as "third world"—with the Europeans and Japanese as "second world" and the United States and the Soviet Union as "superpowers"—in their view of the world that consists of three interdependent but contending three parts) [5]. The ill-fated 1973 "Year of Europe" was a belated recognition by American policymakers that détente could not be a three party affair, but must include all of the North Atlantic. Moreover, Japan, in transition to this first order of the hierarchy, will achieve this level within the next quarter of a century.

Second order powers do not appear randomly. Instead they emerge from the context of the geopolitical region, which provides their action setting, just as the geostrategic region is the action setting for the first order powers. The international system that we are describing—the geostrategic and the geopolitical—is predicated on the national state. It is therefore a subsystem thereof, the hierarchy presupposing the geopolitical region to be subordinate to the geostrategic world militarily and ideologically, but not necessarily economically and in most cases not politically. Even within the same geopolitical region, a first order power is not likely to be able to exercise political dominance over a second order nation for a long period of time. Indeed, a form of love—hate relationship between neighbors may hasten the development of the second order power. As the international system imposes greater and greater moral restraints on the exercise of naked political power in relations among states, and as smaller states play off first order powers to free themselves of subordinate status, the second order states and even some of the third order increase their abilities to pursue their own political goals.

A number of factors account for the rise of the second order hierarchy: political modernization; the diffusion of technology (including nuclear); the armaments race; the development of internal geopolitical regional linkages including the possibilities for regional cartels; and the establishment of economic cartels among second order powers from several different geopolitical regions. The issue of globe-spanning cartels for one product versus regionally organized cartels for a variety of products is just coming to the forefront. The United States, with its need for secured and realistically priced petroleum from Venezuela, Canada and Mexico, might accept a regional cartel whereby it would assure its neighbors of higher prices for such minerals as copper, lead, zinc, iron, sulphur and bauxite.

The view of the world that I have previously but forward includes a hierarchical structure of two geostrategic and ten geopolitical regions: the trade-dependent maritime geostrategic world consists of five geopolitical regions (Anglo–America and the Caribbean; South America; Maritime Europe and the Maghreb; Africa south of the Sahara; and Offshore Asia); and the Eurasian continental geostrategic subsumes two geopolitical regions (USSR and Eastern Europe; and East Asia). In addition, one independent geopolitical region (South Asia), and two shatterbelts (the Middle East and Southeast Asia) are postulated. Not all of these ten geopolitical regions contain the capacities to spawn primary powers with global ambitions and capacities. But nearly all are regional environments that can serve as bases for second order powers (Figure 1–1).

Because there are three classes of geopolitical regions—dependent (seven in number), independent (one) and fragmented (two)—the second order hierarchy cannot be a homogeneous order. Some second order power nodes must compete directly with the primary power located within their geopolitical region (e.g., East Germany vies with the USSR); others compete in their own region with first order powers that are located within adjoining regions (e.g., the United States vying with Brazil or the Chinese People's Republic with India); still others face competition within their region from great powers that are far removed (e.g., the United States as a competitor to Egypt in the Middle East; or Common Market Europe to Indonesia in Southeast Asia). The accessibility factor that is so important to the definition of the geopolitical region has clear military and psychological consequences, somewhat less compelling economic influences and the least impact politically.

Since the competition to achieve second order power status is still in its early stages, we may find two or even three national states vying with one another in any single geopolitical region. Thus, up to 20 states are currently active competitors for places within the second level hierarchical order. With such numbers, is there any wonder that in their rage for order, great power statesmen and their

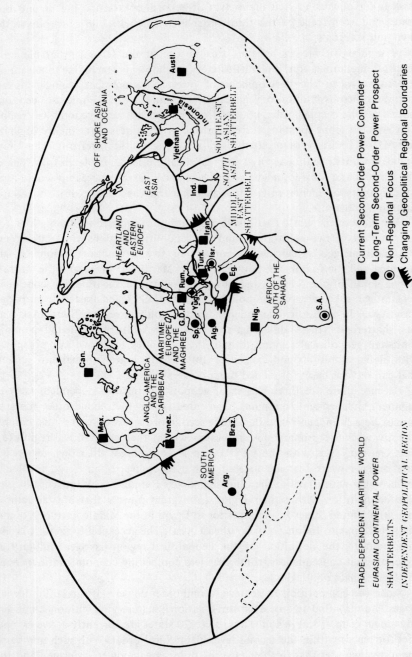

Figure 1-1. Map

publics evince impatience and indeed fear at their failure to be able to cast these contenders in a neat set of boxes? The "poker game" has become much larger; there are both new players and constant redistribution of chips, and the rules are being rewritten. But just as smaller nations of the world have had to sit by in resignation and in fear as the first order powers try to work out their equilibrium state, so must great powers now accept the proposition that they cannot wish instant equilibrium on others. To those who express fear at the proliferation of power bases, especially those with nuclear weapons capabilities, we might simply note that superpowers do not have a monopoly on sane and responsible leadership. Hitler and Stalin exhibited behavior no more responsible than Amin's or Kadaffi's.

Of the cast of about 20 contenders for second order power status, ten to 15 appear to have the potential for achieving this position rapidly within the ten geopolitical regions. Essentially, such status is derived from a combination of the following factors:

1. the success with which a national state can achieve some measure of intraregional superiority;
2. the strength of a state's extraregional ties to other second order powers; and
3. the ability of a state to gain sustenance from one or more great powers without falling into a condition of overwhelming subordination.

GEOPOLITICAL REGIONS AND SECOND ORDER POWERS

In the forces that lead to the emergence of the second order hierarchy, the regional concept plays an especially powerful role. Proximity stimulates the process of national political and economic development, either out of fear, jealousy or hatred, or out of emulation and attraction. The existence both of first order powers and second order aspirants within or near other nations in a particular region nurtures an action-reaction process—a form of dialectic. Far more than mere spatial construct, the region's characteristics stimulate transnational flows (labor, tourism, goods). They also expose to questioning cultural and ideological norms that would otherwise perhaps not be raised, if peoples did not live side by side. As a result, we look to the regional framework to nurture the emergence of more than one second order power, especially when first order powers are located nearby to stimulate the dialectic process.

In the discussion that follows, therefore, most of the regions contain two second order powers. Also, the most powerful of the superpowers, the United States and the USSR, have three and four second order powers within or immediately adjoining their respective regions. In analyzing each of the geopolitical regions, the following conditions characterize the prospects for second order power emergence:

Anglo–America and the Caribbean

Canada is a logical contender for second order power status. It meets the criterion of having ties with other second order powers and of having the ability to gain sustenance from great powers. Economic and military subordination to the United States no longer deprives Canada of freedom of political action, given Canada's close cultural and trade ties with Common Market Europe and growing economic ties with the Chinese People's Republic, Japan and the USSR. Canada's major ties to the Caribbean are her economic and military links to Jamaica. Should a West Indies Federation be reestablished, then the Canadian presence in the Caribbean would be considerably extended. A future force for strengthening the Canadian regional presence may be common oil policies with Mexico. With the forging of such ties, Canada will derive part of its strength from its ability to influence events within the region. Because Canada's nearest neighbors, other than the United States, are Japan, the USSR and Maritime Europe, it does face the liability of peripheral location within its region. However, air and water, as well as cultural and historic links to the Caribbean, can bridge the gap.

Elsewhere within the Anglo–America and the Caribbean geopolitical region, Mexico and Venezuela have second order power characteristics. For Mexico, the challenge is to abandon its general political posture of Caribbean isolation, to continue to expand the development of its raw material base and to bring its population explosion under control. For Venezuela, secondary power status seems well in hand. It is shifting more of its economic ties from the United States to its Latin neighbors, utilizing its vast petroleum revenues to spur regional development by exporting capital, by supporting oil exports to these countries and by otherwise linking them more closely to itself. An example is Venezuela's use of her new-found political leverage to rally Colombia and Costa Rica in support of Panama's claim to sovereignty over the Panama Canal and Canal Zone. It is not implausible to anticipate a future alliance among Mexico, Venezuela and Cuba in which Mexico or Venezuela could emerge as the senior partner.

South America

Here, geopolitical secondary power status is dramatically shifting to Brazil, with its overwhelming weight of population, economic growth record and spirit of modernization. At present Brazil has more significant economic and political links with other South American countries than do any of its neighbors, despite its non–Spanish traditions. It also has powerful links outside the region, with the United States, with Japan and with Europe. Argentina, once with strong aspirations for regional hegemony, still seeks to compete with Brazil. To this end, it has developed a population and economic strategy of internal growth that aims at strength through self-sufficiency. However, it seems to be too unstable politically, too weakened by inflation and too isolated from its Latin American neighbors to compete effectively with Brazil.

Maritime Europe and the Maghreb

The emergence of the Common Market of the Nine suggests that this primary power is now too indivisible to be treated as anything but one unit. This is said despite the cracks in unity that have occurred since the Arab oil boycott of 1973–1974 and the friction over Common Market policies towards depressed areas. The referendum in Britain that affirmed membership in the EEC, despite formal Labour party opposition and in the face of re-emergence of strong economic nationalism, may prove the high-water mark of the threat to European unity.

If we treat Common Market Europe as a first order power we must then seek second order powers elsewhere, in Scandinavia or in the Mediterranean. Northern Europe's subordination to the EEC, with few alternative regional or extraregional links, makes emergence of any of its states as second order powers unlikely. The petroleum base of Algeria, the minerals in Morocco and the industrial energies of Spain would, combined, constitute secondary power potential. But there is little that can bind these three ideologically and culturally. The ideological strength that Algeria draws from her role as revolutionary leader of the Arab world draws her attention strongly to the Arab world to the east. Moreover, these three countries are individually dominated by Common Market policies, Algeria and Morocco being highly subordinate to France economically. Still Algeria, because of her oil wealth, her ideological influence in the Arab world, and her retention of cultural and technological ties with France; and Spain, because of her population and resource base, her rapid rise to modernization, and her extraregional ties to Latin America, have some potential for secondary power status.

Offshore Asia

Australia, with her vast resource base, including and especially food; her modern economy; her support ties to the United States and the United Kingdom; and her growing relationship to the other nations of Offshore and Southeast Asia, is a second order power. Japan's position is already that of a near–first order power, her dependence upon imported energy resources being the only limitation to attaining first rank immediately. Extension of petroleum development in nearby Southeast Asia (Indonesia, offshore of Vietnam) and stabilization of a unified energy front among the major oil consuming countries would overcome this Japanese limitation. Taiwan, economically dependent upon the United States and Japan and militarily dependent on the United States, is not likely to achieve secondary power status, despite its remarkable economic efforts.

Africa south of the Sahara

Only two states appear to possess the capacity for secondary power status in the region: Nigeria and the Republic of South Africa. The former's population and petroleum resource base has, to date, been countered by political instability and isolation from its neighbors. But the gap between the impoverished, drought-stricken, food-short African world and an increasingly prosperous Nigeria is likely to hasten the dominance of Nigeria over her neighbors.

South Africa's military and economic power, her economic dominance over her nearest neighbors and her strategic control of the trans-Cape shipping route should accord her secondary power status. But growing political isolation from the black states of the region and uncertainty as to whether a Bantustan-based multiracial society can work cast some long term doubt upon South Africa's possibilities of becoming a secondary power within the geopolitical regional context.

The USSR and Eastern Europe

The rise of the German Democratic Republic as an industrial power is remarkable in the Eastern European context and overshadowed only in comparison with the unprecedented resurgence of the German Federal Republic. East Germany's economic ties with her Eastern European neighbors and with the Soviet Union cast her in a substantially more dominant economic role within the region than such countries as Poland or Romania. The GDR's physical isolation from the USSR and her historic and cultural links to West Germany strengthen her claim to a greater potential for secondary power status than other Eastern European neighbors of the USSR.

Romania is the other candidate for secondary power status in Eastern Europe. She has established herself as capable of conducting an independent economic policy within COMECON and outside the region, following her success in carving out a unique national communist position. On the other hand, the price that Yugoslavia has paid for independence from the USSR, regional isolation, limits second order status for her, as is the case for other states that lack regional dominance of their hinterlands.

East Asia

Only a United Korea under communist dominance could hope to achieve secondary power status in this region, and this only provided that there were a total recasting of geopolitical regional lines (including the split-off of the Far Eastern provinces from Soviet Siberia). At present, as skillful as North Korean leadership has been in playing off Chinese and Soviet rivalries, the state, even if it should be united, has no arena or hinterland for operating within its region.

South Asia

India's dominance within South Asia has been demonstrated recently in a variety of ways: her nuclear weapons capacity; her military defeat of Pakistan; her guiding of Bangladesh to statehood; her ability to balance off Chinese great power pressures by forging alliances with the USSR; and her success in expelling the United States from a South Asia power base by undermining the US–Pakistani alliance. Neither Pakistan nor Bangladesh possesses the resources base for competing with India for secondary power status within the region. Ceylon and Afghanistan are weak, small and peripheral. Moreover, Afghanistan could shift to the Middle Eastern region, especially if it were to realize its claims for part of Baluchistan and acquire a corridor to the sea as a satellite of Iran.

The Middle East
Four regional powers play a role in this shatterbelt: Iran, Turkey, Egypt and Israel. Israel and Turkey have demonstrated their military capacities; Egypt her political ability to lead the Arab world; and Iran her potential to dominate the Persian Gulf and to use her petroleum resources and large population as the basis for unprecedented economic and military buildup. As vast as is the petroleum that is being exploited under the sands of the Middle East, the Persian Gulf's underwater deposits are even more extensive. Competition between Iran and the Arab Gulf states over control of these resources may well erupt in large scale conflict. The prognosis is for Iran's emergence as the strongest regional power with substantial extraregional and global economic leverage. We doubt whether Iranian pretensions to supremacy over the Indian Ocean are likely to overcome Indian and Australian competition or to encourage them to join her in partnership. But primacy in the eastern half of the Middle East is well within the grasp of Iran, which may extend its orbit eastward to include Afghanistan and Pakistan. Size, cohesiveness, ability to build an élite and to support development through training of an indigenous labor force to handle high technology support Iran's ambitions.

Should Egypt succeed in taking over the oil wealth of Libya, her leadership over most of the Arab world will surely be enhanced. Until there is a resolution of the Arab—Israeli conflict, however, Egypt will lack the military and economic capacities to compete with Iran as a second order power. Israel's position for second power status is analogous to that of the Republic of South Africa—she has no proximate regional ties, her links with Iran and Turkey being highly tenuous because of the Islamic character of these two countries. Turkey, Iran's major competitor for secondary power status, lacks Iran's petroleum wealth leverage but has a population base, strategic location, regional hegemony traditions and the military machine to warrant such secondary status. Recent success in Cyprus and friction with the United States over arms sales and the status of American-controlled military bases in Turkey have lessened subordination to the United States by forcing Turkey to take strides towards greater independence.

Southeast Asia
The politically fragmented and less-developed nature of the peninsular part of this shatterbelt suggest little possibility for emergence of a secondary core, unless it emerges from a politically united and economically rebuilt Vietnam—an eventuality that can be anticipated from the ashes of the recent conflict, but the speed of which will depend upon the extent of outside aid and rehabilitation. A

more likely prospect for the near future is Indonesia, which is strengthening its ties with other parts of the region, especially Malaysia, Singapore and the Philippines. Indonesia's population-to-resources ratio, currently a source of weakness, is likely to improve when the already broad petroleum base is extended through offshore activities. In many ways, Indonesia might be likened to the formerly united Pakistan—both beset by overpopulation, poverty, divisiveness. But Indonesian national unity is favored by the relative strength, centrality and dominance of the Javanese center; by the richness of its soils and water base; by the absence of a nearby hostile competitor like India; and by petroleum export revenues. Thus, secondary power status in its geopolitical region is feasible in contrast to the absence of such status for Pakistan.

CRITERIA FOR SECOND ORDER POWER STATUS

A number of second order power possibilities have been suggested: Canada, Mexico, Venezuela, Brazil, Argentina, Algeria, Spain, Australia, Nigeria, East Germany, Romania, India, Iran, Turkey, Egypt, Vietnam and Indonesia (cf. Map 21), Yugoslavia, Israel and the Republic of South Africa are in anamolous positions vis-à-vis their regions. Of these, Canada, Venezuela, Brazil, Australia, Nigeria, East Germany, India, Iran and Turkey all possess immediate potential for becoming part of a second level hierarchy. They are nearly all well endowed with a raw material base, and nearly all have a substantial population base and have begun to interact with one another. This interaction is in terms of: common economic policies (e.g., cartels in food or mineral resources to raise the floor of raw materials prices and exchange these commodities for those from advanced industrial nations on a complementary basis); ease of intercommunication by water; and a balanced rural–urban setting (necessitating ideological attention to achieving harmony between traditional farming cultures and modern urban cultures). Moreover, most have the ability to fashion a nuclear capacity, either through importing the technology or through mobilizing national capacities. Besides their comparative strength within their own regions, second order powers derive influence from links to others within the order. None of these states can hope to achieve significant extraregional links with all of the others in the secondary hierarchical level, but each can hope to achieve meaningful economic, cultural and political links with a minimum number (three to six) of other members of the group. Thus, in addition to intraregional superiority, we have suggested the significance of extraregional links as a criterion. To search out such links, we have projected forward from data generated from an analysis of interdependent trade links (1 percent or more of total international trade in each direction) and interdependent political links as the basis to measure this degree of interrelation [6].

The third criterion cited for second order status is relationship to the first order power. This relationship is, as has already been noted, ambiguous—with generally high degrees of military and lesser economic subordination, but with considerable political (cultural) independence. Of the states that have been discussed, only East Germany does not enjoy a considerable measure of political independence from a major power. This is balanced off, however, by the GDR's greater degree of economic autonomy, evidenced by its increasing trade ties with West Germany. Moreover, with the formal division of Germany now acknowledged by the West, the ability of East Germany to forge a posture more independent of the USSR (along the Romanian model) is bound to increase. This is reinforced by the cultural distinction between the two states. In most cases, the relationship of the second order to the great powers is still to one, and not to two or more (exceptions are Australia, oriented to three powers—the USA, the EEC and Japan; and Algeria, oriented to the EEC and the Soviet Union).

Nuclear weaponry has not been specified as a separate criterion. Surely nuclear weapons capacity can secure second order power status for those states that fulfill the three criteria that have been noted. But it is possible for a nation (or a group of terrorists) to gain possession of the bomb without having a credible nuclear force, using the weapon for pre-emption or retaliation. This is different from having second-strike capacity which, for second order powers, would be on a paired basis (against another secondary power).

SOME GENERAL PRINCIPLES AND CONCLUSIONS

Our references to geopolitical systems build upon an analysis of the interrelationship of agents (in this case nations) and environments (in this case the objects that constitute geopolitical and geostrategic regions). What the agent-nation knows, feels and values is the basis for selecting the objects that constitute environment [7]. My references, therefore, to geostrategic and geopolitical regions are not to systems that are imbedded in nature, but rather to ones that derive from certain values espoused by men or the agents in interaction with those objects that constitute the environment of the system. The agent's values may be freedom, or justice, or economic well-being, or military security, or systems maintenance or openness—or combinations thereof. The objects that constitute the environment may be water or minerals, or other cultures and other social groups.

Changing Regional Boundaries

In a world dominated by the highest power orders, geopolitical regions were initially fashioned by these great powers, either as regions in which the great powers were themselves situated or as regions that lay exposed to the impact of these power nodes. But neither geostrategic nor, especially, geopolitical regions are fixed or static. As second order powers emerge, they begin to fashion geopolitical regions in their own images. While often the regional boundaries are the same as those shaped by the impact of great power extent and competition, sometimes they are shaped by the action of second level nations. Moreover, two competing second order powers located within the same geopolitical region may shape the boundaries differently. The principle of regional boundedness, then, is not of a frame drawn first, within which objects of the system are then filled in. Instead, the boundary is drawn last, after the relationship between the agent and objects is established. In effect, the geopolitical region is *nodal*, with a boundary that is zonal, not *uniform,* with a linear boundary.

To expand on the above, the boundaries of Anglo–America and the Caribbean are, for the United States, the Orinoco and the Amazon regions. For Venezuela, however, the region is in the process of being extended southward along the Andes and the Pacific to include Colombia and Ecuador (a return to an historically unified North Andean region). And while for Iran the Middle East may well extend eastward to include Afghanistan and Pakistan, for Egypt the region's reach southward may now include Somalia and Ethiopia—i.e., the zone of Arab–African confrontation. Should Egypt ultimately absorb Libya, the weight of adding petroleum to its population base would direct its vision of the Middle East westward to include much of the Maghreb, and southward into Chad.

We continue to assume that geopolitical regions are significant insofar as the relations among states are strongly influenced by distance, proximity and accessibility—all of which are enhanced by the contiguity (land and water) that the region provides. However, improved transportation and telecommunications, and changing exogeneous economic or cultural conditions, make it obvious that for a few agents the region is beginning to lose some of its potency. Therefore, we would not dare to suggest that a quarter of a century from now the regional framework that has been offered, especially its boundaries, will continue to provide the basis for international connectedness. Moreover, foreign policy reorientation by second order powers may shift the status of a geopolitical region from dependence, to shatterbeltdom or independence.

The Automated, Highly Managed Societal Revolution

The impact of innovation in science and technology upon international relations is being increasingly felt. One example is through modern weaponry. Until very recently, the major international impact of modern armaments systems was on the first order nuclear powers. Thus détente has been achieved through the balance of terror and stand-off that resulted from the nuclear capacities of the USA, the USSR, the Chinese PPR and two of the Maritime European countries (France and the UK). That other states possess or have the capacity for developing nuclear weapons (India, Canada, South Africa, Israel, Iran, Egypt, West Germany, Japan, Sweden, Brazil, Argentina) inspires considerable fear among the superpowers that the spread of nuclear weapons will make peace maintenance more difficult. But just as détente was the logical outcome of the balance of nuclear terror for great powers, who is to say whether the spread of nuclear weapons, and consequently the spread of balance of terror, may not extend détente among the second order nations by reducing conventional warfare options?

Moreover, a new balance of terror is in the process of being struck by the development of non–nuclear, (but nuclear capable) electronically and laser-controlled land, sea and air weapons. The diffusion of these weapons deprives some nations of the advantages held by skilled, well-organized sophisticated soldiery. Even though the massive arms race and sales by superpowers to others (approximately 10 percent of total world GNP of $250 billion is devoted to defense) may be motivated by callous economic policy to recycle petrodollars or as a calculated risk to exercise power through other nations, the net result may have the unforeseen consequences of developing a new stand-off between traditional enemies through a new balance of terror.

If the perceived result of pushbutton warfare is "no win" warfare; if manpower skills and morale advantages can be offset by remote control weaponry which cannot miss their targets—then a new détente may emerge in many of the world's current trouble spots: Southeast Asia, the Middle East, Africa, South Asia. This would drastically change attitudes among certain smaller nations which have fought wars under the umbrella of superpowers (e.g., Egypt, Syria, Laos, Cambodia, South Vietnam) on the assumption that they could afford to risk and lose battles, and even wars, without losing the conflict. Under such circumstances, we may find ourselves in a world in which powers will contend with one another in economic and cultural, not military, terms. Certainly wars that will have to be fought with modern space age gadgetry leave us not only with a sense of dread, but with a sense of futility. Will we, then, return to an age where knights on horseback, gladiators or hero-wrestlers will act as surrogates for combat with citizen armies?

High Technology

To achieve second order power status, a nation must have the ability to develop and to sustain high technology. This is critical, not merely in establishing nuclear weapons, nuclear power and nuclear-based industry, but in organizing non–nuclear industry and an independent military system. With sophisticated economic and military sinews, a state can compete effectively with other second order powers; it can exercise dominance over lesser nations; and it can maintain some degree of independence of action vis-à-vis the superpowers.

High technology can be purchased. Thus Saudi Arabia can contract for complete telecommunication and desalinization systems, and a steel plant; and Iran can buy several nuclear energy plants, and arrange for outside contractors to build a port, a petrochemical industry and a major new city. Both states can buy naval and air fleets, and tank and antitank arsenals. The basic question is not whether technology can be purchased; it is, rather, whether imported high technology eventually can be maintained by local energies—scientific and engineering skills, national acceptance, long term capital support. Some countries are seeking to achieve technological self-sufficiency by exporting students for advanced training en masse (e.g., Venezuela, Iran) or by importing universities (e.g., Saudi Arabia). If a nation's manpower is inadequate to the task of maintaining and building upon high technology, then it will remain in a state of dependence, no matter how broad its fiscal base may be. And if a nation is dependent on money alone, then the resource from which the capital is generated may lose its value or be depleted (e.g., most oil producers cannot count on this resource beyond from ten to 25 years).

The issue of the import of nuclear energy plants is a case in point. Whether a nation has the capacity to make its own nuclear bombs relates to its ability to develop a reprocessing plant that can produce pure plutonium. Some nations, with a strong national effort, can surely construct such a plant—they only have to be willing to pay the price. Others, however, are dependent upon outside forces to carry through such a project. As long as this dependence condition exists, secondary power status may be unattainable. A major criterion, then, for such status might well be the capacity to design and maintain a reprocessing plant, whether or not the capacity is actually exercised.

The foregoing has been an attempt to sketch out the ongoing developmental process that characterizes our global geopolitical system. The emergence of a second order in the hierarchy is part of the process of reordering of dynamic equilibrium. There is utility to a world view that is neither exclusively one of superpowers nor one of a bewildering mélange of over 140 national states, but rather of regionally framed power cores of different orders. This concept of global geopolitical equilibrium is the result of specific collective searches by man for ordering the relationships among states. It suggests that there can be a check and balance system in which disfunctional superpower behavior must adapt to secondary power concerns.

The environments that man includes in this system are environments that may eventually be excluded, or that may change in their value or durability. As long as we recognize that our deep psychic search for order is of our own creation and that constructs of these orders will vary according to perceived national needs; as long as we do not ascribe to geopolitical regions lives of their own, but see them as useful to today's technology and today's culture—then searching out a world organized hierarchically in regionally framed units has value and utility.

NOTES TO CHAPTER ONE

1. Press interview reported in the *New York Times*, October 13, 1974, pp. 35–36.
2. P. G. Kuntz ed., *The Concept of Order* (Seattle: University of Washington, 1968).
3. Ludwig von Bertalanffy, *General Systems Theory* (New York: George Braziller, 1968).

4. Saul B. Cohen, *Geography and Politics in a World Divided,* 2nd ed. (New York: Oxford University Press, 1973).

5. Transcript excerpts, address by Deputy Premier Teng Hsiao-ping, head of the Chinese People's Republic delegation to the General Assembly of the United Nations, the *New York Times,* April 12, 1974, p. 12.

6. J. Fonseca, "Linkage Development in the International System" (Doctoral dissertation, Graduate School of Geography, Clark University, June 1974).

7. B. Kaplan, S. Wapner, and S. Cohen, "An Organismic-Developmental Perspective for Understanding Transactions of Men and Environments," *Environment and Behavior,* vol. 5, no. 3, September 1973.

Chapter Two

The United States–Soviet Arms Race: SALT and Nuclear Proliferation

*Stefan H. Leader and
Barry R. Schneider*

INTRODUCTION

There seems to be growing recognition in the United States that Americans and Russians are not the only ones who pay close attention to the pace and scope of the US–Soviet strategic arms race. Secretary of Defense Schlesinger has belabored the importance of third country perceptions of the current state of the US–Soviet strategic weapons balance. His concern is not misplaced, though we think he overestimates the importance of relatively small differences in highly esoteric measures of strategic power. Non-nuclear countries do pay close attention to the US–Soviet arms race because the US–Soviet competition in strategic as well as conventional weapons sets the pace for the development of military technology worldwide. The US–Soviet weapons race is the cutting edge of advanced military technology and military thinking. Moreover, non-nuclear states are concerned that the nuclear stockpiles of the nuclear "haves" may be turned against them one day. Because we and the Soviets are watched closely, what we do influences military policy in many countries. It is obvious that some relation exists between so-called "vertical proliferation" of nuclear weapons in the US–Soviet nuclear arms race and the tendency toward "horizontal" proliferation—the propensity of non-nuclear states to join the nuclear club.

In this chapter we would like to explore these and related problems. The chapter is divided into two parts. In part one, we would like to explore the state of the US–USSR strategic balance and the impact of the SALT negotiations and agreements on that balance. We shall deal with three sets of questions:

1. What is the state of the US–USSR strategic arms race? Who leads? Who lags? How do you measure strategic power?

The authors would like to thank Bob Berman and David Johnson of the Center for Defense Information staff and Sally Anderson, formerly of the Center for Defense Information staff, for their assistance in preparing this paper.

2. How does US and Soviet quest for strategic parity affect the "vertical" proliferation of nuclear weapons in the hands of these superpowers?
3. What impact has SALT had upon US–USSR nuclear weapons proliferation? SALT I? The July 1974 Moscow Agreements? The Vladivostok accords?

In the second section we explore the impact of the US–Soviet arms competition (vertical proliferation) on nuclear aspirants and near-nuclear countries, and on the prospects for "horizontal" nuclear proliferation. In particular we will look at three correlates of the US–Soviet weapons race and political competition: the diffusion of nuclear expertise and technological know-how; the spread of advanced military technology in the form of nuclear-capable delivery systems, i.e., advanced aircraft and missiles; and increased opportunities for nuclear theft. We shall examine several questions here:

1. What is the relationship between the pace and scope of US–Soviet strategic arms race on the availability of nuclear expertise and technology worldwide?
2. What is the relationship between the superpower arms competition and the spread of nuclear-capable delivery systems to possible nuclear powers?
3. What is the impact of the dispersion of US and Soviet nuclear weapons on the chances of nuclear proliferation through theft and seizure?

VERTICAL PROLIFERATION—THE US–SOVIET STRATEGIC WEAPONS RACE

Since the Nonproliferation Treaty was signed five years ago there has been a significant increase in the number of nuclear weapons held by the United States and the Soviet Union (see Table 2–1). Twice as many US and Soviet strategic weapons exist in 1975 as existed in 1970, when the SALT talks began and NPT was about to be signed. The United States has roughly 30,500 nuclear weapons as of mid-1975. Eighty-five hundred of those are strategic nuclear weapons, and 22,000 are tactical nuclear weapons. The Soviet Union has some 2,800 strategic nuclear weapons and an undetermined number of tactical nuclear weapons, perhaps as many as 6,000 [1].

This vertical proliferation can be expected to continue as the US and the Soviet Union deploy new generations of strategic weapons. The present 8,500 US strategic nuclear weapon arsenal could reach 11,800 by 1980 and 21,000 to

Table 2–1. US–Soviet Nuclear Weapons Race—Strategic Nuclear Weapons

Years	1970	1971	1972	1973	1974	1975
USSR	1800	2100	2200	2200	2500	2800
USA	4000	4600	5700	6800	7650	8500

Source: DOD figures taken from the Annual Defense Department Reports to Congress.

30,000 by 1985 as systems under development—Trident submarines with MIRV missiles, the B—1 strategic bomber, additional MIRV warheads for Minuteman missiles and long range cruise missiles—are introduced. The Soviet strategic nuclear arsenal could reach 8,000 to 15,000 weapons by 1985 as new land-based missiles with MIRV warheads and new sea-based missiles are deployed.

The impact of this kind of proliferation on near-nuclear countries is controversial. At the very least it undermines the force of US and Soviet arguments about the dangers of horizontal proliferation. At worst it encourages the development of nuclear weapons by other states.

One new US weapon program—the long range cruise missile—may have an especially bad impact on the horizontal proliferation of nuclear weapons. Kosta Tsipis has recently pointed out that the availability of a relatively inexpensive and effective delivery system such as the long range cruise missile may encourage near-nuclear countries now deterred from acquiring nuclear weapons due to the prohibitive cost of long range delivery systems such as bombers or ballistic missiles to move ahead with the development of nuclear weapons [2].

Vertical Proliferation and Competing Measures of Strategic Parity

One engine that drives the Soviet—United States nuclear weapons race is the inability of policymakers to agree on a definition of what constitutes nuclear parity or sufficiency. For each of the past few years, the release of the annual report of the secretary of defense and the budget of the Department of Defense has provoked debate on the Pentagon's request for funds for strategic weapons and on the relative strategic nuclear power of the US and the Soviet Union. The secretary of defense has argued the importance of "essential equivalence" and "perceived equality" in strategic forces, pointed to new Soviet initiatives and called for corresponding American initiatives to keep pace. Critics have argued that the US already has a more than ample strategic deterrent, that both the US and the USSR are in the realm of nuclear plenty and surplus overkill capacity, and that more weapons are simply not needed. At the heart of the debate is the issue of how best to measure equivalence given the basic dissimilarities between US and Soviet strategic forces.

It is assumed that both the United States and Soviet governments want to have at least comparable strategic forces, so that neither is put at a disadvantage in a dispute, crisis or conflict. Neither can accept marked inferiority. But how is the strategic balance to be measured? How is strategic parity to be defined? Unless a common answer can be found, both within the US government and between the US and the Soviet Union, the nuclear arms race is certain to continue.

Measures of Strategic Power

At least six different measures of strategic power have been used at one time or another in the recent strategic debate and this makes comparisons difficult.

Any comprehensive assessment of the current strategic balance between the US and the Soviet Union requires multiple measures since no one variable by itself is sufficient to describe all dimensions of the US–Soviet balance. This is also a standard approach to social science measurement problems.

Numbers of Launchers. This is perhaps the oldest, most well-known measure of strategic power and also the one incorporated in both SALT I and the Vladivostok agreements. It is also the least significant in and of itself. The Soviets have a slight edge by this measure, in quantity if not in quality. US strategic bombers and missiles have been modernized and replaced by new models at a far more rapid pace than Soviet strategic delivery vehicles. In the 1946 to 1974 period the US has scrapped and mothballed more bombers and missiles than the Soviets have built in total [3]. This does not count the US bombers and missiles currently deployed. The count for the US and the Soviet Union projected to mid–1975 is summarized in Table 2–2. Marginal differences in numbers of delivery vehicles have no substantial impact on the strategic balance.

Force Loadings—Nuclear Weapons. A second measure of strategic power is what the Pentagon euphemistically calls "force loadings." These are independently targetable warheads or nuclear bombs. The significance of this measure of strategic power was given a boost by Secretary of State Henry Kissinger's statement that "one is hit by warheads, not launchers" [4]. Mr. Kissinger's statement seems to have prompted General George Brown, chairman of the joint chiefs of staff, to acknowledge in his FY 1976 posture statement that "the number of

Table 2–2. Strategic Weapon Launchers *(March 1975)*

United States		*Soviet Union*	
ICBM		ICBM	
Minuteman II	450	SS–7	190
Minuteman III	550	SS–8	19
Titan II	54	SS–9/18	288
		SS–11/19	1030
		SS–13	60
SLBM		SLBM	
Polaris	208	SSN–5	24
Poseidon	448	SSN–6	544
		SSN–8	132
Bombers		Bombers	
B–52 G&H	425	Bear	100
FB–111	73	Backfire B	60
Total	2208	*Total*	2447

Source: Center for Defense Information.

strategic warheads and bombs is a very significant measure of the balance" [5]. John Newhouse, counselor of the Arms Control and Disarmament Agency, pointed out in a recent speech that "[i]t is difficult to overstate the importance of this kind of advantage; missile warheads—the actual weapons—not missile launchers—the means of delivery—represent the more critical measure of overall strategic power" [6].

The US, as is apparent from Table 2-3, has almost a three-to-one advantage in nuclear weapons. This is a function of the fact that by mid-1975 the US will have 832 launchers equipped with MIRV warheads while the Soviets are expected by the most optimistic projections to have only a few MIRV warhead missiles deployed [7]. Secretary of State Kissinger has pointed out that "we will be substantially ahead in MIRVs for five to six years" [8].

Accuracy. This is one of the most significant measures of the capability of strategic forces available. As John Newhouse put it, "Although we have fewer ICBMs, ours are more accurate; and accuracy is a more dominant quality than yield" [9]. The standard measure of accuracy is circular error probability (CEP). CEP is defined as the size of the circle in which 50 percent of one's warheads can be expected to land. The ability of a missile to destroy a hard target depends on several factors, including numbers of warheads, yield, reliability and most important of all, accuracy. "The better the accuracy, that is, the smaller the CEP, the larger will be the 'K' (the lethality of the warhead to a hard target) and therefore the probability that the warhead can destroy the silo" [10].

The significance of this is that hard target kill capability is more sensitive to the numbers of warheads in a country's arsenal and to their accuracy than to their size or yield. This has been reaffirmed many times by US defense officials. Admiral Moorer pointed out in 1971 that ". . . accuracy is much more important than weapon yield" [11]. In 1974 Secretary of Defense Richardson made the same point: "Although our warheads are relatively small in terms of megatonage, with respect to target destruction, numbers of warheads and accuracy are much more sensitive measures of destructive power than weight" [12]. US advantages in these areas give the US a significant edge in counterforce capability and in strategic power.

Throw Weight. One of the more recent as well as most controversial measures of strategic power is throw weight. Some have argued that it "is the most useful verifiable measure of relative missile capability . . ." [13]. We disagree and

Table 2-3. Force Loadings—Nuclear Weapons *(mid-1975)*

United States	Soviet Union
8500	2800

Source: James R. Schlesinger, *Annual Defense Department Report, FY1976 and FY1977.*

Table 2-4. Accuracy

United States		Soviet Union	
Missile	Re-entry Vehicle Accuracy (CEP) Nautical Miles	Missile	Re-entry Vehicle Accuracy (CEP) Nautical Miles
Minuteman III	0.2	SS-9	(1)*
Minuteman II	0.3	SS-11, SS-13	(1)
Titan	0.5	SS-N-6	(1-2)
Poseidon	0.3	SS-N-8	(1-2)
Polaris	0.5	SS-7, SS-8	(1.5)

*( ) = estimate.
Source: SIPRI (Stockholm International Peace Research Institute)

will have more to say about this below. Throw weight is defined as the weight of that part of a missile beyond the last boost stage and thus is related to the size and number of warheads that a booster can carry [14]. All things being equal, a missile with larger throw weight will be able to carry a greater explosive punch and more numerous MIRV warheads. This is the basis for the Pentagon's argument about the danger of the US being overtaken in numbers of nuclear weapons by 1979 and endangered by a Soviet counterforce attack. Generally, Soviet missiles carry larger payloads than US missiles. According to Joint Chiefs of Staff Chairman Brown, their missiles currently have an advantage of two to one in payload [15]. Overall, however, the throw weights of US and Soviet strategic forces are about even.

Nevertheless, Secretary of Defense Schlesinger is concerned that if the Soviets are able to marry accurate MIRV warheads to their larger missiles the Soviet force will be able to attack a larger number of targets, including military ones, and improve hard target kill capability [16]. The secretary of defense sees the danger that with this capability the Soviets would have the ability to launch limited counterforce attacks against US land-based missile silos. He has argued that without a comparable capability, the US, in the event of such an attack, would be faced with a choice of surrender to Soviet blackmail or suicide (because it would provoke retaliation in kind) by attacking Soviet cities.

Several comments on throw weight and the Pentagon's scenarios are in order. First, Secretary of State Kissinger has pointed out in a recent background briefing: "The throw weight problem in my judgement, ... is a bit of a phoney, because nobody asked us to design small missiles, that was our choice" [17].

A second objection concerns the significance of the Soviet throw weight advantage. As we pointed out above, *all other things being equal,* a missile with a larger throw weight will be able to carry more and larger warheads. As is usually the case, all other things are not equal. We know from what has been seen of Soviet space technology that Soviet electronics are more primitive than American: that the Soviets still rely heavily on vacuum tubes and are far behind the US in miniaturized solid state circuitry and in computers and minicomputers [18]. CIA Director William Colby pointed out in testimony in 1974 that the technological gap between the Soviets and the US "is an across-the-board-one from ICBM systems to electric razors..." [19]. In short, the Soviets can do much less with their throw weight than we can with ours (lower "throw weight productivity") [20].

A third objection relates to the Pentagon's limited counterforce attack scenarios. The notion that the Soviets would risk a surprise nuclear attack, even a limited one, on our land-based missiles, leaving our formidable (4,500 nuclear weapons) sea-based nuclear forces untouched, makes little or no sense. Secretary of State Kissinger has made the very same point. "There is no way the Soviets can threaten our land based missiles even theoretically for four or five years... it is just one hell of a risk.... They would have to be crazy to do this. What would they achieve with this?" [21] Jeremy Stone has expanded on the same argument.

> Our sea-based forces could respond against any Soviet targets they wish, issuing a counter ultimatum—that full scale attacks on US cities would result in full scale attack on Soviet cities.
> Soviet attacks on our land-based forces would inevitably cause widespread fallout and many millions of casualties. No Soviet planner could assume that we would carefully and restrainedly calculate after that. Nor could he be sure that we could distinguish his attack from an all-out attack. Nor could he be sure that we could restrain our sea-based forces with suitable communications once the crisis began or our airborne bombers.

"The entire scenario," Dr. Stone goes on to say, "is bizarre—enormous risks for no point" [22].

Over the longer run—ten to 15 years—some danger to our land-based missiles is likely to develop as the Soviets improve accuracy of their missiles. Secretary Kissinger has acknowledged this.

> Over a ten year period no matter what you do, land-based systems are going to become highly vulnerable. There is no escaping it. And over a ten year period I think the composition of our force, in our judgement, is a safer one than the Soviets... [23].

The point here is that a much smaller percentage of US total throw weight is derived from land-based missiles than in the Soviet force. Bombers and submarine-based missiles contribute a larger share of total US throw weight than do Soviet bombers and sea-based missiles to their strategic arsenal. In short, because of the US lead in accuracy, Soviet strategic forces are in greater danger of becoming obsolete, and much sooner, than is the American force.

There is a final point that needs to be made with regard to throw weight. It has been acknowledged by Pentagon officials that by MIRVing their ICBMs the Soviets will reduce their yield-to-weight ratios (the ratio of explosive yield to throw weight). In testimony before the Senate Appropriations Committee in 1974, General Harold Collins, air force assistant deputy chief of staff for research and development, pointed out that MIRVed Soviet missiles can carry only a fraction of the explosive payload that they could with single warheads. This degradation of yield-to-weight ratios is due to the added propulsion, hardware and electronics components needed to deliver separate warheads on their targets [24]. All of this raises fundamental and as yet unanswered questions about the significance of throw weight as a measure of strategic power. What is clear, however, is that by itself, throw weight is not a very useful measure of strategic power.

Megatonnage and Equivalent Megatonnage. Still another measure of the state of the US–USSR strategic nuclear balance is explosive power measured in megaton equivalents (i.e., millions of tons of TNT equivalents). In sheer explosive power the Soviet Union maintains an edge because the warheads on their ICBMs and the explosive yields of some of their nuclear bombs are greater than their US counterparts (see Tables 2–5 and 2–6).

These raw megatonnage figures are, however, somewhat misleading since beyond a certain point larger bombs do proportionately far less additional damage than smaller weapons to nonhardened targets. Moreover, beyond a certain point, no more explosive yield is needed to destroy a given target. A better measure of explosive power is "Equivalent Megatonnage" (EMT). This EMT is the basic ability of a warhead to destroy a target and is estimated to be proportional to the two-thirds power of its megatonnage if the yield is less than one megaton, and the square root of the megatonnage if the yield is over one megaton [25].

Several smaller nuclear weapons can accomplish the same destruction as one very large weapon. Five one megaton explosives can inflict damage equivalent to one 25 megaton explosive; four one megaton bombs equal one 16 megaton bomb, and so on. By this measure (EMT) the US and USSR arsenals contain roughly the same destructive power (see Figure 2–1).

In 1975 the US "triad" of strategic bombers, ballistic missile submarines and ICBMs carry 3436 EMT whereas the Soviet Union's strategic launchers are capable of hurtling an EMT of 3847. Thus, in equivalent megatonnage, the Soviet Union maintains a slight, if insignificant, advantage.* Note the breakdown of megatonnage (MT) and equivalent megatonnage (EMT) by US and USSR delivery systems excluding the newly deployed Soviet SS–16, 17, 18 or 19 series which are only now being moved beyond the testing phase.

*Based on unpublished research by Barry Schneider at the Center for Defense Information, Washington, D.C.

Table 2–5. United States Megatonnage (1975)

Type	Number	Warheads Each	Estimated Raw Megatonnage (MT) Per Warhead	Total Raw Megatonnage (MT)	Total Equivalent Megatonnage (EMT)
Sea-Based Strategic Missiles:					
Polaris A–2	96	1	0.8	76.8	83.8
Polaris A–3	176	3	0.2	105.6	180.6
Poseidon C–3	384	10	0.4	153.6	45.0
				336.0	308.0
Intercontinental Ballistic Missiles:					
Titan II	54	1	10.0	540	170
Minuteman I	100	1	1.0	100	100
Minuteman II	500	1	1.5	750	610
Minuteman III	400	3	0.16	192	26
				1582	906
Long Range Bombers (and ASM)					
B–52 A/F	146	4	1.0	584	584
B–52 G/H	274	4	1.0	1096	1096
FB 111	76	2	1.0	152	152
SRAM	1140	1	0.2	228	390
				2060	2222
Summary:					
SLBMs				336	308
ICBMs				1582	906
Long Range Bombers				2060	2222
Totals				3978	3436

Table 2-6. Soviet Strategic Megatonnage (1975)

Type	Number	Warheads Each	Estimated Raw Megatonnage (MT) Per Warhead	Total Raw Megatonnage (MT)	Total Equivalent Megatonnage (EMT)
Sea-Launched Ballistic Missiles:					
SS-N-6	528	1	1	528	528
SS-N-8	108	1	1	108	108
	636			636	636
Intercontinental Ballistic Missiles:					
SS-7	139	1	5.0	695	307
SS-8	70	1	5.0	350	158
SS-9	288	1	20.0	5760	1282
SS-11 Mod I	970	1	1.0	970	970
SS-13	60	1	1.0	60	60
SS-11 Mod III	40	3 (MRV)	0.5	60	76
				7895	2853
Long Range Bombers					
TU20 "Bear"	90	1	5	450	199
Mya 4 "Bison"	36	2	5	360	159
				810	358
Summary:					
SLBMs				636	636
ICBMs				7895	2853
Long Range Bombers				810	358
Totals				9341	3847

Figure 2–1. Comparative Effectiveness in Terms of Destructive Area

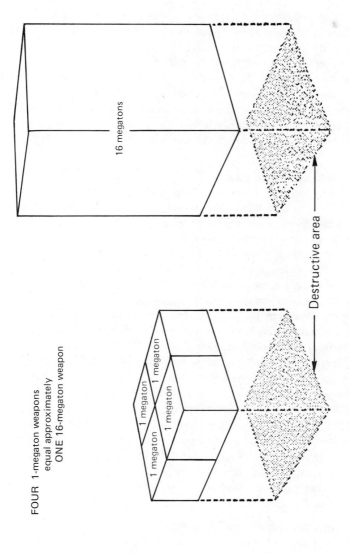

Note: The United States has more, smaller weapons; the Soviet Union has fewer, larger weapons. This gives the USSR a lead in megatonnage, but as this chart shows, it is not the total megatonnage that counts. It is the effectiveness—best measured in "equivalent megatonnage." By this measure, the US and the USSR are roughly comparable.
This chart was released by the Department of State on August 1, 1972.

Reliability. Overall weapon reliability is a composite of several individual variables, including missiles available, communications reliability, launch reliability, boost reliability, re-entry vehicle separation reliability, penetration reliability and detonation reliability. Figures on most of these components are hard to arrive at from the public literature though some evidence is available. For example, availability of US strategic forces has been estimated at about 95 percent to about 75 percent for the Soviets.** Overall, however, US and Soviet missile reliability are roughly the same. The overall reliability of Minuteman II and III is between 75 percent and 80 percent. This can be compared to estimated reliability of Soviet SS-9 of about 75 percent and for SS-11 of about 70 percent. These two missiles comprise the bulk of Soviet ICBMs. The navy hopes to achieve reliability of about 80 percent for its Poseidon missiles by 1977 [26]. By comparison, Soviet sea-based missiles are thought to have somewhat lower reliability—about 65 percent for the SS-N-6 and 70 percent for the SS-N-7 and SS-N-8.

US Has Marked Strategic Edge

If one had to attempt to reach a composite judgement about relative US and Soviet strategic strength from the foregoing analysis, the conclusion is enescapable that the US currently has a marked edge.

The US maintains three times more strategic weapons than the Soviet Union. Its missiles are more accurate. The US also has a better hard target kill capability than the Soviet Union.

The Soviet force, on the other hand, contains more delivery vehicles than the US and Soviet missiles enjoy a throw weight advantage over US missiles. This throw weight advantage might in several years permit the Soviets to reach approximate equality with the US in numbers of warheads assuming they master MIRV technology and deploy large numbers of MIRVed missiles.

At the same time, however, it is also clear that this US edge is of little significance politically. As Secretary of Defense Schlesinger pointed out, ". . . we have a good second strike deterrent, but so does the Soviet Union. Although the two forces differ in a number of respects, no one doubts they are in approximate balance" [27]. The point is not that they are in precise balance in terms of every variable by which one measures strategic power but that they are in balance in terms of their political effect—each force deters the other. Since each can deliver unacceptable damage to the other in a retaliatory blow, real strategic equivalence has been achieved and is unlikely to be upset by incremental changes in any of the six measures discussed here.

SALT—Arms Catalyst or Arms Restrainer?

SALT has produced mixed arms control results. It has succeeded in restraining some arms competition but has failed to halt the strategic arms race

---

**All of these estimates unless otherwise noted are based on unpublished research at the Center for Defense Information.

overall. Indeed, it has acted merely to rechannel it and, in some degree, it has served as a catalyst for further arms production and deployment.

SALT I, signed in May 1972, froze ICBM launcher numbers, SLBM launch tube numbers, ballistic missile submarine numbers and the number of ABM missiles per site as well as the number of ABM sites. The agreement had the virtue of freezing arms building at levels then current and of halting the quantitative arms race in these categories of strategic weaponry. Unfortunately, the agreement left unregulated the number of nuclear warheads on each side—one of the most important measures of strategic power. Also, two significant loopholes were left in the agreement—strategic long range bombers and forward-based nuclear systems were excluded. Thus, a significant portion of the nuclear delivery systems were left unregulated. Furthermore, the qualitative arms race, the process by which new, more efficient and modern generations of strategic weapons continue to replace older systems (e.g., ICBMs and SLBMs and MIRV), was left unregulated.

Nevertheless, the SALT I treaty was a useful step forward in stabilizing the arms race. The ABM treaty and the freeze on offensive missiles lessened the chances of either the US or the USSR being able to launch a pre-emptive nuclear strike without being destroyed in return. The accords thus tended to make nuclear war less likely, although they did nothing to discourage the continued "modernization" of strategic forces. SALT I was a limited victory for those who sought to slow (vertical) nuclear proliferation by the Soviet and US governments.

The 1974 Agreements—Arms Control in Reverse?

The SALT I agreements have been followed by two more summit agreements between the US and the USSR in the summer and winter of 1974. The Nixon–Brezhnev agreement at Moscow in July accomplished little of substance. It established a threshold of 150 kilotons, which underground test explosions were not to exceed. This was virtually meaningless since less than a half dozen such tests had occurred in the previous five years, and fewer still were scheduled for the future.

The July 1974 summit also reduced from two to one the number of sites where ABMs would be permitted. This was a constructive step, although it was far from certain that it prohibited anything either government planned to do, as both had completed only one ABM site at the time.

These agreements, and the November 1974 Vladivostok accords, have led some analysts to suggest that the SALT negotiations have in some ways backfired on those interested in limiting strategic weapons. In their view, SALT has recently been perverted from its initial goals of arresting the US–USSR strategic arms race to the point where strategic arms control negotiations have virtually become a part of the weapons acquisition process.

Figure 2–2. SALT I at a Glance (signed May 26, 1972)

US		USSR	
ICBM LAUNCHERS			
Deployed	1054	Deployed and under construction	1618
Recent construction rate	None	Recent construction rate	250 per year
Planned for 1977 without SALT	1054	Possible 1977 projection without SALT	2000
SALT ceiling	1000–1054*	SALT ceiling	1408–1618*
SLBM LAUNCH TUBES			
Present	656	Present	580
Recent construction rate	None	Recent construction rate	128 per year
Planned for 1977 without SALT	656	Possible 1977 projection without SALT	1200
SALT ceiling	710**	SALT ceiling	950**
BALLISTIC MISSILE SUBMARINES			
Deployed and under conversion	41	Deployed and under construction	52 (approx.
Recent construction rate	None	Recent construction rate	7–9 per year
Planned for 1977 without SALT	41	Possible 1977 projection without SALT	80–90
SALT ceiling	44**	SALT ceiling	62**
ABMs			
SALT ceiling	2 sites (100 missiles each)	SALT ceiling	2 sites (100 missiles each)

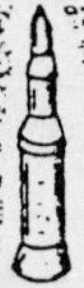
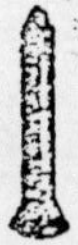
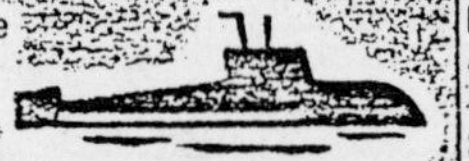
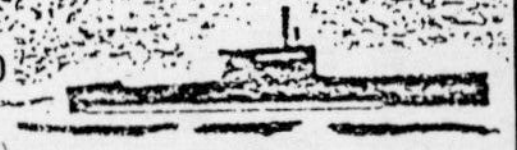
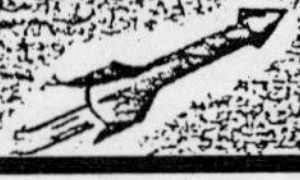
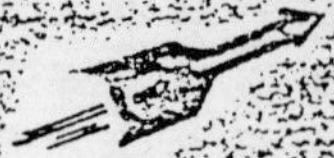

*Depending on whether old ICBMs are dismantled and replaced by SLBMs.
**To reach these levels US would have to dismantle 54 old Titan ICBMs; USSR would have to dismantle 210 old SS 7 and SS 8 ICBMs.

Source: Chart reproduced from *The Defense Monitor* (Washington, D.C.: Center for Defense Information, Vol. I, no. 3, July 1972).

SALT: Three Spurs to Vertical Proliferation

SALT has spurred arms competition in three ways. First, the armed services have gained an argument in the annual budget battles that helps convince skeptics to fund new weapons. This is the "bargaining chip" argument. The Pentagon has insisted that we should continue to fund weapons like the B–1 bomber and the Trident submarine, and should continue MIRVing US strategic missiles—all in order to strengthen our bargaining position with the Soviets. It is not unlikely that similar arguments are being presented inside the Kremlin by representatives of the Soviet military services. Jack Ruina has put the problem very well:

> "Finding a suitable bargaining chip is a fundamental problem. It is hard to think of an arms program that simultaneously is good enough to worry an opponent and bad enough for the military to be willing to give it up in negotiations" [28].

What has happened, of course, is that no "bargaining chips" have been bargained away.

Arms control pacts like the Outer Space Treaty, the Seabeds Treaty and the Biological Convention of 1972 all prohibited the production of future weapons. Only in the ABM Treaty of 1972 has an existing weapon system been actually cut back once deployment had begun. In no other cases have arms control agreements resulted in *disarmament*.

Second, the Strategic Arms Limitation Talks have spurred further weapons development activities in the US. Soon after the July 1974 Threshold Test Ban Treaty was signed, the AEC asked for $58 million more to accelerate the testing of nuclear weapons before the cutoff, which was scheduled for 21 months later. Congress granted $22 million of this request.

Third, agreements like the one negotiated in November 1974 in Vladivostok allowed weapon numbers higher than exist in current inventories. These high limits on MIRVs and strategic delivery vehicles may serve as "lures," rather than as limits, to further arms production. Pentagon spokesmen argue that since the Soviets will build to such limits, the US must also. The Soviet military, quite likely, argue in the same manner.

The Ford–Brezhnev SALT II Agreement at Vladivostok in late November 1974 was advertised by the parties as the long-awaited "cap on the strategic arms race." In this Vladivostok agreement both the United States and Soviet Union agreed in principle to limit until 1985:

1. the number of their strategic bombers and missile launchers (either sea- or land-based) to 2,400 for each country; and
2. the number of multiple independently targeted re-entry vehicles (MIRV), either sea- or land-based, to 1,320 for each country.

Vladivostok Virtues. Defenders of the Vladivostok agreement argue that it helps maintain the atmosphere of détente in US–Soviet relations and keeps the momentum of SALT going in a positive direction. Secretary of State Kissinger has argued that the new ten year pact relieved unfavorable time pressures and the now-or-never quality that the SALT negotiations had taken on in 1974. The new limits established at Vladivostok can be used as base points for further lower negotiated limits and as a basis for planning by defense analysts. A great deal of uncertainty about the future is removed for both the military planner and the arms control negotiator.

While the US agreed to include strategic bombers as a concession to the USSR, it can be argued that the Soviets have given up more than the United States in this particular agreement. The omission of US forward-based systems, even though they constitute a substantial nuclear force in Europe, Asia and on aircraft carriers, was a significant concession in view of the threat that these weapons pose to the Soviet Union. The Vladivostok accords, by stipulating numerical equality in MIRVed missiles and launchers, can help to channel the US–Soviet strategic arms race in a direction where US military advantages are pronounced. The Soviets will no longer be able to compensate for their technological inferiority by having more missiles. It is likely that the US will continue to enjoy an advantage in weapons technology for some time to come.

Lastly, defenders of the Vladivostok SALT accord point out that it "caps" the arms race at some finite point and acts to channel and restrict weapons production at lower future levels than might have been the case with no limits at all.

Vladivostok Vices. Despite all the cited advantages for the Vladivostok pact this "cap on the arms race" may well turn out to be something of a "dunce cap." The limits are so permissive and high that the agreement may be seen more as a license to continue the strategic arms race for another ten years than as an agreement which favors arms restraint.

It is useful to review what President Ford and General Secretary Brezhnev did *not* accomplish at Vladivostok in order to put the agreement into proper perspective. First, the accord imposed no direct limit on the number of nuclear weapons allowed each side. This is perhaps the most serious defect in the accord. The limit of 1,320 MIRVable launchers limits no current weapon or warhead deployment plans of the US. Prior to the November 1974 summit, the US had nearly completed its program to MIRV 550 ICBMs and 496 Poseidon SLBMs. The Vladivostok pact now permits and encourages the MIRVing of 274 additional missiles beyond the 1,046 planned. The result may be the proliferation of far more nuclear weapons in the US arsenal than were initially planned. Indeed, President Ford tells us that "we do have an obligation to step up to that ceiling" (2,400 strategic delivery vehicles, 1,320 MIRVed missiles), and he promises that "the budget that I recommend will keep our strategic forces either up to or aimed at that objective" [29]. The MIRV limits obviously do not cut into Soviet plans as the first operational Soviet MIRVed missile had yet to be deployed at the time of the agreement. MIRVing 1320 may be the work of a decade for the Soviets.

A second failure of the Vladivostok agreement is that it cuts back no current US arms program and imposes no substantial reduction in any Soviet weapons program. Indeed, it will probably increase, not reduce, military spending in both countries.

A third major flaw in SALT I was that no truly inhibiting limits were placed upon the qualitative arms race. MIRV limits were set so high as to constitute no limits at all in the next three to five years. No limits were placed on the number of warheads a MIRVed missile may carry. No restrictions were placed on improvements in accuracy, the creation of mobile ICBMs, improved ASW technology or on MARV[†]—the logical extension of MIRV technology. This is a major loophole that begs for plugging in future negotiations.

Finally, there is no indication that the agreement directs the strategic competition along channels that enhance the national security of either the United States or the Soviet Union. The strategic world permitted by Vladivostok until 1985 looks as threatening as that likely in the absence of an agreement, perhaps more so. We are likely to see greater numbers of weapons, all producing no discernable gain in security.

What the Vladivostok agreement *does* may be as harmful as what it does *not* do. First, it acts to undo the 1972 US–Soviet interim agreement which froze ICBMs and SLBMs then deployed or under construction at 1972 levels. Vladivostok permits any number of ICBMs and SLBMs up to a combined total of 2,400 strategic launchers. The US was limited to a maximum of 1,710 strategic missiles in 1972; the Soviet Union, 2,358. Vladivostok enables the military on both sides to argue for building to the agreed high limits because the rival will also be doing so.

The agreement at Vladivostok, if not followed by further far more restrictive agreements, permits and encourages the building and deployment by each superpower of as many as three to four times as many nuclear weapons as are in each superpower's arsenal now. Mid-1975 figures place combined US–USSR strategic weapons totals at 11,300 (8,500 US; 2,800 USSR). Under the Vladivostok agreement, 29,000 Soviet and American strategic nuclear weapons might well be deployed by 1985 or before.* In addition, the US and Soviets have currently in their arsenals an estimated total of 28,000 tactical nuclear weapons.**

[†]A highly accurate, maneuvering warhead, each one of which could destroy a missile silo, thus destabilizing the nuclear balance by providing great incentives to strike first.

*This assumes production of the new US Trident, B–1, and MIRV to the 2,400 launcher and 1,320 MIRV limits allowed by Vladivostok. This also assumes Soviet MIRVing and deployment of a new family of ICBMs. It is conservative in that it allows for no other new strategic qualitative innovations which might well occur. The conservative projection is that US nuclear weapons will grow in number from 7,650 in mid–1974 to 21,000 by 1985.

THE SUPERPOWERS AND NUCLAR SPREAD

Having examined and assessed the status, pace and direction of the superpowers' nuclear arms race, the question remains: What is the impact of the superpowers' arms race on the nuclear aspirants? How has US—USSR vertical proliferation affected horizontal "Nth country" proliferation?

We will argue that continued Soviet—US nuclear weapons competition has increased the possibilities for membership in the nuclear club at least three ways.

First, US—Soviet nuclear weapons building sets a bad example for the rest of the world. Indeed, it violates a commitment both made in Article six of the Nonproliferation Treaty to reduce their own nuclear weapon stockpiles. Second, the same forces that help spur the nuclear arms race also help spur the diffusion of military-related nuclear technology through the sharing by the superpowers of "peaceful" nuclear technology. (This will be discussed at length by Theodore Taylor in Chapter Five.) Third, nuclear proliferation is facilitated by the transfer of nuclear-capable delivery systems to many countries. Fourth, the basing of nuclear weapons worldwide increases the chances for theft and seizure.

Supplying Nth Countries With Delivery Vehicles

A significant and dangerous trend is the increasing availability and diffusion of nuclear-capable delivery systems to nuclear aspirants. It has become increasingly obvious in recent years that the resources and technology necessary to develop and build reliable nuclear-capable delivery systems—primarily missiles and advanced jet aircraft—with their sophisticated electronics, are far more demanding than those necessary to develop and build nuclear weapons. The extensive trade in conventional arms in recent years has made available to nuclear aspirants, in increasing numbers, the advanced missiles and aircraft necessary to carry nuclear weapons. Credible delivery systems are an essential ingredient for turning a test nuclear explosion into a meaningful military threat.

Both the US and the Soviets have transferred to both allies and clients large numbers of nuclear-capable delivery systems. Literally thousands of nuclear-capable aircraft, ships and guided missiles have been given or sold to foreign countries as part of the grant military assistance programs of the US and the military assistance programs of the Soviet Union. Since 1950 the US alone has transferred nearly 18,000 nuclear-capable guided missiles of all types—ground-to-air, ground-to-ground and air-to-ground—to foreign countries. During the same period the US has transferred about 8,400 nuclear-capable aircraft of all

Soviet weapons, conservatively estimated, will grow to just under 8,000. A worst case estimate of Soviet growth assuming many new military technology innovations would place Soviet totals at 14,500.

**The US is estimated to have in 1975 22,000 tactical nuclear weapons. Estimates of Soviet tactical nuclear weapons are less precise. They are estimated to have about 3,500 tactical nuclear weapons in Europe and perhaps another 2,500 or more elsewhere in the USSR.

Table 2—7. US Nuclear-Capable Guided Missiles, Aircraft and Ships Given and Sold to Foreign Governments, 1950—1973

Name	Assistance	Sales	Total
Surface-to-surface:			
Corporal	113	–	113
Jupiter	63	–	63
Pershing	–	133	133
Sergeant	38	195	233
Honest John*	3,566	2,688	6,254
Subsurface-to-surface:			
Polaris (A—3)	–	102	102
Air-to-surface:			
Bullpup	3,768	1,603	5,371
Walleye	–	104	104
Surface-to-air:			
Nike	3,460	1,766	5,226
Terrier	114	172	286
Total missiles	11,122	6,763	17,885
Aircraft:			
Antisubmarine	160	52	212
Attack	178	581	759
Bombers	113	–	113
Fighters	7,767	1,509	9,276
Patrol	186	102	288
Total aircraft	8,404	2,244	10,648
Ships			
Aircraft Carriers	3	2	5
Cruisers	–	6	6
Cargo Ships	4	–	4
Destroyers	38	62	100
Destroyer Escorts	62	23	85
Submarines	24	33	57
Sub-Rescue	1	1	2
Sub-Tender	1	1	2
Total ships	133	128	261

*1950—1972

Source: This chart is based on unpublished research done by Sally Anderson at the Center for Defense Information, Washington, D.C.

types to foreign countries. US-supplied nuclear-capable missiles are currently deployed by Spain, Taiwan and the Republic of Korea as well as by NATO countries. Both Taiwan and South Korea are believed to be interested in developing nuclear weapons and have in operation, or under construction, nuclear reactors capable of producing militarily significant quantities of plutonium [30].

Soviet nuclear-capable missiles are deployed by Warsaw Pact countries as well as Egypt, Syria, Iraq, North Korea and Cuba. Massive US and Soviet expenditures in recent years on highly sophisticated delivery systems have obscured the fact that nuclear weapons can be delivered with far simpler, less sophisticated weapons than the ICBMs and SLBMs in the US and Soviet arsenals. The only nuclear weapons actually used in war were delivered by aircraft that even the most underdeveloped country would not consider worth buying today. Many nuclear aspirants could acquire a credible nuclear deterrent against one or more of the current nuclear powers with a relatively small number of F-4 Phantom fighter-bombers or comparable aircraft equipped with nuclear weapons. It is worth nothing in this connection that the US's so-called "fourth strategic force," forward-based in Europe, Asia and on aircraft carriers, consists of precisely these kinds of delivery systems—fighter-bombers such as the F-4s, F-8s, F-14s, F-111s, A-4s, A-6s and A-7s as well as relatively short range surface-to-surface missiles such as the Lance, Pershing, Sergeant and Honest John.

Superpower Nuclear Dispersion—Theft and Seizure Dangers

There is another way the superpowers contribute to horizontal nuclear proliferation, at least potentially. The dispersion of US and USSR nuclear weapons across the world has led to the danger that either terrorist organizations or "allies" might steal or seize US or Soviet nuclear weapons. This might well prove easier to accomplish than the production of a nuclear weapon.

Nuclear Spread To Terrorists. The risk of war with the Soviet Union has long been regarded as the primary threat to the security of the United States. In hopes of averting such a catastrophe, thousands of man-years have been spent by analysts exploring how such a war is likely to erupt.

The research and contingency plans designed to control escalation of crises and limited wars, to prevent accidental war or unauthorized attacks, and to deter surprise attacks may be ignoring a likelier danger to US national security—the nuclear terrorist.

A half dozen terrorists with a homemade or hijacked nuclear weapon could cause thousands of deaths in a city like New York. Yet thousands of nuclear weapons are deployed by United States, Soviet, French and British forces in many different locations, some with questionable security precautions. If a terrorist group stole and detonated one "small" 10 kiloton nuclear weapon in New York City, the explosion could cause nearly 100,000 deaths—more than all the US battle deaths incurred together in the Vietnam and Korean Wars. One of the hundreds of US nuclear bombs in the one megaton range, if exploded in Manhattan Island, would inflict casualties exceeding the combined totals of the war dead from the American Revolution, the War of 1812, the Mexican War, the US Civil War, the Spanish-American War, the First World War, World War II, the Korean War and the Vietnam War. Nearly one-and-a-half million people would perish in such an event [31].

More than 40 major terrorist groups are reported to exist worldwide. Urban guerrilla activities, Olympic murders, airplane hijackings, terror bombings and airport massacres are all well known.

How safe are nuclear weapons from theft by terrorist groups? Not very, according to the few indicators available in this highly classified area. US Army Special Forces exercises have shown that nuclear weapons storage areas can be penetrated successfully without detection, despite guards, fences and sensors. Their example could obviously be followed by a daring and well-organized terrorist organization.

The United States, its allies and the Soviet Union have now deployed thousands of tactical nuclear weapons across the world, each in an effort to bolster its own national security. In doing so, these governments have made their societies more vulnerable to the nuclear terrorist, and increase the chances of nuclear proliferation in this form.

Nuclear Spread to Allies. More than half of all US nuclear weapons are stationed abroad or on the high seas. Countries where US nuclear weapons are reportedly stationed include [32]:

Federal Republic of Germany	Greece
United Kingdom	Turkey
Netherlands	Spain
Belgium	Portugal
Italy	Philippines
Iceland	Republic of Korea

Nuclear-capable ships of both US and Soviet navies visit numerous ports in additional countries. Once the US deploys nuclear weapons to an allied country, we put in jeopardy our control of those weapons if that nation ever becomes unfriendly to the United States. Allied seizures of US weapons could result in the United States having to fight its way into an allied country in order to rescue its own weapons.

This scenario takes on more plausibility in the light of the recent Greek and Turkish fighting on Cyprus. Both Greece and Turkey are host countries for US tactical nuclear weapons and US security forces were put on maximum alert guarding the weapons compounds during the short war between these two NATO allies.

Too many of the allied host countries that permit US nuclear bases are dictatorships with oppressive regimes that spark dissent. Franco of Spain and Park Chung Hee of Korea are vintage examples of this genre. Some allies who permit US nuclear weapons already have domestic insurrections on their hands. The Philippines and the United Kingdom are current examples. In Greece and Portugal recent coups d'état have changed the complexion of the ruling groups. In countries plagued by civil wars and coups, US tactical nuclear bases may not be safe from our "allies." During internal political disruptions in allied countries, US bases might find themselves caught in the middle of a firefight. One side or the other might find it advantageous or even necessary to seize US tactical nuclear weapons to gain the upper hand in the local struggle. Nor can be discount the possibility that allies such as the Republic of Korea might seize US weapons in order to defend themselves from—or possibly to attack—their antagonists.

In a recent book, *The Superwarriors,* James Canan points out that "The sobering thing about weapons—especially nuclear weapons—is that they have a life of their own, quite independent of the ups and downs of cold war diplomacy. They just keep growing. Their technologies, once set in motion, become tyrannical" [33]. This is apparent to governments all around the world who generally follow Attorney General John Mitchell's suggestion of a few years ago to pay attention to our actions not our words. Our actions suggest that nuclear weapons are very important, that they are a source of political influence and prestige, and that they are highly desirable. When this is added to the pressures of regional rivalries and power balances it is hard to escape the conclusion that in the years to come we will confront an international environment made far more dangerous by the spread of nuclear weapons to a great many more countries.

NOTES TO CHAPTER TWO

1. Barry R. Schneider ed., "30,000 U.S. Nuclear Weapons," *The Defense Monitor,* vol. 4, no. 2, February 1975 (Washington, D.C.: Center for Defense Information), p. 1.
2. Kosta Tsipis, "The Long Range Cruise Missile," *Bulletin of Atomic Scientists,* April 1975, p. 24.
3. Barry R. Schneider, ed., "Arms Restraint: Too Weak a Voice," *The Defense Monitor,* vol. 3, no. 7, p. 3. August 1974.
4. News Conference, April 26, 1974.
5. Chr., Joint Chiefs of Staff, General George Brown, *U.S. Military Posture for FY 1976,* Department of Defense, 1975, p. 37.
6. John Newhouse, Speech delivered at UCLA—Westwood, January 29, 1975, p. 8.
7. James R. Schlesinger, *Annual Defense Department Report, FY 76 and FY 76T,* Washington: Department of Defense, 1975, pp. 11—12.
8. Henry Kissinger, Department of State. SALT Backgrounder given on Tokyo—Peking flight, November 25, 1974.
9. Newhouse, p. 8.
10. SIPRI, *Offensive Missiles,* Stockholm Paper 5 (Uppsalla, 1974), p. 16. Also Hearings *Nuclear Weapons and Foreign Policy,* Committee on Foreign Relations, U.S. Senate, 1974, p. 170.
11. Testimony of Admiral Thomas Moorer, Senate Armed Services Committee, *Hearings on FY 1972 Defense Department Authorization,* March 15, 1971, p. 19.

12. Testimony of Secretary of Defense Elliot Richardson, House Armed Services Committee, *Hearings on FY 74 Department of Defense Authorization*, pt. I, April 11, 1973, p. 482.

13. Paul Nitze, *Aviation Week and Space Technology*, February 24, 1975, p. 66.

14. Brown, p. 37.

15. *Ibid.*

16. Schlesinger, p. 11–14.

17. Henry Kissinger, Department of State. SALT Backgrounder press briefing. November 25, 1974.

18. See *Aviation Week and Space Technology*, October 7, 1974, pp. 12–14; January 27, 1975, p. 61; February 3, 1973, p. 19. Also see SIPRI, *Offensive Missiles*, pp. 24n, 25n; and Dr. Malcolm Currie, "Department of Defense Program of R.D.T. and E.", Department of Defense, *Report for FY 1976*, Washington, 1975, p. 11–17.

19. Testimony of CIA Director William Colby, U.S. Congress, Joint Economic Committee, *Hearings on Allocation of Resources in the Soviet Union and China*, April 12, 1974, p. 18.

20. Edgar Ulsamer, "The Soviet Drive for Aerospace Superiority," *Air Force Magazine*, March 1975, p. 46.

21. Kissinger, SALT Backgrounder.

22. Jeremy Stone, *F.A.S. Public Interest Report*, Washington: Federation of American Scientists, February 1974, p. 8.

23. Kissinger.

24. Testimony of General Harold Collins, US Air Force assistant deputy chief of staff for research and development, Senate Appropriations Committee, *Hearings on Department of Defense Appropriations FY 75*, pt. 4, May 2, 1975, p. 704.

25. Testimony of David Packard, Senate Foreign Relations Committee, *Arms Control Implications of Current Defense Budget FY 72*, June 1, 1971, p. 179.

26. House Armed Services Committee, *Hearings on Department of Defense Authorization*, pt. IV, p. 3992.

27. Schlesinger, p. 1–16.

28. Jack Ruina, "SALT in a MAD World", *New York Times Magazine*, June 30, 1974, p. 42.

29. Transcript of President Ford's News Conference, *New York Times*, December 3, 1974, p. 28.

30. Hearings before the U.S. Congress. House Committee on Appropriations. Subcommittee on Foreign Operations and Related Agencies. *Foreign Assistance and Related Agencies, Appropriations, FY 1975*, March 1974. p. 1383.

31. Schneider, "30,000 U.S. Nuclear Weapons," 4, 2 p. 8.

32. *Ibid.*

33. James Canan, *The Superwarriors*, (New York: Weybright and Talley, 1975), p. 133.

Chapter Three

The Nonproliferation Treaty and the Nuclear Aspirants: The Strategic Context of the Indian Ocean

Robert M. Lawrence

INTRODUCTION

Six of the Indian Ocean nations are found somewhere on almost all lists of potential nuclear weapons states. These are Australia, India, Indonesia, Iran, Pakistan and the Republic of South Africa. The placement of these states on such lists results from the fact that they now possess, or are thought likely to obtain in the future, the necessary technical infrastructure to support at least a first generation nuclear weapons program; and because the motivation is thought to exist in greater or lesser degree to move each under certain circumstances toward the development of nuclear weapons.

To date three of these nations—India, Pakistan and the Republic of South Africa—have refused to be bound by the Treaty on the Nonproliferation of Nuclear Weapons.

The purposes of this chapter are to inquire into the reasons why three of the Indian Ocean nations have rejected the NPT; to examine the strategic actions of the two superpowers in the Indian Ocean region with the view of establishing which actions are likely, if continued, to operate in a way to reinforce the three nonratifying nations in their refusal to accept the NPT, and to cause those now adhering to the treaty to consider abrogation; and to speculate upon what it all means in terms of future US interests.

WHAT FACTORS CAUSED THREE INDIAN OCEAN STATES TO REJECT THE NPT?

The logic of the NPT seems to be, to paraphrase the old saying, in the eye of the beholder. Thus it may be stated that the rationale of the NPT appears to be one thing to the three nuclear weapons states which sponsored it, and something quite else to other nations in different situations and holding different interests.

To put it another way, the logic of the NPT seems to be a relative matter depending upon the circumstances in which an individual nation finds itself. Taking a cue from this perspective, the analysis which follows will attempt to establish the logic of the NPT as seen by the three nuclear weapons states which sponsor the treaty; will note the different perception of the NPT rationale which has been expressed by the two nuclear weapons states which have not signed the treaty; and will conclude by exploring the view of the NPT logic taken by the three Indian Ocean nuclear aspirant states which have to date refused to accept the treaty.

Understanding the differing perceptions regarding the logic of the NPT should assist in comprehending the current failure of the NPT to gain full acceptance among the Indian Ocean nuclear aspirants. Understanding the differing perceptions of the NPT logic may also suggest modifications in the treaty, and in the behavior of the superpowers in the Indian Ocean region, which could encourage acceptance by states now rejecting the treaty, and continued adherence by those currently associated with the treaty.*

The NPT Logic According to the Nuclear Weapons State Sponsors

Documents of record concerning the negotiations which culminated in the NPT suggest that from Washington's perspective the logic of the NPT may be stated in the form of four basic assumptions [1]:

1. There is a nearly universal dread of nuclear proliferation beyond the US, USSR, UK, PRC and France; and there is a correlative belief that the greater the number of nations acquiring nuclear weapons the greater will be the danger of nuclear weapon detonation, either by accident or design, with the result that substantial, possibly even catastrophic damage will be done to civilization.

2. The motivation to acquire nuclear weapons on the part of states not possessing them can largely be removed or attenuated by actions taken or pledged by the sponsors of the NPT. These actions or pledges include:

a. the offering of security guarantees to those nations agreeing not to develop nuclear weapons. The first such guarantee was issued by President Lyndon Johnson shortly after the first nuclear test by the PRC in the fall of 1964. At that time the American president stated that nations agreeing not to develop nuclear weapons could expect US assistance against nuclear blackmail. The Soviet Union's position on guarantees differed from the one adopted by the

---

*The author should state his bias in favor of the proposition that more success in limiting further national nuclear proliferation can be expected from efforts designed to remove or reduce incentives for proliferation, than from efforts designed to control the availability of fissionable materials and delivery systems. Due to technological diffusion the latter appears to have the most validity as an antiproliferation strategy only on a relatively short-run basis. On the other hand, efforts to prevent fissionable materials from falling into the possession of terrorists, criminals and deranged persons may represent the preferred approach indefinitely.

US. In 1966 Premier Kosygin suggested a statement be added to the pending NPT which would prohibit the use of nuclear weapons against a non-nuclear weapons state signatory of the treaty. The result of the debate among those hammering out the NPT following the American action and the Soviet suggestion was to drop the matter of protective guarantees from the NPT and its seek inclusion in a UN resolution. This was done on June 19, 1968. The resolution reads in part:

> That aggression with nuclear weapons or the threat of such aggression against a non-nuclear-weapon State would create a situation in which the Security Council, and above all its nuclear-weapon State permanent members, would have to act immediately in accordance with their obligations under the United Nations Charter [2].

b. the offering of assurances that the treaty is not a device to prevent commercial exploitation of civil nuclear energy by the non-nuclear weapons states nor an arrangement which would facilitate nuclear expionage by the treaty's nuclear weapons state sponsors. Articles III, IV and V attempt to deal with these matters. In the first paragraph of Article IV the non-nuclear weapons signatories of the NPT are assured by the sponsors that the former's peaceful nuclear programs will not be obstructed by international inspection. In Article III the emphasis regarding what is to be inspected is upon "source and special fissionable materials" rather than upon facilities or reactors. Thus in the sponsor's view the danger of industrial espionage regarding facilities and reactor types is greatly reduced.

c. offers non-nuclear weapons signatories of the treaty "plowshare" type civil nuclear explosives on a nondiscriminatory basis, with the statement, "that the charge to such Parties for explosive devices used will be as low as possible and exclude any charge for research and development."

3. The sponsors of the treaty are generally considered to be responsible and trustworthy stewards of nuclear destructive power and thus pose no threat to nations which remain unarmed in a nuclear sense; and that non-nuclear weapons signatories will accept as sincere the statement that the nuclear weapon state sponsors of the NPT will move toward their own arms control and disarmament plan for nuclear weapons, as well as working for "General and Complete Disarmament." To secure this objective Article VI states:

> Each of the Parties to the Treaty undertakes to pursue negotiations in good faith on effective measures relating to cessation of the nuclear arms race at an early date and to nuclear disarmament, and on a treaty on general and complete disarmament under strict and effective international control.

4. The International Atomic Energy Agency inspection procedures and associated safeguards techniques will prove adequate to prevent the clandestine

evasion of the Treaty's prohibitions against the illegal manufacture of nuclear weapons.

Not articulated in public by the American government, nor by the governments of the United Kingdom and the Soviet Union, is the additional assumption that in international affairs the interests of the nuclear weapons states are advanced in a context which features the fact that many nations with whom they must interact remain unarmed in regard to nuclear weapons.

The record concerning the negotiations which culminated in the NPT suggests that in general the Soviets were in accord with the assumptions stated above as being held by the Americans. Thus one may presume that the logic of the NPT perceived in Moscow is similar to that held in Washington. There are two exceptions however. First is the special concern of the Soviet Union over the necessity to prevent West Germany from acquiring nuclear weapons. This objective appears stronger for obvious reasons in the Soviet Union than in the United States. Then there is the matter of the US forward-based systems and the nuclear sharing doctrine between the US and some NATO allies. During the negotiations for the NPT the Soviet Union sought for a time to instill in the NPT prohibitions against the stationing of US nuclear forces in other nations and against any nuclear sharing between Washington and its allies. At one time the Soviet delegate Mr. Kuznetsov criticized the Irish proposal—the first initiative toward limiting the spread of nuclear weapons which evolved to become the NPT—as being deficient because it did not, "even mention prohibiting states from having nuclear weapons outside their territory" [2]. The Soviet Union failed in securing the removal of US forward-based systems in the NPT. It should be noted that to date Moscow has also failed in securing that objective in either the SALT I accords, or by means of the Vladivostok guidelines for negotiation of a SALT II treaty for the limitation of strategic nuclear weapons for the next ten years.

French and Chinese Views of the NPT Logic

More than a decade ago the two nuclear weapons states which have not adopted the NPT saw in that document a different logic than that perceived in Moscow and Washington. While many Americans will disagree with the perceptions of Paris and Peking, it is useful to note these views because the French and Chinese have constructed arguments against the NPT which may be appealing in either the original versions, or in altered form, to some of the Indian Ocean nations.* In this regard it should be recalled that in general the Indian Ocean aspirants with which this chapter is concerned are probably closer in outlook to France and China regarding the proliferation of nuclear weapons than they are to either the United States or the Soviet Union.

---

*It can be argued that French and Chinese views of the NPT expressed some years ago are not relevant to the future. One can also note the point made by Albert Wohlstetter that after the passage of time the latest nation to join the nuclear club may tend to accept the view that no others need apply [4]. However the earlier views of France and China are still important, even if they are receiving less support from those who originated them, because their appeal may fit the current needs of potential nuclear weapons states.

Much of the logic perceived in the NPT by the French and the Chinese is associated with the general point that it is dangerous to entrust nuclear weapons solely to the two superpowers because both tend to be imperialistic and interventionist states. In such a context it follows that considerations of prudence and equity may justify other states being accorded the option of acquiring nuclear weapons which will serve the purpose of constraining the harmful proclivities of the US and USSR. Because of the possible importance of the French and Chinese views to some of the Indian Ocean aspirants a closer examination of the specifics in the French and Chinese positions is warranted.

Seven years before the NPT was signed the French General Pierre Gallois noted that the proliferation of nuclear weapons beyond the superpowers, most specifically the development of such weapons by a France independent of either the United States or England, would exercise a positive influence upon the chances for peace. His argument was that nuclear weapons in the possession of additional nations would create further constraints upon those already possessing them, and that newcomers to the nuclear weapons club would by the fact they had obtained nuclear weapons status have greater influence in working for arms control and disarmament than when they were unarmed in a nuclear sense [5].

A number of Gaullist writers advanced the proposition that the two superpowers wanted to prevent others from obtaining nuclear weapons as a means of maintaining their worldwide hegemony. American predominance in NATO and Soviet preeminence in the Warsaw Pact served to confirm French suspicions concerning the hegemonic tendencies of the United States and the Soviet Union. To many Gaullists one means by which to terminate, or at least reduce, superpower domination was to break up the US—USSR nuclear weapons monopoly by the development of a French nuclear force. It was argued by proponents of this view that such a force would directly provide Paris more independence of action, and would operate indirectly to afford greater flexibility to other states which found themselves in either the American or Soviet orbit [6].

The French advanced other reasons to justify the development of nuclear weapons which may be appealing to some of the Indian Ocean nuclear aspirants. For example, there was the argument that the credibility of the US coming to the defense of France was being eroded due to the increasing capabilities of the Soviet Union to mount a strategic attack upon the US should the Americans seek to protect Europe by threatening a nuclear strike upon Soviet cities. The point of this argument was that the French must look to their own deterrent rather than depend upon what was considered to be an uncertain deterrent in the hands of the US. Further, there was the argument made popular by the late General Ailleret that France required nuclear forces because France might be in combat with similarly equipped forces, presumably the Soviets but possibly others.

Variations of the French criticism of the NPT logic have often appeared in the publications emanating from the People's Republic of China. Even before the first Chinese nuclear test, Peking issued the following statement as part of an attack upon the then new Partial Nuclear Test Ban Treaty:

> With regard to preventing nuclear proliferation, the Chinese Government has always maintained that the arguments of the U.S. imperialists must not be echoed, but that a class analysis be made. Whether or not nuclear weapons help peace depends on who possesses them. It is detrimental to peace if they are in the hands of imperialist countries; it helps peace if they are in the hands of socialist countries. It must not be said undiscriminatingly that the danger of nuclear war increases along with the increase in the number of nuclear powers [7].

The general theme that "ideologically selective proliferation" is in the interests of the world's masses was continued after the first Chinese nuclear explosion conducted in the fall of 1964. The next year Foreign Minister Chen Yi stated the hope that the "Afro-Asian countries will be able to make atom bombs for themselves" [8]. In 1967 Peking expressed the view that non-nuclear states had the right to develop nuclear weapons so as to be able to resist what had become the twin evils threatening the world in China's eyes—US and Soviet imperialism [9]. Still later, in 1971, it was stated that ". . . other countries [have] just rights to develop nuclear weapons and resist nuclear threats posed by the superpowers" [10].

For more than a decade the PRC has in parallel with its verbal support for selected proliferation also called for the complete prohibition of nuclear weapons. This position was restated in 1973 when Peking criticized the superpowers for their continuation of the arms race [11]. The Chinese explain away the seeming contradiction between "ideologically selective proliferation" and the prohibition of nuclear weapons this way:

> Regrettably, however, the two nuclear powers have thus far failed to make a positive response to the Chinese nuclear prohibition suggestion. Instead [they] have concocted the Partial Nuclear Test Ban Treaty, the Treaty on the Non-Proliferation of Nuclear Weapons, etc. These agreements . . . are in essence a camouflage for their own nuclear arms expansion in the name of nuclear disarmament, a means for consolidating their nuclear monopoly. . . . In the absence of the complete prohibition and thorough destruction of nuclear weapons, it is impossible to expect the other countries, which are subjected to the threat of the two nuclear powers, not to develop nuclear weapons for the purpose of self-defense [12].

It should be noted that the People's Republic of China has apparently not backed its pro-proliferation rhetoric with substantive assistance to any nation regarding actual nuclear weapons development. However, Peking has developed a set of arguments supporting proliferation which could be persuasive to some of the Indian Ocean aspirants with which this study is concerned.

Indian, Pakistani and South African Views of the NPT.

India's Dominant Role. Of the three Indian Ocean nuclear aspirants which have not ratified the NPT, India has the greatest number of outspoken critics of the NPT logic as set forth by the superpowers. For this reason—and because of the considerable quantity of information regarding the NPT in the public arena in India; because India saw fit to detonate what New Delhi termed a "nuclear device" on May 18, 1974; and because India's actions concerning the NPT during the next ten years may be crucial in regard to the position other regional nuclear aspirants may take on the NPT, particularly Pakistan and Iran—more space will be devoted to analyzing Indian perceptions of the NPT than those of the other two states.

One is safe to contend that in general Indians perceive the logic of the NPT to be largely the reverse of that perceived in Washington and Moscow. For example, in regard to the view that the further spread of nuclear weapons will be dangerous, one of the most outspoken and articulate Indians advances the proposition that the basic danger to world peace is found in the motivations and actions of those possessing nuclear weapons, particularly the US. On this point K. Subrahmanyam, director of the Indian Institute for Defense Studies and Analyses has written that it may be more dangerous for the current five nuclear weapons states to hold a monopoly on nuclear force than to have a situation in which some "middle range" nations such as India acquire nuclear weapons. His argument is that the nuclear weapons nations have a history of intervention into the affairs of other states, hence threatening the peace. According to Subrahmanyam the possession of nuclear weapons by "middle range" powers could serve as a check upon the interventionist tendencies of the nuclear weapons states, thus contributing to peace. He does not think realistic the fears of some that "middle range" states possessing nuclear weapons would themselves become interventionist [13].

The Indians in general have not been impressed by the efforts of the superpowers to (1) guarantee the protection of non-nuclear weapons signatories to the treaty; (2) guarantee no industrial espionage via the inspection system while promising the inspection arrangements will not hamper civil nuclear developments; and (3) offer PNEs to nations not developing nuclear weapons. The Indian explosion of a nuclear device in May of 1974, and the discussion in official sources of additional nuclear explosive experiments in the future, certainly constitutes a categorical rejection of the last point.

Many Indians have serious reservations regarding the premise of the NPT that the superpowers are responsible and trustworthy stewards of nuclear destructive power. The following quotations emphasize that point:

> In a world in which naked power unfortunately dominates the resolution of vital issues involving the security of nations the case for India producing such weapons [nuclear ones] is extremely strong [14].

It may sound strange to some but it is true that limited nuclear armament has now become an inescapable requirement for the preservation of our real independence which constitutes the core of our non-alignment [15].

Prime Minister Gandhi has expressed herself regarding the lack of trust in the responsible actions of the US thus (the statement was made in the aftermath of the Indian-Pakistani war of 1973 and the so-called American "tilt toward Pakistan"):

A great power must take into account the existence not only of countries with comparative power, but of the multitude of others who are no longer willing to be pawns on a global chessboard. Above all, the United States has yet to resolve the inner contradiction between the traditions of the founding fathers and of Lincoln and the external image it gives of a superpower pursuing the cold logic of power politics [16].

The dominant view in India regarding the superpower pledge to work for a reduction in their strategic nuclear stockpiles (Article VI of the NPT) is one of cynicism and distrust. What particularly concerns Indians is the fact that the superpowers continue under SALT I and the Vladivostok guidelines what the Indians term "vertical proliferation" while attempting to ban by means of the NPT what is called "horizontal proliferation"—i.e., the superpowers may continue to amass nuclear weapons but others may not. This point of view has been popularized in India by the former Ambassador from New Delhi to the Eighteen Nation Disarmament Committee V. C. Trivedi [17]. This cynicism is expressed another way in the following statement by C. R. Gharekhan, a former deputy secretary in the External Affairs Ministry:

It is conceivable that whatever limited agreement might be reached in the SALT, negotiations would be bandied about by the Super-Powers as fulfillment of their obligation under Article VI of the NPT. First of all one can be fairly certain that the U.S. and the U.S.S.R. will not agree on any measure of vertical [that is, decrease in their own stockpiles] nonproliferation [18].

While India has not emphasized the point that the NPT does not offer particularly impressive safeguards against the clandestine manufacture of nuclear weapons, the issue has been raised [19]. Recently the Indian government has tightened security on fissionable materials.

Perhaps surpassing in importance the above objections is the widespread belief in India that the basic logic of the NPT is an effort on the part of the superpowers to enforce by treaty an unjust and inequitable arrangement whereby nations such as India are kept in an inferior status. This view of the NPT logic has been frequently expressed. For example, in 1974 the following statement was made by Dr. H. N. Sethna, director of the Indian AEC: "We have to consider that the NPT is an unequal legal instrument: we would object to it so long as it remains discriminatory in character" [20]. K. Subrahmanyam put the matter this way:

> One [objection] is a widespread suspicion that the treaty is a means of keeping the nuclear-weapons club limited. It is believed this position has the support of the two superpowers particularly, with the other nuclear-weapon states in general sympathy. Success in this matter would mean placing India in a perpetual second-class status [21].

It should be noted that the basic inequities seen in the NPT by Indians are: (1) the fact the superpowers and the other three nuclear weapons states may increase their nuclear stockpiles while others are denied any such weapons (Article II); (2) the requirement that non–nuclear weapons states signatories shall be inspected to prevent clandestine diversion of fissionable materials to weapons production while the five nuclear weapons states are not subject to any type of inspection (Article III); and (3) the prohibition against the non–nuclear weapons states developing their own PNEs while the five nuclear weapons states may develop and use such devices (Article II).

To a degree, what the Indians are saying about the NPT and its inequity is somewhat analogous to what American blacks said about "Jim Crow" laws. A basic problem in both contexts is how a reasonable amount of equality can be obtained for both Indians and blacks when they are inferior to, respectively, the superpowers and whites in terms of wealth and political power. History suggests that those with superior power do not readily grant equal rights to the less powerful, and the current situation in regard to the nuclear weapons superpowers is no exception.

There may be another analogue to the developing situation in India in the history of the black movement in the US. This would be the evolution from what Hamilton and Carmichael have termed the "politics of deference" to adoption of "Black Power" [22]. This shift in black strategy in the US, from showing deference to whites in the hopes of obtaining some degree of better treatment to employing economic and political power in bargaining for concessions regarding civil rights from the white majority, is being mirrored today in India.* In India there is considerable movement away from the practice of making appeals to more powerful states on the grounds of morality and ethics, and toward a cautious adoption of a strategy of moderate realpolitik. A major philosophical obstacle to this development is the legacy of Gandhi found in a

---

*The authors of *Black Power* reserve the use of physical force by blacks against whites only for those situations wherein black lives and property are in danger of an immediate sort from White violence (see pp. 52–53).

strict constructionist interpretation of the Mahatma's teachings. Those who support the development of a moderate realpolitik approach to international politics are making a flexible interpretation of Gandhi's writings. If successful in this effort, a philosophy will be available which will permit India to be both Gandhian and a modest practioner of realpolitik. In contrast to the strict constructionist interpretation which emphasizes only the nonviolent facets of Gandhi's writings, the flexible constructionists emphasize that portion of Gandhi's works which make the point that nonviolence is the preferred method of resisting evil, but that the main point is that evil must be resisted [23]. The following quotation from Gandhi's *My Religion* is being used by flexible constructionists to bolster their case that under some conditions the use of violence in a just cause is in accord with Gandhian philosophy:

> I am a member of an institution which holds a few acres of land whose crops are in imminent peril from monkeys. I believe in the sacredness of all life and hence I regard it a breach of Ahimsa (the principle that injury should not be done to any living thing) to inflict injury on the monkeys in order to save the crops. I would like to avoid this evil. I can avoid it by leaving or breaking up the institution. I do not do so because I do not expect to be able to find a society where there will be no agriculture and therefore no destruction of some life. In fear and trembling, in humility and penance, I therefore participate in the injury inflicted on the monkeys, hoping someday to find a way out [24].

It is the author's impression that the flexible constructionist school of thought is growing in India, encouraged by a series of external international events which suggest that for the weak and meek to plead ethics to the strong is often a disappointing business.*

It should be noted as an aside at this point, that a useful study would be one in which the interpretations of Gandhian thought be closely examined in the context of the relationship between such thought and possible modifications in the international behavior of India.**

Some Speculative Conclusions Regarding the Crucial Role of India Relative to the NPT and other Indian Ocean Nuclear Aspirants. The preceding analysis of

---

*Some care must be taken in projecting the growth in the flexible constructionist school of Gandhian thought due to the recent rise in public interest in India in the nonviolent teachings of Gandhi follower Jayaprakash Narayan. This individual is leading nonviolent demonstrations in opposition to certain practices charged against the government of India.

**In fairness to Indians who argue for a modest amount of realpolitik in Indian foreign policy, in conjunction with the claim that they practice Gandhian thought, it should be pointed out that they say nuclear armaments would be employed by New Delhi to bargain for nuclear disarmament on the part of the current nuclear weapons states—not for Indian self-aggrandizement. In fairness to skeptics it should be noted that nations before have sought power claiming to be just in its use.

the Indian perspective toward the NPT indicates that New Delhi generally rejects the logic of the NPT advanced by the superpower sponsors, while extrapolating from much of the logic previously imputed to the NPT by France and the People's Republic of China. New Delhi's perspective is not likely to alter, assuming continued viability of the nation, unless the NPT is altered and the superpowers alter some of the behavior which serves to intensify Indian suspicions of Washington and Moscow. All of this is important because it is suggested that during the next ten to 15 years India's perspective of the NPT logic will probably be crucial in terms of the decisions some of the Indian Ocean nuclear aspirants take regarding the NPT. There are several reasons for this speculative conclusion.

Should India continue to reject the logic of the NPT as currently constituted, and move incrementally beyond its current possession of a "political nuclear weapon," Pakistan cannot realistically be expected to accept the NPT. Then there is the matter of Iran. In another decade Iran may have the technical infrastructure to commence the development of nuclear weapons. Iran could use the presence of an India not a signatory to the NPT and gradually moving toward a "military nuclear weapon" as justification for abrogation of the NPT, which Teheran accepted in 1970.* Further, Iran could view development of nuclear weapons by India as a threat to a fellow Moslem state, Pakistan, which could require NPT abrogation on the grounds that Indian action constituted an "extraordinary event" which "jeopardized the supreme interests" of Iran—the basis stated in Article X for a withdrawal from the NPT upon 90 days notice. Of course the shah could merely use the presence of a non-NPT-supporting India, which was seeming to move beyond a "political nuclear weapons," as an excuse to justify what could be a quest for nuclear weapons independent of Indian actions.

On the other hand, should India accept either a current or modified NPT, and also give the appearance of not purposely moving toward a "military nuclear weapon" capability, considerable pressure would be placed upon Pakistan to follow suit. Further, the context would be established which would be conducive to Iran continuing adherence to the treaty, unless relations between Teheran and the Arab states, the USSR or the US dictated other options.

Because the Pakistani government does not encourage the free discussion of political matters there is much less material which can be analyzed on the proliferation question from that country than from India. Available information indicates that the NPT per se is not as important to Prime Minister Bhutto as the presence of several threats to which development of nuclear weapons might constitute a Pakistani response. At this time the most serious threat perceived by Pakistan is that from India. While this general fear has concerned the Pakistanis

*It seems clear that the confluence of Indian nuclear explosive experimentation and peaceful missile experimentation will be the technical capability to produce at least IRBMs with fission warheads sometime in the late 1970s or early 1980s.

for some years, the Indian detonation of May 1974 served to heighten the concern. This worry is illustrated by recalling the newspaper headlines in Pakistan after the Indian nuclear test which asserted (rather than raising the question): "A Nuclear Bomb Directed at Us." It should be noted that even if India would accept the NPT, Pakistan would still be faced with superior Indian conventional forces, to which for some in Pakistan nuclear weapons might be a suitable counter.

There are other considerations more important to the decisionmaking processes of Australia, Indonesia and the Republic of South Africa than whether India ratifies the NPT. However, Indian acceptance of the treaty would seem to provide some minimum support for NPT acceptance by these other nuclear aspirants.

In general the Republic of South Africa perceives the same kind of disutilitarian logic in the NPT as does India. However the expression of displeasure by South Africans has to date been muted relative to statements made by Indians. Perhaps the most irksome aspect of the current NPT from Pretoria's perspective concerns the commercial disadvantages for non-nuclear weapons signatories of the treaty. Delegates from South Africa engaging in debate before the United Nations were rather frank on this point when they stated that the inspection arrangements would encourage commercial espionage to the extent that "hard earned technological advances in such innocent activities as refining processes, uranium oxide extraction processes which have given some small commercial and economic advantages will be lost to the non-nuclear weapon states" [25]. The South African sensitivity on the matter of commercial espionage may be explained by the fact that Pretoria claims to have developed a revolutionary method of enriching U-235 at much less cost than contemporary enrichment methods [26].

Pretoria has also voiced displeasure over the discriminatory aspects of the NPT in regard to the fact that some may have nuclear weapons while other states may not. Further, the South African government seems to hold little faith in the guarantees of the UN resolution to assist those faced with nuclear blackmail. This skepticism stems in part from the poor relations South Africa has experienced in the past with the United Nations because of the racial policies of Pretoria.

Despite official rejection of the NPT for the reasons noted, one may suspect there is a more fundamental reason for South Africa refusing to be bound by the restrictions of the treaty. That reason is the unwillingness to signal the giving up of the trump card of nuclear weapons until such time as the republic's internal and external racial problems are closer to a solution acceptable to the white minority and its government.

One observer of South African nuclear decisionmaking has recently written that while the republic will not likely accept the treaty, neither will Pretoria develop nuclear weapons in the near future. According to the analysis of John Spence, South Africa faces no immediate threat requiring her to exercise the nuclear option, while it does face the possibility that nuclear weapons development would conjure up considerable worldwide concern over its racial intentions, and would damage relations with Western Europe and the US—a kind of damage Pretoria wishes to avoid [27].

The Three Indian Ocean Nuclear Aspirants Which Accepted the NPT

During the 1960s there was considerable public and government discussion in Australia concerning the advantages and disadvantages of developing nuclear weapons. One aspect of the matter which was not much at issue was whether Australia had the technological capability to produce such weapons. Given the deposits of uranium in the land, and the technological infrastructure, one must agree with the general Australian assessment that with considerable expense the Australians could indeed develop at least atomic weapons. Why Canberra has accepted the NPT provisions despite this opinion results from a combination of factors which indicate Australia perceives its current status as considerably different from, say, India, thus requiring different policies on the NPT.

In the first place Australia is a member of the ANZUS Treaty Organization, and as such is directly tied to the United States. Beyond the formal ties, Australia has enjoyed generally warm relations with the US since the days of World War II cooperation. Further, the US maintains military facilities in Australia, of which the most important is probably the communication base on the Northwest Cape, used to communicate with US submarines capable of operating in the Indian Ocean. These ties with America suggest that Australia does not stand alone in the world dependent entirely upon its own resources.

Since the end of World War II to the beginnings of the 1970s portions of the Australian population have been fearful of external threats, often associated with a generalized concern over Asian communism. For example, there was concern over the Malayan emergency, the rise of the PRC, the Korean War and the Vietnam War. There were lesser dangers perceived in association with lesser insurgencies in the islands to the north of Australia, and periodic worries over Indonesian nationalistic expansion. However, within the past few years these fears have subsided, although specific events could no doubt bring them to the fore again. Thus, despite some doubts about American credibility arising from the US withdrawal from Southeast Asia, Australia generally perceives little in the way of external threats to which nuclear weapons acquisition would be an appropriate response. Further, Australia does not see serious commercial disadvantages in the NPT although Canberra has set on the record some mild reservations concerning the commercial aspects of the NPT.

At this time Australia seemingly has few if any dreams of becoming a major nation residing at the level just below that of the superpowers. In other words, most in Australia seem content to live in a very modest nation, unadorned with the most prestigious trappings of great states—nuclear weapons.

Of course all of this could change with the complete withdrawal of the US from the Indian Ocean, Asia and parts of the Southwest Pacific, coupled with a rise in the perception of an external threat. However, as noted in the Australian debate on the NPT, should those kinds of events occur Canberra can always give three months notice and remove itself from the constraints of the treaty while its scientists convert their civilian nuclear knowledge into military nuclear applications.

Nine years ago a research study of proliferation possibilities in Asia contained the following statement in regard to Indonesia:

> However, since Indonesia does possess several aircraft capable of carrying atomic bombs for short distances, it is the inability to produce quantities of fissionable material that constitutes the primary physical barrier to an Indonesian nuclear force [28].

Much the same statement can be made today, because little has changed in Indonesia in regard to facilities such as reactors to produce Pu–239 or facilities to enrich U–235. In time Indonesia may, like many other nations, make use of centrifuge or laser technology to enrich U–235 to weapons-grade levels. However, such efforts are not apparent at this time.

The paucity of technical infrastructure, including fissionable material, plus the requirements for other types of investment in the Indonesian economy, help to explain Indonesia's acceptance of the NPT. Another explanation is the current lack of external threats for which a nuclear weapons response would appear reasonable. Thus it seems that at least for the 1970s Indonesia is not likely to abrogate the NPT.

It is difficult to evaluate the position taken by Iran regarding the NPT, other than to note that Iran has ratified the treaty. The difficulty arises from two circumstances. One is that since Iran is an authoritarian nation, there is no public discussion of public policy except for announcements of decisions taken by the shah. Thus there is practically no material to examine concerning the reasoning behind Iran's acceptance of the NPT. One is therefore left with either the hypothesis that Iran accepted the NPT because the shah agreed with the logic of the treaty as set forth by the superpowers sponsors; or that he accepted it for other reasons best known to himself, but probably linked with his view of Iranian self-interest. The other difficulty in evaluating Iran's perception of the NPT is that one familiar with the shah's statements concerning his intention to build Iran into "an England or a France" of the region, and with his ambitious military and industrialization programs,* may be suspicious whether a man with

---

*Early in 1975 it was announced that Iran hoped to add eight US nuclear reactors to the two ordered from France and Germany to be built in the 1970s. During the past two years the shah is estimated to have purchased between $4 and $6 billion in arms from the US.

these dreams can stop short of seeking the capstone achievement of great nation status. In particular, students of international power politics and scholars who have studied the US efforts to contain the Soviet Union by various means may have trouble viewing Iran's acceptance of the NPT at face value. The rapid drive for national power and prestige on the part of the shah, and the interest of the US in supporting Iran as a major power block to Soviet pressures, can be interpreted to mean that Iran's association with the NPT is only a temporary matter. On the other hand, it may be possible to satisfy the US requirement for an increasingly strong anti-Soviet force in the Persian Gulf region with a conventionally armed state, and to satisfy the shah's aspirations for Iran short of nuclear weapons acquisition.

Thus one is led to the conclusion regarding Iran and the NPT that a difficult task for the superpowers in the future will be keeping Iran from abrogating the treaty. Initially at least it appears that this task might be undertaken both by modifying the NPT to reduce inequities between nuclear weapon and non-nuclear weapon states and, more importantly perhaps, by the two superpowers taking actions to reduce the threats to Iran which could result in Iran seeking to produce nuclear weapons as a response.

In any event it should take approximately a decade for the construction of the currently projected reactors which must be in operation to produce indigenous Pu–239 for an Iranian fission bomb program. This means that time remains for the superpowers to remove incentives for proliferation to the extent that such action is within their power. Of course it may be that nothing will do but for the shah to lead a second Persia into the world's spotlight armed with nuclear weapons.

HOW MIGHT AN EXTRAPOLATION OF CURRENT SUPERPOWER ACTIVITIES IN THE INDIAN OCEAN INFLUENCE THE POSITION TAKEN REGARDING THE NPT BY INDIAN OCEAN ASPIRANTS?

It appears that the US, and to a lesser extent the Soviet Union, have undertaken several types of action which if continued will probably reinforce the reasons given by at least one Indian Ocean nuclear aspirant for not accepting the NPT, and may undercut adherence to the NPT by at least one other nuclear aspirant.

American activities in the Indian Ocean, presumably taken for strategic reasons, probably exercise the greatest influence upon India. If continued they will likely increase the pressure upon India not to accept the NPT and possibly to move closer to "military nuclear weapons." There are four basic types of US activity which seem to carry with them the probability of intensifying Indian hostility toward the NPT. These are:

1. continued and possibly increased presence of US naval forces in the Indian Ocean, coupled with the expansion of fixed facilities to service both naval

forces and possibly Strategic Air Command bombers and Military Airlift Command cargo jets. Examples of this type of activity would be the periodic visits to the Indian Ocean by a US aircraft carrier and escorting ships, and plans by the Department of Defense to enhance basing facilities on the island of Diego Garcia in the Indian Ocean some 800 miles south of the tip of India [29]. More recently the US is reported to have asked for use of the Royal Air Force field on the island of Masirah, in the Arabian Sea off the coast of Oman.
2. the lifting of the US arms embargo to Pakistan announced late in February of 1975. It should be noted that this support for Pakistan may be partially offset should the US sell military equipment to India as some have speculated will be the case.
3. military aid and sales to Iran of such a character and in such an amount that the shah could extend Iranian forces out from the Persian Gulf, into the Arabian Sea and on into the Indian Ocean. Should some US military equipment to Iran be used to assist Pakistan in support of the shah's statement that under no circumstances would he permit Pakistan to be further fragmented, the pro-proliferation influence upon India would be heightened [30].
4. overt and covert US military presence in Southeast Asia.

There is a complicated, and difficult to assess, factor in evaluating the consequences of US activity in the Indian Ocean in the context of possible stimulation of proliferation tendencies on the part of India. That is the fact that US military presence in the Indian Ocean probably serves to retard proliferation in the Republic of South Africa, Iran, Pakistan and Australia because these nations feel less threatened with US forces in the area than if they were withdrawn, at least for the immediate future. Thus the US is in the difficult position of contributing to proliferation pressures whether American forces stay in the Indian Ocean region or are withdrawn.

Soviet naval presence in the Indian Ocean, which to date has not equaled in strength that of American forces, would seem to have the greatest proliferatory influence upon Iran. If continued and increased Soviet presence could cause the shah to abrogate the NPT as a self-defense measure, particularly if the US military presence is reduced in the region. Soviet naval presence in the Indian Ocean would also seem to contribute to proliferation pressures in Pakistan and the Republic of South Africa, and possibly in Australia.* Again the point should be made that such a result would be intensified should US forces be withdrawn.

The military activity of the two superpowers in the Indian Ocean is no doubt considered by the respective military establishments of each nation as being justified in terms of strategic requirements. However, the case can be made that

---

*To date the Indian government has not expressed concern over the presence of Soviet naval forces in the Indian Ocean. One explanation given by Indians is that the Soviet presence is merely a response to the American presence.

such activities should be evaluated in terms of the costs which may be associated with them in the form of continued and increasing pressures upon certain Indian Ocean nations to develop nuclear weapons. Perhaps such costs are more than offset for the superpowers by the advantages each feels it can obtain by continuing and increasing its military presence in the Indian Ocean, but perhaps not. In any event the matter should be subjected to intensive analysis to determine the various trade-offs involved.

HOW MIGHT POSSIBLE FUTURE ACTIVITIES OF THE SUPERPOWERS IN THE INDIAN OCEAN INFLUENCE THE POSITION OF THE NPT ADHERENTS AND NONADHERENTS?

There are a number of possible future activities, generally associated with the US, which could occur and which would seemingly reinforce the anti-NPT perspective of India. In several instances such US action might operate to force NPT abrogation on the part of Iran. These possible actions are:

1. although increasingly unlikely, reintroduction of large US military forces in Southeast Asia. The Indians would likely view such activity as further evidence of US imperialism which would require steady movement toward "military nuclear weapons" as a hedge against further imperialism, and possibly as a bargaining gambit.
2. the initiation of US military sales or assistance to the People's Republic of China. This kind of activity could be taken in the context of building China as a counter to the Soviet Union, or in the interests of supporting an anti-Soviet faction in the post–Mao death period. Such activity on the part of the US would substantially increase the pressure upon and within the Indian government to move toward "military nuclear weapons."
3. US military moves to increase the leverage Washington could bring to focus on the Middle East oil-producing nations. In this event New Delhi would see US imperialism on the rise.
4. the actual use of force by the US against one or several of the oil producing states, or in support of Israel. Such action would have substantial consequences in terms of generating strong pressures in India to offset what would be viewed as imperialism by obtaining a nuclear weapons capability.

Depending upon the exact circumstances, the kinds of activities mentioned under (3) and (4) above could also trigger abrogation of the NPT by Iran and the beginning of a military nuclear program by the shah—assuming the technical ability to mount such a program had been acquired.

CONCLUSIONS

The superpowers are partially responsible for the fact that three Indian Ocean nations have not accepted the NPT, and for the possibility that Iran may in time abrogate the treaty.* Part of the responsibility borne by Washington and Moscow is found in the substance of the NPT and in actions which appear to run counter to their pledge that the superpowers would undertake disarmament of nuclear weapons. Part of the responsibility is also found in the form of actions taken in the Indian Ocean which can be interpreted by at least India and Iran to mean that the US and the USSR respectively are untrustworthy and imperialistic.

If the two superpowers wish to decrease their contribution to nuclear proliferation in the Indian Ocean region they should consider two types of remedial action. First, they need to cooperate in devising ways to enhance the attractiveness of the NPT. This would involve the need to establish greater equality in treatment between non–nuclear weapons and nuclear weapons states, and the need for the superpowers to engage in nuclear disarmament. Second, they need to cooperate in undertaking to lower their military presence in the Indian Ocean.

The necessity for mutual action by the two superpowers needs to be underscored. This is because it will be insufficient for only one to withdraw from the Indian Ocean, for in that situation the remaining superpower will be much more menacing to some Indian Ocean nuclear aspirants. Similarly, it will be insufficient for only one superpower to engage in nuclear disarmament because in that situation the armed posture of the other superpower will appear more threatening than before.

Should the superpowers decide against substantially altering the terms of the NPT; if they refrain from making substantial nuclear disarmament moves; and should they decide their interests preclude a lowering of their military presence in the Indian Ocean; they must seriously consider the possibility of more "political" and "military" nuclear weapons in the possession of Indian Ocean nations. Since the chances seem relatively low for mutual American–Soviet efforts designed to remove the incentives for proliferation, both nations should undertake the necessary contingency planning to protect their interests given the possibility of additional proliferation in the Indian Ocean region.

Fortunately there is time for either American–Soviet cooperation aimed at removing proliferation incentives, or, failing that, for planning to live with proliferation in the Indian Ocean area. This is because all the Indian Ocean nuclear aspirants will require considerable time to develop nuclear weapons capabilities. As noted previously, the Indians should not have an IRBM capability until late in the 1970s or early 1980s, and an ICBM would follow that achievement. Iran will require eight to ten years before indigenously produced Pu–239 will be

---

*Of course there are many other factors which account for the refusal of some nations to accept the NPT, and for the possibility that others may abrogate the treaty.

available and machined for nuclear weapons. While Pakistan could illegally divert Pu–239 from its currently operating commercial reactor, there is no chemical separations plant in operation, and it would take three or so years to build one. A Pakistani-built reactor would seem eight to ten years in the future. Because of the need to build plutonium-producing reactors and chemical separations plants, Australia, Indonesia and the Republic of South Africa all seem to be a decade away from possession of the technical infrastructure and fissionable materials necessary for nuclear weapons fabrication. Of course some technology which would enable U–235 to be enriched to weapons grade levels may develop rapidly among some of the Indian Ocean nations. This would substantially alter the time frame suggested above as being required for nuclear weapons to be indigenously produced in the Indian Ocean area. However there is little evidence now that this will be the case.

Thus the superpowers should, with deliberate speed, but no sense of panic, sort out their priorities in the Indian Ocean as between some proliferation and other goals.

NOTES TO CHAPTER THREE

1. For an examination of the debate on the NPT see, Robert M. Lawrence and Joel Larus, "A Historical Review of Nuclear Weapons Proliferation and the Development of the NPT," in Robert M. Lawrence and Joel Larus eds., *Nuclear Proliferation: Phase II* (Lawrence: University Press of Kansas, 1974), pp. 1–29.

2. The complete text of Security Council Resolution 255 (1968) is found in the *Yearbook of the United Nations 1968*, p. 28.

3. UN Document A/4680.

4. Albert Wohlstetter, "Nuclear Sharing: NATO and the N + 1 Country," *Foreign Affairs,* April 1961, pp. 355–87.

5. Pierre Gallois, *The Balance of Terror: Strategy for the Nuclear Age* (Boston: Houghton Mifflin Co., 1961), p. 113.

6. See for examples of this perspective Edward A. Kolodziej, "Revolt and Revisionism in the Gaullist Global Vision: An Analysis of French Strategic Policy," *Journal of Politics,* May 1971, pp. 448–477.

7. Quoted in William Griffith, *The Sino–Soviet Rift* (Cambridge: The MIT Press, 1964), p. 3.

8. Chen Yi, "China is Determined to Make All Necessary Sacrifices for the Defeat of U.S. Imperialism," *Peking Review,* October 8, 1965, p. 8.

9. "Nuclear Hoax Cannot Save U.S. Imperialism and Soviet Revisionism," *Peking Review,* September 8, 1967, p. 34.

10. Chen Chu in the *New China News Agency,* December 10, 1971.

11. "Chinese Comment on Strategic Policy and Arms Limitation," Permanent Subcommittee on Investigations, *Committee on Government Operations,* Committee Report, United States Senate, (Washington, D.C.: Government Printing Office, 1974), pp. 5–6.

12. "Chiao Kuan-hua Explains Chinese Government's Principled Stand," *Peking Review*, December 3, 1971, pp. 14–16.

13. K. Subrahmanyam, "India: Keeping The Option Open," in Lawrence and Larus eds., *Nuclear Proliferation: Phase II* (Lawrence: University Press of Kansas, 1974), pp. 140–142.

14. Ravi Kaul, *India's Strategic Spectrum* (New Delhi: Chanakya Publishing House, 1969), p. 200.

15. Raj Krishna, "India and the Bomb," *India Quarterly*, April–June 1965, p. 135.

16. Indira Gandhi, "India and the World," *Foreign Affairs*, October 1972, p. 75.

17. V. C. Trivedi, "Vertical Versus Horizontal Proliferation: An Indian View," in James E. Dougherty and J. F. Lehman, Jr. eds., *Arms Control for the Late Sixties*, (Princeton: D. Van Nostrand, 1967).

18. C. R. Gharekhan, "Strategic Arms Limitation–II," *India Quarterly*, October–December 1970, p. 389.

19. See *India's Position on the Proposed Draft Treaty on Non–Proliferation of Nuclear Weapons* (New Delhi: India Information Service, 1968); and *India News*, May 24, 1968.

20. H. N. Sethna, "PNE Technology Vitally Important for India," *Nuclear India*, October 1974, p. 6.

21. Subrahmanyam, p. 140.

22. Stokely Carmichael and Charles V. Hamilton, *Black Power* (New York: Vintage Books, 1967).

23. Sampooran Singh, *India and the Nuclear Bomb* (New Delhi: S. Chand and Co., 1971).

24. Quoted in Ibid., pp. 104–105.

25. UN Document, General Assembly, A/C.1/PV.1571, pp. 59–60.

26. South Africa, Parliament, House of Assembly, *Debates*, vol. 25, cols. 57/8, July 20, 1970.

27. John Spence, "The Republic of South Africa: Proliferation and the Politics of Outward Movement," in Lawrence and Larus eds., *Nuclear Proliferation: Phase II* (Lawrence: University Press of Kansas, 1974), pp. 232–233.

28. William Van Cleave, Harold Rood, and Judith Pettenger, *Nth Country Threat Analysis: The Asiatic-Pacific Area* (Menlo Park, Cal.: Stanford Research Institute, 1966), p. 135.

29. For a recent statement on US military presence in the Indian Ocean see James R. Schlesinger, *Annual Defense Department Report, FY 1976 and FY 1977* (Washington, D.C.: Government Printing Office, 1975), p. III 23.

Discussion Essay: Proliferation and the International Strategic System

Decisions made by leaders of near-nuclear countries to "go nuclear" hinge more upon their assessments of the international strategic system than on any formal requests which might be made by the present nuclear countries to forego nuclear weapons development. The conclusion of the preceding chapters is that the strategic system is moving in a direction which makes it highly likely that several additional countries will develop nuclear weapons in the not too distant future. Why this is and what proliferation means in terms of international security relate to several factors—among them are the superpowers' continuing nuclear arms race, regional conflicts and shifts in the geopolitical structure of the international system.

THE RACE BETWEEN THE SUPERPOWERS

The 1970 Nonproliferation Treaty stipulated that the existing nuclear powers were to progress toward disarming their nuclear military establishments concurrently with contributing to developing peaceful nuclear energy capacities in the non-nuclear states which adhered to the treaty. Although the treaty's terms are very general, the clear intent of its authors was that an obligation had been assumed by the Soviet Union and the United States to decrease the weapons gap between themselves and the non-nuclear states through mutual force reductions.

Because of the treaty's assumption of reciprocity, a critical question prior to the treaty review conference in May 1975 was one of military trends, specifically the state of the superpower arms race and of nuclear military technological developments. Chapter Two (Leader and Schneider) concludes that vertical proliferation between the United States and the Soviet Union has continued at a high rate throughout the years between the drafting of the initial treaty and the review conference.

The present position of the superpowers regarding strategic arms may be even more distant from limitations that the figures in Chapter Two suggest, although those figures are much higher than those usually quoted in official circles (Leitenberg). Inadequate assessment methods are responsible for continual underrepresentation of the actual numbers of nuclear weapons and missiles. For example, the United States estimates the numbers of nuclear missiles in the Soviet Union on the basis of the number of missile launchers, although the ratio of missiles to launchers varies. Furthermore, tactical nuclear weapons are not included in the estimates of nuclear missiles and warheads. The actual number of nuclear weapons in the two countries may be reasonably estimated to be about 30 percent higher than official figures. Lastly, looking toward the future, the "thresshholds" to which the superpowers agreed in SALT II were nothing more than acknowledgments of the respective weapons planning processes in the United States and the Soviet Union, not actual curbs on present weapons development.

If they are to portray the strategic system adequately, the weapons figures should also be supplemented by including civil defense arrangements which contribute to proliferation (Taylor); deployment trends (Subrahmanyam); and the persistence of the cold war doctrine with its implications for nuclear weapons use (Subrahmanyam). In sum, the evidence of the failure of the superpower signatories to make significant progress toward narrowing the weapons gap is substantial.

While the superpowers have confronted a nuclear stalemate in the past, the future of the balance is somewhat uncertain, at least in the calculations of policymakers in the two countries. This uncertainty provides a poor basis for either power to limit investment in nuclear weapons production, research and development. On the other hand, as long as the stalemate continues, the potential for "mutual assured destruction" means that the cost to a nuclear state of guaranteeing protection to a non-nuclear ally continues to be high (the leaky nuclear umbrella). The non-nuclear states are encouraged to exercise nuclear options on both these counts.

One approach to the current stand-off is for the superpowers to recognize the imbalance of military capabilities between themselves and the second order states and to deal with it at least indirectly by such means as:

1. establishing nuclear-free zones and guaranteeing them by surveillance
2. promulgating nonuse pacts, guaranteeing that nuclear weapons will not be used against non-nuclear countries;
3. reducing stocks of tactical nuclear weapons, which seem to be most incriminated in stimulating proliferation;
4. controlling the distribution of nuclear materials and technology (Schoettle).

The first three of these proposals attempt to provide alternatives to proliferation by removing some of the incentives to proliferate. Yet, to some discussants, the major problem is not met by these preventive measures. Despite the high potential for change in the international strategic system, the superpowers' national security policies and strategic doctrines do not appear to have changed (Leitenberg). The superpowers' assumptions about the international strategic system are that it continues to be bipolar and that the major events are those which involved superpower confrontations.

REGIONAL ISSUES

The manner in which the superpowers view the international strategic system is very different from the views held by the various non-nuclear or "lesser" nuclear states. The different viewpoints suggested to several discussants that the basic shortcoming of the NPT was not that the superpowers had failed to disarm but that regional strategic issues fuel the horizontal nuclear arms race (Kemp). The superpowers' nuclear umbrellas, on the other hand, are only responsive to threats to the global system.

A strong conventional military force might be sufficient for Iran to "maintain security" in the Persian Gulf region without the development of a nuclear capacity (Chubin). However, India has been engaged in armed conflict with one nuclear weapons state (China) and is confronted with nuclear weapons deployed in the Indian Ocean by the United States (an uncertain dimension) and potential nuclear weapons in Pakistan in the future (Subrahmanyam).

Whether or not Iran would continue to rely upon conventional weapons if proliferation were to continue in the Indian Ocean region is uncertain. What occurs in the Indian Ocean region in the near future from the regional perspective of Iranian policymakers will depend upon developments in the Indian nuclear establishment and its relations with Iraq just as much as it will upon Iran's reaction to the superpowers' arms race.

Whichever approach is adopted, the importance of regional issues to second order states and the concurrent regional arms races are supported by several developments. A major factor has been the change in the structure of the international economic system (Kemp). Shifts in control of nation states over economic resources both encourage higher levels of arms and nuclear reactor sales (Subrahmanyam) and greater internal political insecurity—garrison state reactions—in several of the near-nuclear countries (Kemp). In this environment, policies to limit the transfer of nuclear technology and materials may have been superseded already by the nuclear powers' willingness to sell both nuclear technology (as in the United States' Atoms for Peace program) and aircraft which can deliver nuclear weapons (Leitenberg).

Observers disagree about the incentives which have led to burgeoning conventional arms sales. Non-nuclear states may bargain for them by threatening to go nuclear—non-nuclear blackmail (Cahn). The military needs of individual second order states are highly variable. This line of reasoning may be appropriate for some recipient states, but not others.

An elementary classification of nuclear powers would distinguish among (1) those states with a second-strike capability both against lesser powers and against the two superpowers; (2) those with a second-strike capability against lesser powers; (3) those with the capacity to wreak unacceptable destruction on a first-strike basis; and (4) non-national actors (Kemp). A region in which conflict potential existed among three states, each with the same level of second-strike nuclear capability, would present a different tactical situation from that in which one maintained a first-strike capability and the other two strong conventional forces (two probable futures for India, Pakistan and Iran).

Nonetheless, whatever the incentive, the contemporary scene is one in which the weapons gap may have, in fact, been narrowed from the "bottom up." The maritime world has lost control of the oceans, the superpowers face a nuclear stalemate and new economic centers of power have developed (Kemp). In the changing environment described in Chapter One (Cohen), treaties to stabilize the status quo are not likely to be popular with the new "second order" states (Subrahmanyam). In this setting, such a treaty is much more likely to be met with the argument that deterrence is more reliable than abstinence. Realistic counterforce threats, not treaties, explain restraint in the use of biological weapons (Schoettle).

Despite the attractiveness of indigenous deterrents against regional hostilities, some second order states may still be persuaded against them by superpower guarantees. For example, potential reactions to continued economic upheavals could occur in Japan and Germany which would bolster the position of nuclear weapons advocates (Kemp). But for these states, a responsive US posture could be critical in heading off proliferation—a US posture of continuing nuclear standoff vis-à-vis the Soviet Union.

This position is essentially a third view of the strategic system, that of the second order, near-nuclear state which has close ties to one of the superpowers. It also brings a dilemma into focus—the dilemma of a multifaceted strategic system in which areas of overlapping interests are not yet clear.

Part Two

Chapter Four

Risks of the Nuclear Fuel Cycle and the Developing Countries

Christoph Hohenemser

Abstract: It is shown that the spread of nuclear weapons to additional countries has been far slower than anticipated from capability estimates made 17 years ago. This can in part be accounted for by the relatively slow development of nuclear power; it must, however, also be attributed to the decision by many countries to forego independent national development of nuclear weapons. Today a possibly greater risk than that arising from independent national weapons programs grows from the proliferation of the mature nuclear fuel cycle. To this end, a preliminary review of risk assessment research on the nuclear fuel cycle is presented. The conclusion reached from this is that risk assessment is to a large extent incomplete, and in many cases is concentrated on conceivable consequences without discussion of event probabilities that lead to these consequences. For the developing countries, this means that caution should be exercised in adopting the mature nuclear fuel cycle for nuclear electric power production. As one interim solution, the adoption of a truncated fuel cycle similar to Canada's is suggested.

INTRODUCTION

In 1956, 14 years before the signing of the Nuclear Nonproliferation Treaty, at a time when the IAEA was just being established and the world found itself in the middle of the cold war, the problem of nuclear proliferation loomed large. Known in those days as the "Nth country problem," it formed, with nuclear fallout and increasing nuclear weapons stockpiles, one of the prime motivations for a nuclear test ban.

From a political point of view, many feared a world in which a fanatical ruler of some small, unimportant country would attempt nuclear blackmail in order to achieve his ends; or what is worse, trigger a nuclear war between the superpowers.

From a technical point of view, it was beginning to be recognized that with the spread of nuclear know-how to many countries, significant quantities of plutonium would become available outside the boundaries of the nuclear weapons countries. To produce nuclear weapons in this context, a country needed to do only the following: (1) divert sufficient spent fuel from a nuclear power station; (2) build a small chemical separation facility to extract the plutonium from the spent fuel; (3) learn how to employ chemical explosives so as to compress 5–10 kg of plutonium to about two-thirds its normal volume. Based on the available scientific, economic and industrial resources, Davidon, Kalkstein and Hohenemser [1] estimated in 1958 that 12 countries could mount a completely independent weapons program, from mining through bomb assembly, for a cost of about $130 million. A summary of these results appears in Table 4–1. The essential correctness of these results was accepted by the Atomic Energy Commission in 1960 [2].

Mitigating this situation was the fact that for a credible military threat, an "Nth country" would have to *deliver* as well as manufacture nuclear weapons. Against the then available air defenses of many countries, this was considered far more difficult and expensive than nuclear weapons production itself [3].

Today, nearly 20 years after the "Nth country problem" became part of our doomsday baggage, some significant positive changes in perception must be recorded.

1. The growth of civilian nuclear power has been far slower than originally anticipated, as can be seen in Table 4–2 [4].
2. Only three additional countries have acquired nuclear weapons since 1956, and the political effect of these new entries has not been notably destabilizing.
3. The crucial ingredients for independent acquisition of nuclear weapons—i.e., chemical separation plants or isotope separation plants—have been much slower to proliferate than reactors. Outside the six nuclear weapons countries, isotope separation plants of significant magnitude are found only in the Netherlands, West Germany and South Africa [5]. Similarly, fuel reprocessing plants of significance are restricted to Belgium, West Germany, Italy and Japan [6].
4. The Nuclear Nonproliferation Treaty has been in effect for five years, and at the very least serves as a restraint against, and perhaps as an early warning of, further nuclear weapons proliferation.

If one wishes to be optimistic, it is possible to argue that the "classic" nuclear proliferation problem, with nuclear weapons in the arsenal of many lesser powers, has not materialized. Even powers who clearly have the capability to develop weapons have chosen not to. Thus the difference between capability and intent, which was all too often minimized in the early discussions of nuclear proliferation, is very significant.

Table 4-1. Capital and Annual Costs of a Nominal Nuclear Weapons Program as Estimated in 1960

Stage in the Development	Capital Cost ($ million)		Annual Operating Cost ($ million)	
	Plants Emphasizing Power	Plants Emphasizing Weapons-grade Pu	Plants Emphasizing Power	Plants Emphasizing Weapons-grade Pu
Uranium (ore to metal)	0.5	1	0.5	4
Fuel element fabrication	0.5	1	0.5	1
Moderator preparation (heavy water)	3.0	3	1.0	1
Reactor	24.0	16	1.0	3
Chemical processing	6.0	18	0.2	2
Waste disposal	3.0	3	1.0	1
Bomb assembly	10.0	10	5.0	5
Total	47.0	52	9.0	17

Source: W. Davidon, M. Kalkstein and C. Hohenemser, *The Nth Country Problem and Arms Control* (Washington, D.C.: National Planning Association, 1960). p.

Table 4-2. Plutonium Production from Civilian Nuclear Power—Projection versus Reality

Country	Plutonium Production** (kg/year)		Percent of Projection Achieved
	1960 Estimate for 1975	1975 Actual	
Belgium	1500	~ 200	13
Canada	?	600	–
China*	?	?	–
Czechoslovakia	5000	~ 200	4
France*	8000	~ 600	7
Germany, E.	3000+	~ 100	3
Germany, W.	6000	1000	18
India*	?	200	–
Italy	500+	200	40
Japan	7000	1000	14
Netherlands	3000	~ 100	3
Norway	?	?	–
Poland	1800	?	–
Rumania	500	?	–
Spain	1800	400	22
Sweden	2000	500	25
Switzerland	?	200	–
Soviet Union	?	~1000	–
United Kingdom*	6000+	~2000	33
United States*	?	~5000	–

*Nuclear weapons countries. The US, USSR, France and UK have all produced additional plutonium in military reactors. China and India may have done so.

**To obtain maximum number of nominal weapons, divide by 5. (the critical mass of Pu is ~5).

Sources: W. Davidon, M. Kalbstein and C. Hohenemser, *The Nth Country Problem and Arms Control* (Washington, D.C.: National Planning Association, 1960); M. Willrich, ed., *International Safeguards and the Nuclear Industry* (Baltimore: Johns Hopkins University Press, 1973).

At the same time, with the coming of age of nuclear electric power, the problem of catastrophic failure within the mature nuclear fuel cycle has become an increasingly strongly felt concern. Conceivable reactor accidents involve fission product inventories that are 1,000 times greater than those of individual nuclear bombs and are thus capable of producing radiological catastrophes that far exceed the effect of a single nuclear weapon. This suggests that it is prudent to examine all nuclear fuel cycle risks, and in connection with developing countries, to concern ourselves with these as well as independent nuclear weapons production.

Table 4-3 illustrates currently projected nuclear power growth [7]. In the context of the traditional nuclear proliferation problem, this indicates a great deal of plutonium which could get into the wrong hands. It is time we looked at such a table also as a measure of possible risks associated with the fuel cycle, quite independent of weapons production. In contrast to past projections, which we saw in Table 4-2 were far too high, I would expect present projections to be more accurate, since nuclear power has acquired such a marked economic advantage over fossil fuel.

In this chapter I will, therefore, survey the risks of the mature nuclear fuel cycle and try in a brief space to indicate what is known and what is not known. It is worth stating at the outset that despite 33 years of reactor development, the study of catastrophic risks in the nuclear fuel cycle is still in its infancy. It may therefore be anticipated that the reader will be left with more questions than answers. After surveying the risks, I will turn to some limited comments on implications for developing countries.

THE NUCLEAR FUEL CYCLE

Reactor Types

There are a number of different reactors in commercial use. For the present discussion, in order to illustrate the mature fuel cycle, I will limit myself to the most common type, the light water moderated reactor (LWR). This is in wide use in the United States and also plays a major role in export trade. LWRs require slightly enriched uranium, in which the abundance of U–235 is about 4 percent. Alternatively, LWRs may be fueled by natural U mixed with an appropriate quantity of recycled Pu.

All US LWRs utilize a massive pressure vessel inside of which the fuel is stacked vertically and through which primary coolant is circulated. In one type of LWR, the so-called pressurized water reactor (PWR), the water in the pressure vessel is at $315°C$ and 150 atmospheres, and does not boil; turbine steam is made in a heat exchanger operating at somewhat lower pressure and temperature, as shown in Figure 4–1. In a second type of LWR, known as a boiling water reactor (BWR), the water in the pressure vessel is at $315°C$ and 67 atmospheres, and boils during normal reactor operation; turbine steam is made directly, without intervention of a separate heat exchanger, as shown in Figure 4–2. The principal design differences between the PWR and the BWR represent a trade-off between two types of safety consideration. In the PWR, there are two barriers between the primary coolant and the environment; in the BWR there is but one. This means that dissolved fission products in the primary coolant of the BWR are more likely to get out and produce low-level environmental contamination. In the BWR, on the other hand, a transient increase in reactivity produces a rapid increase in bubble formation which in turn serves as an important negative feedback mechanism preventing runaway; the PWR also has a negative void coefficient, but responds more slowly to transients because its primary coolant does not normally boil.

Stages of the Fuel Cycle

The mature LWR fuel cycle is illustrated in Figure 4–3. For convenience of discussion, the fuel cycle is divided into six (vertical) stages, which may be briefly described as follows.

Table 4–3. Installed Nuclear Capacity in 1974 and 1980 (showing a four-fold increase over this period)

Country	Total Nuclear Capacity 1974 (MWe)	Number of Power Reactors (over 20 MWe)	Total Nuclear Capacity 1980 (MWe)	Number of Power Reactors (over 20 MWe)
Argentina	320	1	920	2
Austria	—	—	700	1
Belgium	400	1	1,700	3
Brazil	—	—	600	1
Bulgaria	440	1	1,800	4
Canada	2,500	7	6,100	12
Czechoslovakia	110	1	1,800	5
Finland	—	—	1,500	3
France	2,900	10	15,000	23
FR Germany	4,200	10	22,000	28
German DR	430	2	800	3
Hungary	—	—	440	1
India	780	4	1,600	8
Italy	600	3	3,400	7
Japan	5,000	10	19,000	29
Korea, South	—	—	1,200	2
Mexico	—	—	1,300	2
Netherlands	530	2	530	2
Pakistan	120	1	120	1
Spain	1,070	3	8,600	11
Sweden	2,600	4	8,300	11
Switzerland	1,000	3	5,700	8
Taiwan	—	—	3,000	4
Thailand	—	—	500	1
USSR	3,500	16	10,000	24
UK	5,800	31	11,000	39
USA	40,500	60	138,000	156
Yugoslavia	—	—	1,400	2

Table 4-3. continued

Country	Total Nuclear Capacity 1974 (MWe)	Number of Power Reactors (over 20 MWe)	Total Nuclear Capacity 1980 (MWe)	Number of Power Reactors (over 20 MWe)
TOTALS	1974		1980	
Countries	19		28	
Reactors	170		393	
Capacity (MWe)	72,800		270,000	

Source: *Preventing Nuclear Weapon Proliferation* (Stockholm: International Peace Research Institute, 1975).

Figure 4-1. Pressurized Water Reactor Fuel Cycle

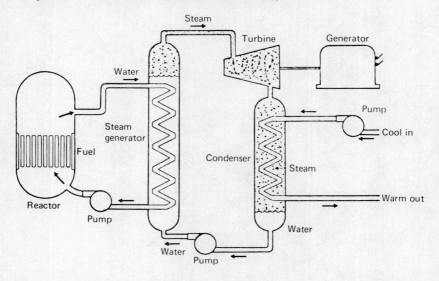

Source: Adapted from D.R. Inglis, *Nuclear Energy, Its Physics and Its Social Challenge* Reading, Mass.: Addison–Wesley, 1973), p.

Figure 4-2. Boiling Water Reactor Fuel Cycle

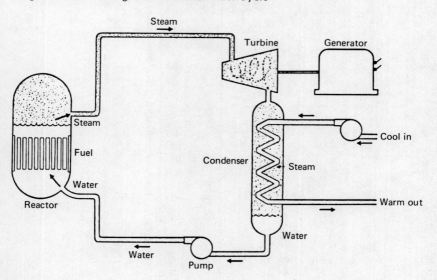

Source: Adapted from D.R. Inglis, *Nuclear Energy, Its Physics and Its Social Challenge* (Reading, Mass.: Addison–Wesley, 1973), p.

Risks of the Nuclear Fuel Cycle and the Developing Countries 93

Figure 4-3. Light Water Reactor Fuel Cycle *(Six Stages)*

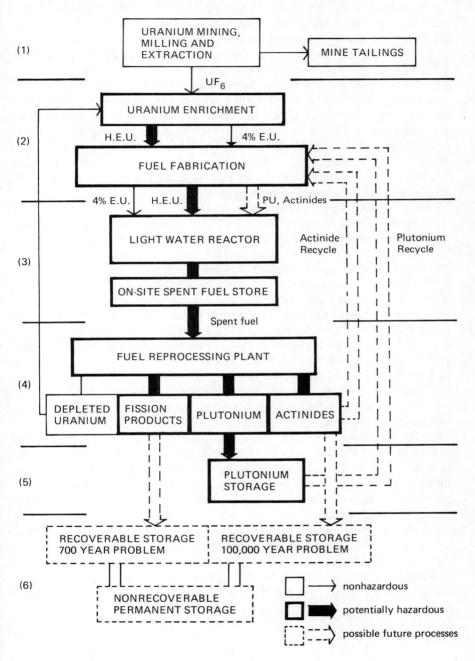

1. *Uranium mining, milling and extraction.* This is an activity which is functionally little different from the mining of any metal ore. It involves, however, the exposure of miners to radioactive dust; in addition, mine tailings constitute a well-defined radioactive hazard.

2. *Uranium enrichment and fuel fabrication.* This stage has the purpose of increasing the 0.7 percent natural abundance of U-235 to at least 44 percent. Significant enrichment plants have in the past involved large scale gaseous diffusion installations, though more recently smaller centrifuge plants have been developed [8]. They play a dual role since they may also be used for making weapons grade U (90+ percent U-235. Fuel fabrication is normally performed at some distance from the enrichment plant and requires that the UF_6 first be converted to U_3O_8 or U-metal. Fabrication involves machining and cladding of the fuel elements. With possible future plutonium recycling, the fuel fabrication process may have plutonium inputs as well.

3. *The LWR.* The average reactor is refueled about once a year. At this time, approximately one-third of the fuel inventory is replaced with fresh fuel and spent fuel is removed to an on-site storage area for cooling. Fresh fuel may have the form of 4 percent enriched U, highly enriched U mixed with natural U, or, in the future, Pu with a suitable admixture of natural U.

4. *Fuel reprocessing plant.* After some months of on-site storage, spent fuel is shipped in a cooled and heavily shielded container to the reprocessing plant. Here, after removal of the cladding, the spent fuel is dissolved in acid and Pu and U are extracted. The remnant liquid is highly radioactive, and contains fission products and long-lived actinides. The wastes are stored on the site of the processing plant in refrigerated double walled steel tanks. In some cases, after a period of storage, the liquid wastes are solidified.

5. *Plutonium storage.* Plutonium recovered in reprocessing of spent fuel is presently stored in solution in a specially designed storage area. Uranium extracted in reprocessing is recycled to the enrichment stage.

6. *Permanent waste storage.* Permanent storage of wastes will ultimately be the last stage of the fuel cycle. To date, such storage has not been undertaken because no method of storage has been agreed upon. Because of the vastly different half-lives of fission products and transuranic actinides, waste storage naturally divides into two problems: (a) storage of fission products for 700 years; (b) storage of actinides for one million years or more. As indicated by Rose, one may envision recycle for burnup in the LWR as an alternative to storage [9]. To date, actinides have not been separated from fission products at the fuel reprocessing plant.

Readers desiring additional, largely nonspecialized, information about the nuclear fuel cycle are referred to the excellent review of the subject by Inglis [10].

RISKS OF THE NUCLEAR FUEL CYCLE

Routine and Catastrophic Risk

A large number of hazards have been identified within the nuclear fuel cycle. These may be conveniently divided into two classes, which we denote by "routine" and "catastrophic," and which can be described as follows.

Routine risks. Generally characterized by well-defined hazards and consequences, these risks are akin to the risks of automobile driving or cigarette smoking: i.e., they can be characterized by a well-defined probability distribution. Examples are low-level normal radioactive emissions from nuclear installations or radioactive dust encountered in uranium mining.

Catastrophic risks. Characterized by poorly defined or undefined probabilities common to all rare events, these are akin to risks from meteors and earthquakes. Because quantitative statements about them have proven difficult to make, much of the literature is concerned with conceivable consequences alone, as distinct from the probability that such consequences will be experienced.

During most of the 1960s, the dominant public discussion of nuclear safety concerned routine risks. Despite a heated debate on the proper level of radiation standards, the routine risks of the nuclear fuel cycle can be shown to be small compared to many other hazards that we tolerate, including the burning of fossil fuels [11]. Most occur as occupational hazards, and the largest part of them are not related to radiation at all. It is therefore possible to dismiss them as insignificant for present purposes. They simply do not carry any weight in arguments dealing with the proliferation of nuclear power.

In 1970, an increasing concern with catastrophic risk surfaced in the public discussion. One root of the new concern has been the safety of the reactors themselves, particularly the effectiveness of the emergency core cooling systems (ECCS). This discussion was to a large extent initiated by the Union of Concerned Scientists [12] and resulted in extensive rule-making hearings held by the AEC [13]. The result of the hearings was a substantial upgrading of ECCS acceptance criteria. Another root of concern has been the increasingly wide availability of separated Pu in the maturing fuel cycles of advanced countries (see Table 4-2). This concern was stimulated from two independent directions. Taylor [14] worried about theft of weapons grade material and its subsequent use by terrorists. He pointed out that the manufacture of crude Pu bombs, even with strongly Pu-240 denatured material, was much easier than had previously been believed. Speth, Tamplin and Cochran [15] and others saw potential disaster in the toxicity of Pu, particularly through the hypothetical accumulation of hot particles in the lungs.

Stimulated by critics and pressed by the environmentalists in the courts, the Atomic Energy Commission has responded with a major study which considers reactor safety for the first time from the point of view of a systematic statistical treatment of accident initiation and consequences. Known variously as the "Rasmussen Report," Draft WASH 1400," or the AEC Reactor Safety Study [16], the study concludes that the risk of catastrophic accident is so small as to be of little practical concern. Rather than settling the issues previously raised, the AEC study has been subject to continuing debate, with major critiques issuing from the American Physical Society [17], the Environmental Protection Agency and the Union of Concerned Scientists [18].

Typology of Catastrophic Nuclear Risk

In order to determine to what extent the ongoing discussion of catastrophic nuclear risk is inclusive of all potentially significant problems, a typology of risks has been constructed in the form of a 3 × 6 matrix. Though no more than a bookkeeping procedure, it will be seen that such a typology has an immediate and obvious qualitative implication. In a subsequent section, the matrix will be used to illustrate the incompleteness of present research on catastrophic nuclear hazards.

We distinguish between three stages of occurrence in the course of a nuclear catastrophe. A symbolic illustration of the stages appears in Figure 4–4.

1. *Initiating events.* Initiating events are events at the beginning of accident sequences which potentially, but not necessarily, lead to a nuclear catastrophe. Specific kinds of initiating events are quite large in number. For a reactor, they are exemplified by "break in a primary coolant pipe" or "operator error" or "electric power failure." In another part of the fuel cycle, initiating events are exemplified by "theft of weapons grade material," or "sabotage of fission product storage." Most hypothetical accident sequences invoke one or more simultaneous initiating events in order to arrive at a hazardous situation.

2. *Hazards.* Hazards are physical consequences which in themselves involve a direct threat to human life or property. Unlike initiating events, hazards are relatively few in number. We have identified just three types that cover all cases: (a) a nuclear explosion; (b) massive release of plutonium and/or transuranic actinides; and (c) massive release of fission products.

3. *Consequences.* We define consequences as events which directly affect people, such as loss of life, injury or property damage. The course of consequences may be quite different in dissimilar geographical locations, despite equivalent hazards. Like the chain of initiating events, a correct description of consequences is complex and problematical.

With the three stages of catastrophe in mind, the previously mentioned typology of nuclear risk can now be constructed according to the following recipe:

1. List the three hazards as columns of a matrix.
2. List the stages of the fuel cycle as rows of the same matrix.

Risks of the Nuclear Fuel Cycle and the Developing Countries 97

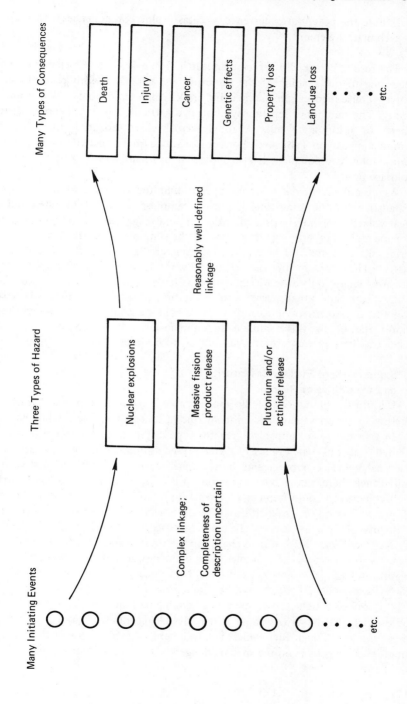

Figure 4–4. Three Stages in the Development of a Nuclear Catastrophe and the Linkages Between Them

3. Denote the potential occurrence of catastrophic consequences by symbols in each matrix element.

The results of the above recipe appear in Table 4–4. Matrix elements, which according to our reading of the literature are associated with potential catastrophic consequences, are marked by an X; elements devoid of these are left empty. The number of separate symbols in a particular element denotes separate *classes* of initiating events. A rough verbal characterization of the matrix symbols appears in Table 4–5. For comparison, a qualitative picture of potential catastrophic consequences is also indicated in Figure 4–3, in this case through boldface lines.

As described above, it seems evident that the typology of nuclear risk is exhaustive. At the same time it must be emphasized that Tables 4–4 and 4–5 treat what are obviously risks of quite different magnitude in an equivalent way. For example, properly evaluated, one of the symbols in matrix element C4 may have risks that are 100 or 1,000 times greater than some other symbol in the matrix. The typology of risks is thus a two dimensional cut of a three dimensional problem: only when each symbol is replaced by a vertical bar (out of the plane of the paper) that measures the magnitude of risk on a commensurate scale will the assessment of nuclear fuel cycle risks be complete. As we will see in more detail below, this complete three dimensional form of nuclear risk assessment is, with some exceptions, considered to be impossible at the present time.

Significance of Fuel Cycle Maturity and the Choice of Reactor

Despite the limitations of the above typology of risk, we may see clearly that fuel cycle maturity plays a significant role in determining the total nuclear risk.

In Canada, for example, the reactor type differs from that employed in the United States in that: (1) heavy water is used as moderator, and (2) natural U is used as fuel. Fuel reprocessing is not done because without enriched U in the initial fuel, there is no economic reason for it (extraction of Pu does not suffice). Consequently, a large number of risks vanish from the matrix, with the result shown in Table 4–6. Even this may be overstated to some extent, since matrix elements 3B and 3C are widely believed to carry a much smaller risk in the CANDU system. The dramatic change in the risk typology for the Canadian fuel cycle is, furthermore, not unique. Most countries do not have fuel reprocessing plants, and several countries use natural U in some of their power reactors. In all such cases, fuel cycle risks would be characterized by Table 4–6.

I therefore conclude that one way of coping with nuclear fuel cycle risks may be through the use of *truncated* fuel cycles. A recent article by B. T. Feld [19] urges the use of such fuel cycles. His argument is based not so much on composite fuel cycle risks as on the specific dangers of nuclear weapons spread.

Table 4-4. Typology of Catastrophic Nuclear Risks—Light Water Reactor Fuel Cycle

Fuel Cycle Stage	Hazard Type		
	(A) Nuclear Explosion	(B) Plutonium and/or Actinide Release	(C) Massive Fission Product Release
(1) Mining, milling and extraction			
(2) Enrichment, conversion and fuel fabrication	X X F	F F	
(3) Light water reactor		X X X X X X	X X X X X X
(4) Fuel reprocessing	X	X X X X	X X X X F F
(5) Plutonium storage	X F	X X X	
(6) Waste disposal		F F	F F

Key: X indicates the presence of one class of initiating event; F indicates the future occurrence of one class of initiating event; a blank indicates there are no initiating events.

Table 4-5. Catastrophic Nuclear Risk Described in Terms of Classes of Initiating Events

A. *Hazard Arising From Weapons Grade Uranium or Plutonium*
  1. Mining, milling and extraction:
     Empty.
  2. Enrichment, conversion and fuel fabrication:
     a. Theft of H.E.U. at output of enrichment plant or in subsequent transport.
     b. Theft of H.E.U. at output of fuel fabrication plant or in subsequent transport.
     c. Theft of mixed oxide fuel elements at output of fuel fabrication plant or in subsequent transport. Occurs only under recycle.
  3. Light water reactor:
     Empty.
  4. Fuel reprocessing:
     a. Theft of plutonium at output of processing plant or in subsequent transport.
  5. Plutonium storage:
     a. Theft of plutonium from storage facility.
     b. Theft of plutonium in transport of same to fuel element fabrication, should "recycle" be instituted. This is not a present danger.
  6. Waste storage (once this is instituted):
     Empty.

B. *Actinide and Plutonium Release*
  1. Mining, milling and extraction:
     Empty.
  2. Enrichment, conversion and fuel fabrication:
     a. Sabotage of fuel fabrication plant, causing release of plutonium to the environment. This is a hazard only if plutonium and/or actinide recycle is instituted.
     b. Unanticipated malfunction of fuel fabrication plant, causing release of plutonium and/or actinides to the environment, as for example, in a fire. This is a hazard only if plutonium and/or actinide recycle is instituted.
  3. Light water reactor:
     a. Sabotage of the reactor, causing core-melting,
     b. Unanticipated malfunction of reactor, causing core-melting, partial vaporization, and breach of containment.
     c. Sabotage of on-site spent fuel storage, resulting in release of material to the environment.
     d. Unanticipated malfunction of on-site spent fuel storage, resulting in release of material to the environment.
     e. Sabotage of spent fuel transport from onsite storage to fuel reprocessing plant.
     f. Unanticipated failure of containment during spent fuel transport from on-site storage to reprocessing plant.
  4. Fuel reprocessing:
     a. Sabotage of reprocessing plant, causing release of plutonium and/or actinides to the environment.
     b. Unanticipated malfunction of reprocessing plant, causing release of plutonium and actinides to the environment.
     c. Sabotage of on-site storage of plutonium and/or actinides, causing release of plutonium and/or actinides to the environment.
     d. Sabotage of transport of plutonium or actinide storage, causing release of same to the environment.

Table 4-5. continued

5. Plutonium storage:
   a. Unanticipated malfunction, causing release of plutonium to the environment.
   b. Sabotage of storage area, causing release of plutonium to the environment.
   c. Sabotage of transport to fuel fabrication plant. This is a hazard only if recycle is instituted.
6. Waste storage (once this is instituted):
   a. Unanticipated malfunction in recoverable storage areas, causing release of actinides to the environment.
   b. Sabotage of recoverable storage areas, causing release of actinides to the environment.

C. *Massive Fission Products Release*
   1. Mining, milling and extraction:
      Empty.
   2. Uranium enrichment and fuel fabrication:
      Empty.
   3. Light water reactor:
      a. Sabotage of reactor, causing core-melting, partial vaporization and breach of containment.
      b. Unanticipated malfunction of reactor, causing core-melting, partial vaporization, and breach of containment.
      c. Sabotage of on-site spent fuel storage, resulting in release of material to the environment.
      d. Unanticipated failure of on-site spent fuel storage, resulting in release of material to the environment.
      e. Sabotage of spent fuel transport from on-site storage to fuel reprocessing plant
      f. Unanticipated failure of containment during spent fuel transport from on-site storage to reprocessing plant.
   4. Fuel reprocessing:
      a. Sabotage of reprocessing plant, causing release of fission products to the environment.
      b. Unanticipated malfunction of reprocessing plant, causing release of fission products to the environment.
      c. Sabotage of on-site storage, causing release of fission products to the environment.
      d. Unanticipated malfunction of on-site storage, causing release of fission products to the environment.
      e. Sabotage of transport to recoverable storage. A hazard only when this is instituted.
      f. Unanticipated failure of transport to recoverable storage. A hazard only when this is instituted.
   5. Plutonium storage:
      Empty.
   6. Waste storage:
      a. Sabotage of recoverable storage, causing release to environment. A hazard only when this is instituted.
      b. Unanticipated failure of recoverable storage, causing release to the environment. A hazard only when this is instituted.

Table 4-6. Typology of Catastrophic Nuclear Risk—Canadian Heavy Water Fuel Cycle

Fuel Cycle Stage	Hazard Type		
	(A) Nuclear Explosion	(B) Plutonium and/or Actinide Release	(C) Massive Fission Product Release
(1) Mining, milling and extraction			
(2) Enrichment, conversion and fuel fabrication			
(3) Reactor		X X X X	X X X X
(4) Fuel reprocessing			
(5) Plutonium storage		F F	F F
(6) Waste disposal			

Key: X indicates the presence of one class of initiating event; F indicates future occurrence of initiating event; Stages (4) and (5) of the fuel cycle are not currently in operation.

RESEARCH ON CATASTROPHIC NUCLEAR RISKS

An Ideal Formulation

Ideally, one would like to have available for catastrophic risks the kind of information that exists for routine risks of most kinds—i.e., a quantitative probability statement, such as "the risk of death in auto accident is 1/5000 per year for an average US resident." As already noted, statements such as these are usually based on a body of relevant statistics, which in turn requires that the human consequence in question (death in our example) has occurred a fair number of times in the past. For automobile deaths, this is no problem: 50,000 per year die in this case. For nuclear catastrophes, on the other hand, we can say at the outset that there will be no analogous empirically based statistics because there have been no nuclear catastrophes so far. This is not to say that there is no information. There have been serious accidents in a few cases, and there have been thousands of abnormal events or near-accidents. These, then, must form the basis for any risk analysis.

In order to discuss the present state of research in the assessment of catastrophic nuclear risk, consider how an ideal formulation of the problem would appear. In this it is convenient to make use of the three stages of an accident defined in the previous section and illustrated in Figure 4–4. We introduce mathematical notation for the sake of conciseness, realizing at the same time that many of the functions introduced have not, and perhaps cannot, be calculated.

Define

$$C_{kj}(q_j) \equiv \text{magnitude of consequences of type } k \text{ due to a hazard of type } j \text{ and magnitude } q_j. \quad (4.1)$$

In general, q_j is measured in physical units, such as kilotons of explosive power, curies of radioactivity and so forth. Calculation of $C_{kj}(q_j)$ depends upon assumptions about population distribution and density, weather and a number of other factors. Given definition (4.1), then

$$\sum_k C_{kj}(q_j) \equiv \text{all consequences due to a hazard of type } j, \text{ magnitude } q_j. \quad (4.2)$$

Define further

$$h_j(q_j)dq_j \equiv \text{the probability that a hazard of type } j, \text{ magnitude in the range } (q_j, q_j + dq_j) \text{ occur in a given year.} \quad (4.3)$$

To express $h_j(q_j)$ in terms of initiating events, we require further probabilities as follows.

$$P_i \equiv \text{probability that an initiating event } i \text{ occur in a given year.} \quad (4.4)$$

$$f_{ij}(q_j)dq_j \equiv \text{probability that an initiating event } i \text{ leads to a hazard } j \text{ of magnitude } (q_j, q_j + dq_j) \text{ in a given year.} \quad (4.5)$$

Then clearly

$$h_j(q_j)dq_j \equiv \sum_i P_i f_{ij}(q_j)dq_j \quad (4.6)$$

and

$$\sum_{ik} \int dq_j P_i f_{ij}(q_j) C_{kj}(q_j) \equiv \text{expected annual consequences of all kinds due to hazard of type } j. \quad (4.7)$$

$$\sum_{ijk} \int dq_j P_i f_{ij}(q_j) C_{kj}(q_j) \equiv \text{expected annual consequences of all kinds.} \quad (4.8)$$

Evaluation of the above relation is clearly what is desired. To obtain a numerical result it is necessary to evaluate two kinds of probabilities and a consequence function, and then to sum over products as indicated. The process is in principle straightforward. The difficulty lies, as we will see, in specifying the required functions.

Two Examples

With the help of the above notation, it is now instructive to examine two "milestone" studies of nuclear reactor safety.

WASH 740. This was the first significant reactor safety study conducted by the U.S. Atomic Energy Commission [20]. Its title is, significantly, "Theoretical Possibilities and Consequences of Major Accidents in Large Nuclear Power Plants." Its purpose was to estimate a range of consequences that might be expected in the case of core meltdown in a 500 Mw_t reactor (about one-sixth the size of present generation reactors). Published in 1957, WASH 740 considered three cases: (1) no release of fission products beyond containment; (2) release of 100 percent of all volatile fission products; (3) release of 50 percent of all fission products, volatile or not. For case (3), the worst of the three, the study predicted 0–3,400 fatalities and 0–43,000 injuries. In effect, the report gives

$$\sum_k C_{kj}(q_j) \equiv \text{all consequences due to hazard } j, \text{ where } j \text{ refers} \quad (4.9)$$
to the massive release of fission products and
$q_j = (0, 0.5 \text{ or } 1) \times$ (volatile fission product inventory).

Other than the fact that q_j is related to the fission product inventory of a reactor, the study really has nothing to do with reactors. It deals in no way with event probabilities. In regard to the "missing" probabilities, the report had the following comments:

> Many outstanding leaders in reactor technology and associated fields were consulted in the course of this study. It is their unanimous opinion that the likelihood of a major reactor accident is low. There is a general reluctance to make quantitative estimates of how low the probability is. There is a common aversion to attachment of quantitative estimates to a phenomenon so vague and uncertain as the probability of occurrence of catastrophic accidents, particularly since such assignment of numerical estimations conveys an erroneous impression of the confidence or firmness of the knowledge constituting the basis for the estimate. Also, some hold a philosophic view that there is no such thing as a numerical value for the probability of occurrence of a catastrophic accident; that such a thing is unknowable.
>
> Thus, many decline to make even order-of-magnitude guesses of the probability of catastrophic reactor accidents. On the other hand, a few have ventured to express their confidence of the extremely low probabilities of occurrence of such accidents by stating numerical, order-of-magnitude estimations [21].

WASH 1400 (Draft). Recently released, this $3.5 million reactor safety study goes considerably beyond WAH 740. It essentially obtains the quantity

$$\sideset{}{'}\sum_{jki} \int P_i f_{ij}(q_j) C_{kj}(q_j) \, dq_j \equiv \text{expected annual losses of all} \quad (4.10)$$
kinds due to 100 light water reactors.

The result, though formally similar to the complete expression (4.8) above, is distinct from (4.8) because the sums over k and i are incomplete. Explicitly omitted in WASH 1400 are (1) events of all kinds referring to fuel cycle sectors other than the light water reactor itself; (2) initiating events involving sabotage. The omissions are symbolized by the primes on the summation symbols in expression (4.10).

The somewhat surprising findings of Draft WASH 1400 are as follows.

1. The probability for core melting is larger than had previously been guessed (1 in 17,000 per reactor year).

2. Given core meltdown, the most probable magnitude of the fission product hazard (quantity of fission products released) is smaller than assumed in WASH 740.
3. As a result of (1) and (2), the annual expected losses for 100 light water reactors are a factor of 100–1000 below commonly experienced catastrophic hazards, such as floods and hurricanes, and a factor of 10,000 below the sum of all man-caused catastrophic hazards.

The publication of Draft WASH 1400 has been widely regarded as a major advance in dealing with nuclear fuel cycle risks. At the same time, critics have listed major shortcomings that go well beyond the already mentioned scope limitation of the study. In the language we have established, two principal criticisms have been as follows.

First, the calculation of probabilities $h_j(q_j)$ has been labeled as "fundamentally flawed" because the method used (fault tree analysis) is in principle not capable of yielding absolute values. To state it another way: there is no way of knowing for sure that the sum in expression (4.6) is inclusive of all relevant terms. The authors of Draft WASH 1400 have repeatedly claimed to have identified most major modes of failure. Others, such as the Union of Concerned Scientists, have argued the contrary [22]. Particular points of contention have been the treatment of common mode failure, the likelihood that emergency core cooling will work as intended and the probability, however small, that the pressure vessel will fail. Both issues would affect the probability of $h_j(q_j)$.

Second, the calculation of the consequence function is regarded as incorrect for two reasons: (1) overly optimistic assumptions have been made about evacuation; and (2) the effects of long-lived radioactivity, particularly Cs–137, have been left out. The former point was first made by the Union of Concerned Scientists [23]; the latter was raised in the recently published American Physical Society Reactor Safety Study [24]. Without any changes in the WASH 1400 probabilities, the effect of the changed consequence functions leads to an increase in losses by factors ranging from 10 to 100. An example of specific results given in Table 4–7 [25]. In congressional testimony, Norman Rasmussen,

Table 4–7. Revised Consequences for LWR Accident Having a Probability of 5×10^{-6} Per Reactor Year

Consequence	Draft Wash 1400	Corrected Result
Prompt fatalities	62	620–990
Lethal cancer	300	10,000–20,000
Genetic defect	300	3,000–20,000
Thyroid nodules	25,000	22,000–350,000

Source: Data is based largely on the American Physical Society Study and was presented by H. Kendall in testimony before the House Committee on Interior and Insular Affairs, Subcommittee on Energy and Environment, April 29, 1975.

director of WASH 1400, has indicated that due to the above mentioned criticisms revisions in the final version of WASH 1400 are planned [26].

Getting All the Terms

Independent of the particular outcome of the debate centered on Draft WASH 1400, it is clear that this most comprehensive reactor study does not get close to describing all the terms that contribute to the nuclear fuel cycle risk. To my knowledge, no one has in fact attempted to give a nominally complete analysis of the problem defined by expression (4.8). Critics like Kendall regard the problem as inherently insoluble at this time [27]. Some argue that the risks associated with the light water reactor dominate all other terms [28], and one individual, B. L. Cohen [30], has given a series of informal arguments that support this view quantitatively.

In my reading of the literature, I have reached the conclusion that most studies dealing with other than reactor-associated matrix elements are either nonexistent or are in the "WASH 740" stage of development: i.e., the discussion deals with conceivable hazards and conceivable consequences, but no estimates are made of event probabilities and no quantitative measure of risk is assigned. A typical example of such a study is the work of Taylor and Willrich on nuclear theft [30], or the section of the recent American Physical Society Study [31] on sabotage. A possible exception to the rule is the literature that deals with transportation accidents.

I conclude, therefore, that the aggregate risks of the nuclear fuel cycle are at best quantifiable in terms of a lower limit, and this is given very roughly by the Draft WASH 1400 results as amended by the critics.

CONCLUSIONS: IMPLICATIONS FOR THE DEVELOPING COUNTRIES

In the preceding discussion I have demonstrated that whole sections of the full matrix of nuclear risks have only been analized in terms of conceivable hazards and conceivable consequences, without a concomitant assessment of event probabilities. This suggests that nuclear aggregate fuel cycle risks may turn out to be larger than commonly anticipated.

In the case of developing countries which are now obtaining nuclear power, largely through importation of technology, the fuel cycle risk may be further increased because of inexperience and the absence of continued monitoring for safe operation. Without quantification of the latter factors, I can perceive a situation in the not too distant future where the greatest threat produced by rapid growth of nuclear power in the developing world is *not* the classic problem of nuclear weapons proliferation through independent national weapons programs, but rather the aggregate risk of the fuel cycle. At this writing, with all the prevailing uncertainty, this is admittedly an incompletely substantiated conjecture that only the future course of political and scientific development can confirm.

At the risk of being presumptuous, here are two recommendations for developing nations that are, in my view, implied by the preceding discussion.

First, it is best to avoid the fully mature nuclear fuel cycle at the present time. This suggests two alternatives: (1) acquiring a natural U–fueled reactor, moderated by heavy water or graphite, and, like Canada, ending the fuel cycle at the point of spent fuel removal from the reactor; (2) buying an enriched U, light water moderated reactor of the US variety and shipping spent fuel to a developed country for reprocessing. The risks associated with the first alternative are smaller, and it has the added advantage of not requiring extensive dependence on outside services.

Second, it is essential that any developing country that introduces nuclear electric power production also make a concomitant major and long term investment in scientific manpower and safety personnel. Given the risks of the nuclear fuel cycle, it is not sufficient to follow the route of "buying the machine and turning it on." Of particular importance is the existence of a well-supported cadre of "critical scientists" who deal with nuclear safety in a local context, from a base independent of the managers of the nuclear technology and in continuing communication with the international scientific community concerned with nuclear safety.

NOTES TO CHAPTER FOUR

1. W. Davidon, M. Kalkstein and C. Hohenemser, *The Nth Country Problem and Arms Control* (Washington, D.C.: National Planning Association, 1960).

2. United States Atomic Energy Commission, *Annual Report to Congress* (Washington, 1960).

3. C. Hohenemser, "The Nth Country Problem Today," in *Disarmament, its Politics and Economics,* S. Melman ed. (Boston: The American Academy of Arts and Sciences, 1962).

4. M. Willrich, ed., *International Safeguards and the Nuclear Industry* (Baltimore: Johns Hopkins University Press, 1973); Davidon, Kalkstein and Hohenemser, *The Nth Country Problem and Arms Control.*

5. P. Boskma, "Uranium Enrichment Technologies and the Demand for Enriched Uranium," in *Nuclear Proliferation Problems* (Cambridge, Mass.: MIT Press, 1974), p. 56.

6. B. M. Jasani, "Nuclear Fuel Reprocessing Plants," in *Nuclear Proliferation Problems* (Cambridge, Mass.: MIT Press, 1974), p. 89.

7. *Preventing Nuclear Weapon Proliferation,* (Stockholm: International Peace Research Institute, 1975).

8. Boskma, p. 56.

9. David J. Rose, "Nuclear Eclectic Power," *Science* 184: 351 (1974).

10. D. R. Inglis, *Nuclear Energy, Its Physics and Its Social Challenge* (Reading, Mass.: Addison–Wesley, 1973).

11. Rose, "Nuclear Eclectic Power."

12. D. F. Ford and H. W. Kendall, "Nuclear Safety," *Environment* (September 1972); D. F. Ford, T. C. Hollocher, H. W. Kendall, J. J. MacKenzie, L. Scheinmen and A. S. Schurgin, *The Nuclear Fuel Cycle* (Cambridge, Mass.: Union of Concerned Scientists, 1973).

13. D. F. Ford and H. W. Kendall, *An Assessment of the Emergency Core Cooling System Rulemaking Hearings* (Cambridge, Mass.: Union of Concerned Scientists, 1973).

14. T. B. Taylor and M. Willrich, *Nuclear Theft: Risks and Safeguards* (Cambridge, Mass.: Ballinger 1974).

15. J. G. Speth, A. R. Tamplin and T. B. Cochran, "Plutonium Recycle, the Fateful Step," *Bulletin of the Atomic Scientists* (November 1974).

16. U.S.A.E.C., Report WASH 1400–Draft, "Reactor Safety Study," (1974).

17. "Report to the American Physical Society by the Study Group on Light Water Reactor Safety," *Reviews of Modern Physics,* vol. 47. suppl. 1 (1975).

18. H. Kendall and S. Moglewar, ed., *Preliminary Review of the AEC Reactor Safety Study,* (Sierra Club–Union of Concerned Scientists, Cambridge, November 1974). See also testimony of H. Kendall before the House Committee on Interior and Insular Affairs, Subcommittee on Energy and Environment, April 29, 1975.

19. B. T. Feld, "Making the World Safe for Plutonium," *Bulletin of the Atomic Scientists,* May 1975, p. 5. Also "Nuclear Proliferation–Thirty Years after Hiroshima," *Physics Today,* July 1975, p. 23.

20. U.S.A.E.C., Report WASH 740, "Theoretical Possibilities and Consequences of Major Accidents in Large Nuclear Power Plants," (1957).

21. Ibid.

22. Kendall and Moglewar, *Preliminary Review.*

23. Ibid.

24. "Report to the American Physical Society," *Reviews of Modern Physics,* vol. 47, suppl. 1 (1975).

25. Kendall, testimony before the House Committee on Interior and Insular Affairs.

26. N. Rasmussen, testimony before the House Committee on Interior and Insular Affairs, Subcommittee on Energy and Environment, April 29, 1975.

27. H. Kendall, private communication.

28. Ralph Lapp, David Rose, private communications.

29. B. L. Cohen, private communication and lecture notes.

30. Taylor and Willrich, *Nuclear Theft.*

31. "Report to the American Physical Society, *Reviews of Modern Physics,* vol. 47, suppl. 1.

32. Marc Ross, "The Possible Release of Cesium in a Spent-fuel Transportation Accident," preprint, Department of Physics, University of Michigan, Ann Arbor Michigan, 1975.

ACKNOWLEDGMENTS

The analysis in the sections "Risks of the Nuclear Fuel Cycle" and "Research on Catastrophic Nuclear Risks" is in part an outgrowth of a research project centered at Clark University, and entitled "Risk and Rare Events: A Three Nation Study of Disaster Prevention in Nuclear Energy Programs." The project involves Anne Kirkby, University of London; Ian Burton, University of Toronto; and Roger Kasperson, Robert Kates, Michael McClintock and the author, all of Clark University. The manuscript was read and commented on by Norman Rasmussen, Ralph Lapp, David Rose and Marc Ross, with the result that several useful corrections were made. Despite collaboration and review, views expressed and errors made here are the sole responsibility of the author.

For initial funding of the work, the support of the Ford Foundation is acknowledged.

Chapter Five

Commercial Nuclear Technology and Nuclear Weapon Proliferation

Theodore B. Taylor

My subject concerns what I believe to be perhaps the most critical task facing humanity—to establish, within a very few years, worldwide controls that will reduce the risks of destructive use of nuclear energy to tolerable levels.

I would first like to present an overview of technological factors that bear on these risks, then focus on several policy issues that I believe should receive special attention in the United States and internationally and, finally, discuss in somewhat more detail a few subjects that bear on these risks and opportunities for dealing with them effectively.

Given the required nuclear materials, any country in the world could design and build nuclear explosives, using information, materials and equipment that are generally available worldwide. How difficult a task this would be depends on the type of explosive wanted. It is much more difficult to make reliable, efficient and lightweight nuclear warheads usable for a variety of military purposes than to make crude, inefficient nuclear explosives with unpredictable yields in the range equivalent to several hundred to several thousand tons of high explosive.

Crude fission bombs that could, under some circumstances, kill 100,000 people or more and destroy billions of dollars worth of property could also be made from stolen nuclear materials by non-national groups of people, such as terrorists. Such devices could conceivably even be designed and built by one person, working alone.

Producing the nuclear materials presently required for nuclear explosives is now generally much more difficult than designing and building the explosives themselves. Three types of materials can now be used as the core materials for nuclear explosives: uranium that is highly enriched in the isotope $U-235$; plutonium, which can be made by capturing neutrons in $U-238$, the abundant isotope in natural uranium; and $U-233$, which can be made by capturing neutrons in natural thorium. None of these so-called "weapon grade" materials exist naturally in significant quantities.

It is also possible that present efforts in the United States and elsewhere to make so-called "pure fusion" nuclear explosives that use only deuterium, the naturally occurring heavy isotope of hydrogen, and tritium, the still heavier, but insignificantly naturally abundant isotope of hydrogen, will succeed. Production of tritium requires some source of neutrons which are then captured in naturally abundant lithium to make the tritium.

A country that wants to acquire the materials necessary for making fission explosives now has several options, at least in principle. I shall mention some of these briefly, not necessarily in order of relative difficulty. Which option would appear most attractive to a particular country would depend on many technological, political, military, economic and other considerations.

1. Build a uranium isotope enrichment facility for converting domestically available natural uranium to highly enriched uranium. There is a widespread belief that it would require a billion dollars to build such a plant, based on currently available technology—that is, the gaseous diffusion or gas centrifuge methods. However, the basic physical principles of a dozen or so different conceivable methods have been described in open technical literature. In particular, the possibility of using high-power lasers to stimulate chemical transitions in U-235, but not in U-238, that would make subsequent chemical separation possible with much less energy than either the gaseous diffusion or gas centrifuge techniques, has been publicly reported in several countries, including the United States, France, the United Kingdom and Australia. I find it quite credible that such a technique may be shown to be relatively easy to use, compared to present gaseous diffusion methods, within ten years or less. It has been reported, further, that South Africa now has the expertise to extract and enrich uranium. I know of no authoritative reports about what method they may be using.

I should also point out that natural uranium exists in commercially extractable concentrations practically everywhere on earth. Although extraction of uranium from ordinary granite, for example, may cost considerably more than from high grade ores, the cost per unit of contained U-235 is well within reason for the production of highly enriched uranium for nuclear explosives.

2. Build a natural uranium reactor, using domestic or imported uranium, domestic or imported heavy water, and high purity graphite or beryllium for slowing down neutrons sufficiently to sustain a chain reaction, and build the reprocessing plant necessary to extract the plutonium from the irradiated fuel. Detailed designs of such types of reactors and separation plant components are widely published. This option is certainly within the capability of at least several dozen nations.

3. Build a high energy charged particle accelerator, similar to some that are now used for basic high energy physics experiments, to produce neutrons, by bombardment of ordinary uranium, that are subsequently captured to make plutonium. This is likely to be somewhat more difficult than building a natural uranium reactor. A high energy proton accelerator somewhat similar to the

so-called "Meson Factory" at Los Alamos Scientific Laboratory could be used to make enough plutonium for at least several fission bombs per year. I mention this possibility because I do not believe such facilities are now subject to safeguards by the International Atomic Energy Agency (IAEA), and because it is an example of a conceivable means for acquiring fissionable materials outside the context of conventional views of how a country might acquire nuclear weapons.

4. Build a reactor that uses nuclear fuel material supplied by another country, and use some of the fresh fuel, if it contains highly enriched uranium, plutonium or U–233 that is not effectively safeguarded. Otherwise, also with the condition that the reactor is not effectively safeguarded, extract plutonium or U–233 from spent fuel at a domestic, unsafeguarded fuel reprocessing plant.

5. Obtain a reactor, or assistance in building it, from another country, and exercise the appropriate one of the two options referred to under point 4. The second of these was apparently the option used by the Indian government.

6. Obtain a nuclear reactor and/or nuclear fuel from another country, and, if the nuclear materials are effectively safeguarded from diversion for use in nuclear explosives, abrogate the agreement at some later time. Depending on the nature of the fuel, the material for nuclear explosives might quickly and easily be obtained from fresh fuel, or might require separation in a domestic fuel reprocessing plant from the highly radioactive fission products in spent fuel. Facilities for doing the latter could be operating within about two years of the start of construction of the reprocessing plant.

7. Arrange for the theft of weapons grade nuclear materials from another country where physical security measures applied to the materials are inadequate.

8. Arrange for the theft of weapons grade nuclear materials from facilities that are within the country and subject to safeguards agreements, but in such a way that the theft appears to be the work of a criminal group without any connections with the government.

9. Arrange for the theft of complete nuclear weapons from another country.

Many other options are conceivable. Those above, and others, may or may not appear significantly credible in a particular situation. They serve to illustrate my main point, however, that there is a wide variety of ways that countries might obtain nuclear weapons.

As I have already suggested, there are also risks that non-national organizations, including terrorists, might covertly divert or openly steal weapons grade nuclear materials from civilian or military nuclear fuel cycles and make their own nuclear explosives, or even steal complete nuclear weapons, in either case from facilities or vehicles that are inadequately protected from armed theft or diversion by "insider" members of such a group. I am especially concerned about this possibility because nuclear threats, by terrorists for example, are not subject to deterrence by the threat of nuclear retaliation, and because physical security safeguards in the United States, and very likely in other countries, are not yet adequate to prevent theft from all places where substantial quantities of weapons grade materials exist.

I hasten to add, however, that this situation is now changing *very* rapidly in the United States as a result of recent actions by the AEC which are being continued under the authority of the Nuclear Regulatory Commission and the Energy Research and Development Administration. I am also convinced that a *very* high level of security for all such materials in the United States can be achieved, at costs that are equivalent to less than 2 percent of the costs of nuclear electric power or military nuclear systems for which these materials are used. By very high security I mean such as to make it highly unlikely that attempts at theft involving resources and skills significantly greater than any used in historical thefts would be successful.

It is important to recognize that the present state of nonmilitary nuclear technology, worldwide, is such that inventories and flow rates of weapons grade nuclear materials (that is nuclear materials not requiring isotope enrichment or separation from highly radioactive fission products) are very small compared with what they are likely to become within the next four or five years. For reasons that I shall present later in this chapter, we appear to have a few years of "relative grace" for accomplishing the immense task I referred to at the beginning of the chapter.

In this connection, I would now like to turn to a number of related policy questions that I believe need resolution, with considerable care, but also with a high sense of urgency. These are the following:

1. Should the US adopt a generally uniform policy with respect to agreements for nuclear assistance to other countries, or should there be substantial differences that depend on special circumstances? For reasons that are connected with information I have just presented, for example, it might be argued that US should insist on fuel being reprocessed outside the recipient country, and our not supplying any plutonium-bearing fresh fuel. The US might offer to pay the recipient country a fair price for plutonium extracted from spent fuel or to provide an equivalent amount of "free" slightly enriched uranium fuel. But such a restriction on nuclear support to a Middle Eastern country, for example, if not also applied to, let us say, Western Germany or Japan, might be considered highly discriminatory and therefore politically unacceptable.

2. Should US agreements with other countries include a provision that would allow strengthening of IAEA safeguards in the future, if this can be done in ways that, objectively, are not discriminatory, costly or generally intrusive? In spite of the difficulty in negotiating such agreements, which may appear too "open ended" to the recipient country, I believe such a provision should be included, since the IAEA safeguards system is still in the process of development.

3. How can the United States and other countries not directly involved in nuclear assistance agreements be assured that the physical protection of pluto-

nium or other nuclear weapon materials in a recipient country is adequate to deal with attempted thefts by non-national organizations? One helpful step in this direction could be to rely more than at present on reports by IAEA inspectors of any evident major deficiencies in physical security that are observed in the course of their carrying out their normal surveillance and materials accounting duties. Beyond this, the U.S. may have to insist on some kind of US control over the security state of materials subject to the agreement, such as using specially designed vehicles, provided by the United States, for transport of fission product-free plutonium within the recipient country's borders.

4. Which entities of our federal government should have the responsibility and authority to assess and formally approve foreign nuclear assistance agreement proposals before they enter the formal negotiation stage? Since such agreements can have an important bearing on the well-being of all of us, and are likely to involve issues that are not amenable to cut and dried cost-benefit analyses, I suggest that at least the substantive guidelines for such agreements should require review and approval by both houses of Congress.

5. Are we going to rely on the IAEA as the primary international institution with responsibility for international control of materials from which nuclear explosives could be made? I strongly urge that we do so. But I also strongly recommend that the United States play a much more aggressive and creative role than it has in the past to assure that the IAEA is able to fulfill this responsibility effectively, and on a worldwide basis. From some direct observation of safeguards-related activities of the IAEA during two years I spent in Vienna as an independent consultant (1966 to 1968), I concluded that they were moving in the right directions but needed a much more definite sign of resolve on the part of member states to see to it that the agency has the manpower, financial resources and international backing to allow it to perform its extremely important safeguarding role effectively. The agency has made significant progress along these lines since that time, but has, I believe, a long way yet to go in a rather short time.

The rest of this chapter is divided into three parts. First, I shall briefly describe some of the basic principles of nuclear explosives technology, to give some idea of the resources required to make various kinds of nuclear explosives. Second, I shall present a brief profile of worldwide nuclear technology as it may be expected to look in 1982, the earliest date at which nuclear power plants that the US decided in 1974 to provide to a foreign country are likely to start full scale operation. Finally, I shall discuss some of the problems and opportunities related to safeguarding nuclear materials and facilities provided to a foreign country by the US against their use by that country for making nuclear explosives and assuring that nuclear materials from which nuclear explosives can be made are not diverted or stolen outright by non-national groups, such as terrorist organizations in that country.

NUCLEAR EXPLOSIVES TECHNOLOGY

Three kinds of nuclear materials that are used in fission power reactors can also be used for making nuclear explosives: U-235, plutonium and U-233. Of these, only U-235 exists in significant quantities in natural deposits, where it represents only about 0.7 percent of the uranium, the rest being the isotope U-238. To be usable as the core material in a practical nuclear explosive, uranium must be enriched to more than 20 percent or so in the isotope U-235. Both plutonium and U-233 in the forms they are made in nuclear reactors (by neutron capture in U-238 or in naturally occurring thorium), however, can be used without isotope enrichment for making nuclear explosives.

The critical masses of spheres of metallic forms of highly enriched uranium (U-235/U-238 greater than 90 percent), plutonium and U-233 inside a good neutron reflector (such as metallic natural uranium) are about 17 kilograms, five to eight kilograms and about six kilograms, respectively. These critical masses are not necessarily equal to the masses required to make a fission bomb. Using highly advanced nuclear weapons technology, practical nuclear explosives can be made with substantially smaller quantities of core materials. For some types of nuclear weapons, on the other hand, substantially greater amounts (initially kept subcritical by physical separation) are required. The physical size of nuclear weapon cores of these materials can range from something smaller than a baseball to something larger than a volleyball. Overall weights of complete nuclear explosives can range from a few tens of pounds to several tons and their minimum diameters can range from a few inches to several feet.

Plutonium made in nuclear reactors by capturing neutrons in the naturally abundant isotope of uranium (U-238) can have an important basic property that distinguishes it from U-235 or U-233 as a material for nuclear weapons. As it is normally made in nuclear power reactors, plutonium consists of significant quantities of several different isotopes; Pu-239, which is highly desirable nuclear weapons material; Pu-240, which is generally undesirable for two reasons—it is relatively difficult to cause to fission at a high rate (and therefore acts as a dilutant) and, more importantly, it spontaneously emits neutrons which, under some conditions, can cause an assembling fission weapon to explode prematurely before it has become sufficiently supercritical to produce an efficient explosion; Pu-241, which is similar to Pu-239; and unimportant quantities of still heavier isotopes of plutonium.

The relative concentration of Pu-240 can be reduced, however, by reducing the amount of time that fuel is left in an operating reactor—i.e., by relatively frequent refueling. One might call plutonium made in this way relatively "high grade," of the type easiest to use in relatively efficient military nuclear weapons with reasonably predictable and reproducible yields.

Nevertheless, nuclear explosives *can* be made with the relatively "low grade" plutonium of the type normally produced in nuclear power plants. The critical mass of such plutonium may be 20 to 30 percent higher than that of "high grade" plutonium. Under some circumstances, the yield of different explosives of the same design may vary by as much as a factor of ten or more, but in all cases be very likely to qualify as a full-fledged nuclear explosion. In short, the *utility* of fission explosives made with "low grade" plutonium will depend both on the expertise of the designers and the purposes for which they are to be used.

Although, generally speaking, the best chemical form of plutonium or uranium to use in fission explosives is metal, reasonably efficient explosives can be made using somewhat larger quantities of the oxides or carbides of these elements. This fact is especially important when considering the risks of theft of materials which are new or are likely to be used in the future in making fuel for power reactors by non-national groups who want to make relatively crude nuclear explosives with a minimum of effort.

Given access to the required amounts of these fissionable materials, the skill, resources and time required to design and build fission explosives depend considerably on what they are to be used for. Under conceivable circumstances, for example, one person who possessed about ten kilograms of plutonium oxide and a substantial amount of high explosive could, within several weeks, design and build a crude fission bomb. By a "crude fission bomb" I mean one that would have a good chance of exploding with a yield equivalent to at least 100 tons of high explosive and be small enough to be carried in an automobile. This could be done using information that is easily accessible to the general public and materials and equipment that are commercially available worldwide. Such a person would require a working knowledge of nuclear physics or engineering, would be innovative and skilled with his hands and would be willing to take a significant risk of killing himself during the construction operation.

Considerably more efficient, reliable and compact fission explosives would require a fairly large team effort by experts. But such a design and construction team could draw on extensive, widely published information that was not known to the nuclear weapon designers who assembled in Los Alamos in 1943 and produced the plutonium implosion bomb that was successfully tested in August 1945. Much of the work required to design a variety of types of militarily useful fission explosives, and build and test their non-nuclear components, could be done before the required nuclear core materials were made available. If it were very important to a nation to do so, it is conceivable that it could successfully test a fission weapon on which such design, construction and testing efforts had been carried out in advance, within a few days of the time when it had access to the required amounts of concentrated plutonium or uranium.

In short, it takes considerably more time and resources to produce the nuclear materials required for fission explosives than it does to design and build the other components of explosives that could satisfy a wide variety of national or subnational purposes.

I would like to turn now to a brief description of what we can expect to be the state of worldwide nuclear technology about eight years from now, when nuclear power plants ordered now are likely to be operating. I shall emphasize aspects of this technology that I believe bear on the likelihood of further nuclear weapon proliferation.

PROFILE OF WORLDWIDE NUCLEAR TECHNOLOGY, 1982

Nuclear Explosives

Aggregate numbers of nuclear weapons in stockpiles of the five nations that have announced nuclear weapon systems will amount to somewhere around 100,000, give or take perhaps a factor of two. Other nations may have openly joined the military nuclear club, with or without violation of any international agreements.

Some large scale peaceful uses of nuclear explosives, for natural gas stimulation or for excavation, for example, may have been proved out, as suggested by US and Soviet studies and experiments, and by the official announcements by the Indian government regarding their recent test. It is possible that continued intensive research and development in this area may yield new types of nuclear explosives and peaceful applications of them by 1982 that will reduce the environmental impact of their use to acceptable levels. Other nations besides India may well have joined the "peaceful" nuclear explosives club.

Nuclear Power

If recent AEC and IAEA forecasts are fulfilled, the following situations are likely to exist in 1982.

1. The worldwide, installed nuclear electric power generating capacity will be about 400,000 megawatts from roughly 400 nuclear power plants situated in perhaps 35 different countries. Nuclear power in the US will probably account for a little less than half this installed capacity and number of power plants. Perhaps 15 nations will be using nuclear power reactors provided by the United States. Such US support now exists or is planned for Western Germany, Italy, Japan, France, Spain, India, Switzerland, Sweden, Korea, Taiwan, Brazil and Mexico and is now being negotiated with Iran.

2. The large majority of these power plants will use so-called light water moderated reactors (LWRs) that now use slightly enriched uranium for fuel. But significant numbers of reactors that use natural uranium for fuel (such as gas cooled, graphite moderated or heavy water moderated reactors) will also be in operation, along with perhaps as many as ten or so helium cooled, graphite moderated (HTGR) reactors that use highly enriched uranium for fuel, and a smaller number of liquid metal cooled fast breeder reactors (LMFBR).

All these reactors, except the HTGR, produce substantial quantities of plutonium. The annual rate of production of plutonium in a 1000 MW(e) power plant that contains the naturally abundant isotope of uranium, U-238, can vary from roughly 400 to 800 kilograms per year, depending on the type of reactor and the frequency of refueling operations. Since some of the produced plutonium is consumed before fuel is extracted for reprocessing, however, the *net* amount of plutonium discharged each year from such a power plant (if not operated primarily as a plutonium production plant) is substantially smaller—about 250 kilograms per year from a 1000 MW(e) LWR reactor.

3. The plutonium contained in used fuel extracted from all the world's nuclear power plants in 1982 will be somewhere in the vicinity of 70,000 kilograms per year, sufficient for more than 10,000 nuclear explosives. The cumulative worldwide *net* production of plutonium by the world's power reactors, through 1982, will be perhaps as much as 400,000 kilograms.

4. Since this plutonium is a valuable nuclear fuel worth roughly $10,000 per kilogram, strong economic incentives will exist for extracting the plutonium at fuel reprocessing plants and recycling it, as a supplement to the U-235 contained in natural or slightly enriched uranium, in *fresh* fuel for reactors. Such recycling is not yet happening on a routine basis, but is very likely to have become routine in many of the world's power reactors by 1982, perhaps as early as the late 1970s. Before this time, plutonium will be either contained in highly radioactive fuel elements awaiting reprocessing or stored after separation from radioactive wastes in spent fuel, awaiting recycling in fresh fuel. After plutonium recycle starts on a routine basis, the rates of flow of plutonium to fuel fabrication plants and nuclear power plants will increase dramatically.

5. Large scale reprocessing facilities for separating plutonium and uranium from spent nuclear fuels will exist in at least the following countries in 1982: the United States, the Soviet Union, the United Kingdom, France, the People's Republic of China, the Federal Republic of Germany, Japan, India and Belgium.

It is also likely, however, that the worldwide capacity to reprocess spent nuclear fuel will be insufficient to handle accumulated backlogs of spent fuel and the rates of fuel discharged from power reactors at that time. In the United States, at least, it is likely to take between eight and ten years between the time a commercial enterprise decides to enter the nuclear fuel reprocessing business and the time it starts full scale operations. No commercial nuclear fuel reprocessing plants have been operating in the United States since mid-1972, when the Nuclear Fuel Services plant near West Valley, New York shut down for renovation and expansion. According to present schedules for operation of fuel reprocessing plants and nuclear power plants, the total US fuel reprocessing capacity in 1982 may be less than one-half the rate at which used fuel will be extracted from US power reactors at that time. In addition, it is likely that a very large backlog of unreprocessed nuclear fuel (perhaps as much as 20,000 tons) will be in storage in the US at that time. Although this kind of mismatch is likely to be much less severe in Western Europe, it is nevertheless likely that, even if plutonium recycle makes use of separated plutonium about as fast as it is produced in the early 1980s, considerable quantities of irradiated nuclear fuel will be "standing in line" in many countries until at leas the mid-1980s.

6. Large scale uranium isotope enrichment facilities capable of producing slightly enriched uranium *or* highly enriched uranium suitable for nuclear explosives are likely to be operating not only in the US, the USSR, the UK, France and the Peoples' Republic of China, but also in the Netherlands (as a multinationally operated facility). It is also quite possible that, by 1982, other countries may have gas centrifuge-type isotope separation facilities capable of separating enough concentrated U-235 for at least several dozen nuclear explosives per year. In addition, current programs in several countries to develop alternative simpler methods for uranium enrichment such as the use of laser-stimulated techniques, may well succeed by 1982. It is therefore conceivable that, by that time, the capability to produce strategically important quantities of highly enriched uranium may become available to any country that has access to quantities of natural uranium in the range of tens of tons or more and that can muster the required technical expertise and other resources. These required resources may well be very much smaller than are now required for either gaseous diffusion or gas centrifuge-type enrichment plants. If and when such relatively simple isotope enrichment techniques are developed, I find it hard to believe that any important process details can be kept secret very long. The basic approaches to the laser-stimulated process, for example, have already been presented in some detail in the open technical literature.

With these observations to help broaden the perspective, I shall now discuss briefly some of the problems and opportunities related to safeguarding those materials and facilities that the United States may be providing to other countries, to prevent their use for destructive purposes.

SAFEGUARDS AGAINST DIVERSION OR THEFT OF NUCLEAR MATERIALS PROVIDED TO OTHER COUNTRIES BY THE UNITED STATES

All nuclear materials and facilities provided by the US to other countries are expected to be subject to International Atomic Energy Agency safeguards, whether or not the recipient country has signed or ratified the Nonproliferation Treaty. In addition, further restrictions and safeguards may be imposed by the US in bilateral agreements with the recipient country. The fundamental question then is: Will these safeguards and restrictions be sufficiently effective to assure that the benefits, to the US and the world generally, of such assistance be worth the residual risks of destructive use, or threats of destructive use, of nuclear materials provided by this assistance?

I do not believe that any group of people now has sufficient information and has explored this question sufficiently to be able to answer it objectively. I am quite sure that I cannot. But perhaps I can help shed some light on some of the issues involved.

I find it convenient to think of these risks in four broad categories:

1. abrogation of agreements by the government of the recipient country;
2. clandestine diversion of nuclear weapon materials by the government of the recipient country;
3. clandestine diversion of nuclear weapon materials by non-national organizations;
4. overt theft of nuclear weapon materials by non-national organizations or by agents of another country.

The first two of these involve violations of agreements by the country to which the nuclear assistance has been provided. The last two do not. Clandestine diversion might be attempted by facility employees intending to sell the materials through a black market; by a non-national organization, such as a group of terrorists, one or more of whom may have infiltrated the facility organization; or, conceivably, by action of a facility management organization that is not effectively controlled by its government. Overt theft of materials might be attempted by terrorists or other types of criminal groups for subsequent sale through a black market, for blackmail threats within the country where the material was stolen or in some other country, or for unannounced acts of terrorism using nuclear explosives.

The application of IAEA or US safeguards offers no direct safeguard against abrogation of agreements by a government, or by a new government that has replaced the one in power at the time when the agreements were made. Assuming that a country had not yet diverted any plutonium from a safeguarded power plant before abrogating its agreements, it could then have several options for using the plant to provide plutonium for nuclear weapons, as follows:

1. If the power plant uses recycled plutonium and a fresh fuel core loading is in storage at the plant, it could easily separate the plutonium in facilities that could be constructed in a few weeks or less. A normal refueling load for a 1000 MW(e) power plant would yield about 250 kilograms of "low grade" plutonium, enough for at least a dozen nuclear explosives.
2. If the country had even a small spent fuel reprocessing plant, it would have two further options: reprocess some or all of the plutonium in the reactor core (about 750 kilograms of low grade plutonium); or replace some of the fuel with domestically fabricated natural uranium rods, remove the irradiated uranium frequently and extract "high grade" plutonium at a maximum rate of a few dozen kilograms a year. If designed in advance of the abrogation of an agreement, such a reprocessing plant could be built in less than two years

if it were small and its cost unimportant. Or it might conceivably have been built clandestinely before the abrogation. Still another possibility is that the country had been operating its own reprocessing facility, subject to safeguards.

3. If the country had its own uranium enrichment plant, based, for example, on new principles developed during the next eight or ten years and either built clandestinely or openly operating under safeguards, it could also separate about 1,000 kilograms of U-235 from stored fresh fuel. If it also had a reprocessing plant, it could extract roughly another 1,000 kilograms of U-235 from fuel in the reactor core. The time required to build a new type of isotope separation plant, however, would likely be at least several years, depending on new developments in this field. I should also point out that such a facility would enable the country to make material for nuclear weapons without using any reactor if it had even relatively small domestic sources of uranium.

Turning now to the possibility of diversion of plutonium from a safeguarded power plant, there are several possibilities. How likely it is that such diversion might go undetected depends on the detailed nature of the IAEA and US safeguards. Present safeguards are specifically designed to deal with all of these possibilities to at least some degree. It is also likely that new, more effective safeguard techniques will be developed and implemented during the forthcoming years.

This is a large and complex subject about which I cannot go into in satisfying detail in this brief study. I can summarize by saying that I am convinced that current types of safeguarding actions can be extended in ways that would make it very likely that diversion of kilogram quantities of plutonium from a large nuclear power plant would be detected, and, perhaps just as important, that this would be evident to operators of the plant.

Diversion of significant amounts of plutonium from a power plant requires removal of fresh fuel elements from storage (if they contain recycled plutonium), or removal of irradiated fuel elements from the core or from storage, or insertion into and subsequent removal from the reactor of units of uranium not normally present in the reactor.

Diversion of fresh fuel would be inherently the easiest, if there were no safeguards. Each fuel assembly for a light water moderated reactor weighs anywhere between about 1,000 and 2,000 pounds, and would contain somewhere between about one and three kilograms of plutonium (*if* it is being recycled) depending on the specific design. Several complete fuel assemblies would therefore have to be diverted to yield enough plutonium for a relatively unsophisticated nuclear explosive. Since IAEA safeguards inspectors could keep careful track of the number of fresh fuel elements, and identify each one with special markings, "dummy" elements containing less or no plutonium would have to be substituted. It is fairly easy to visualize safeguards that would make it extremely difficult to do this without detection, especially if safeguards inspectors could make unlimited inspections of the fresh fuel storage facility.

Safeguards inspectors can insist on being present whenever the reactor pressure vessel is "unbuttoned" to allow refueling or other types of maintenance. Frequent inspections and a variety of types of seals can be used to safeguard against changes in the reactor core not observed directly by inspectors. Another safeguard against unauthorized changes in the reactor or the way it is operated is provided by detailed reviews of the reactor design and inspection by IAEA authorities during construction and through the various stages leading to full scale operation.

Irradiated fuel in storage at the power plant's temporary storage pools is probably more inherently vulnerable to diversion and substitution by "dummy" fuel elements than materials inside the reactor. But frequent inspections, seals to make it difficult to make substitutions for irradiated fuel between inspections without detection and careful inventory of all spent fuel can make clandestine diversion at this stage a risky business.

Safeguards against nongovernmental diversion from a power plant are generally in the best interests of the recipient government. Relatively simple instruments now exist for monitoring access points to the plant to detect unauthorized transfer of even small amounts of plutonium through doors or passageways. It should also be remembered that light water reactor fuel assemblies weigh at least 1,000 pounds or so and yet contain only several kilograms of plutonium at most. With even a rather rudimentary monitoring system irradiated fuel would be extremely difficult to remove without detection. For preventing this type of diversion, heavy reliance could be placed on the recipient government itself, but the US should be assured that the safeguards for this purpose would be effective.

The inherent risks of overt theft by terrorist or other criminal groups in the recipient country are small if the fresh fuel is slightly enriched uranium and the irradiated fuel is not reprocessed within the country. Such risks would be considerably higher, however, if plutonium is recycled in the fresh fuel or irradiated fuel reprocessing is done within the country. Since safeguards against such theft should rely especially on physical security measures rather than materials accounting, and the IAEA currently has no responsibility for providing or even formally requiring such protection, the use of recycled plutonium or the establishment of a fuel reprocessing plant in the recipient country poses some special problems not dealt with by the IAEA safeguards system. These problems have to do with providing the US, as well as other countries that might conceivably be threatened by nuclear explosives made with stolen plutonium, with assurance that the materials are physically protected against any credible thefts. Realistically, I believe such protection would probably have to depend on the effective use of the recipient country's law enforcement or military authorities to deal with theft attempts by large, heavily armed and well-equipped groups.

Effective arrangements of this kind may be difficult to work out, and even more difficult for the US to enforce. This is, in my view, one of the most important issues related to US nuclear assistance to other countries, and should be carefully thought through in drawing up agreements for such assistance.

Chapter Six

India's Nuclear Policy

K. Subrahmanyam

A lot of conventional wisdom has accumulated over the years burying the real issues of nuclear proliferation and it has become difficult for most of the people concerned to isolate the crucial issues from the enormous overburden covering them. The conventional framework is constructed on the following lines. There is a problem of proliferation since the greater the number of countries acquiring nuclear weapons, the higher the probability of their being used. Underlying this axiom is another that nothing can be done to reduce the present number of nuclear weapon powers, and that consequently, by definition, nuclear proliferation concerns only the sixth, seventh, eighth nuclear weapon powers and so on. Further, the powers that have nuclear weapons are responsible and have behaved with restraint, and the assumption is that the new powers acquiring the weapons are not likely to. Lastly, according to this school, pragmatism requires that the peripheral issues are tackled first so that hopefully at some future time the main issues may be dealt with. Most of the debates in the Western world are conducted within this framework. It is the contention here that this framework is inadequate to deal with this issue comprehensively.

Is it realistic to envisage a world where certain powers have a military doctrine based essentially on the use of nuclear weapons and prevent this doctrine from being adopted by other countries? Is it feasible to have an international system in which certain countries derive political and technological benefits by virtue of their possession of nuclear weapons and at the same time prevent other countries which are within the reach of this technology from aspiring for it? How far are present nuclear strategic doctrines realistic and how much do they constitute convenient myths which further the interests of the possessors of nuclear weapons? Lastly, is there something special about nuclear technology which is different from other technologies that in this case alone it is impossible to draw a boundary line between peaceful and nonpeaceful uses, while the same can be done in regard to chemicals and bacteriological agents? Unfortunately, the debate on nuclear issues is carried on generally within like-minded groups in a particular cultural and political milieu with the result that these basic questions are raised very rarely. Analyses of this issue in most Western countries predominately restricted to marginal areas of nuclear proliferation to new countries or to esoteric nuances in regard to balances and advantages in numbers of warheads, launchers, throw weights, accuracies and so on.

In these circumstances, therefore, there is no reason to be surprised that the stand of the government of India evoked considerable amounts of incredulity in the Western world. This credibility gap between the West and India has a very long history. What appeared to the Indians as an oppressive imperialist burden was regarded in the Western world in the nineteenth and the first half of the twentieth century as a civilising mission of the white man. We were told that in spite of the fact that we have a long tradition of civilisation, orderly government and so on, except for the benefits of the British rule, Indians would have cut each others' throats (as a young student in school under the British Raj I had this experience). Those who were trying to brainwash us with this thesis of the white man's burden did not care to reflect on the history of violence in Europe. Similarly, when India proclaimed nonalignment, the credibility of that doctrine, which is today accepted by 90 nations of the world, was low. It was denounced as immoral, leaning towards communism or stooging for imperialism. We have gotten over that phase and John Foster Dulles's successor in office, Henry Kissinger, now acknowledges the contribution the nonalignment doctrine has made toward better understanding of the complexity of the world [1]. We are, therefore, used to living with this issue of credibility. The five nuclear weapon states developed nuclear technology primarily to reach weapon capability and deriving the civil benefits as spin-offs. It is inconceivable to them for a nation which is almost at the bottom of the poverty scale to have developed a strategy of nuclear development of its own, independent of the five nuclear weapon powers. There is an implicit assumption that all nuclear strategic wisdom is confined to the five nuclear weapon powers and those closely associated with them.

It will not be surprising if the participants of this conference are disappointed with not finding answers in this study to the questions they have in mind and if, instead, it raises a new series of questions to which, in my view, there are enormous mental and emotional blocks in this part of the world. It is like the well-known story of the emperor's new clothes. In the court of the Emperor Atom, full of nuclear courtiers, perhaps an unsophisticated mind—like the child in the story—is needed to ask some inconvenient but highly relevant questions.

THIRD WORLD PERCEPTIONS OF NONPROLIFERATION AND THE NPT

What is the problem of proliferation? The United States, the first country to become nuclear, made a decision to explore the possibility of developing nuclear weapons in 1940, found it feasible at the end of 1942 and successfully developed them in 1945. The development of nuclear weapons by the first nation in the world and its reluctance to destroy them at that stage under proper international supervision was the beginning of all proliferation and led to similar efforts on the part of the Soviet Union and Britain to develop nuclear capability. The Soviet Union became a nuclear power in 1949, four years after the United States, and Britain in 1952, seven years after the United States. This was a period of very rapid proliferation. After that, France took its decision sometime in 1956 and became a nuclear power in 1960. China took its decision in 1957 and became a nuclear power in 1964. India conducted its nuclear test ten years after China. It is, therefore, reasonable to conclude that the pace of proliferation of weapon countries is slowing down.

But the pace of nuclear weapon proliferation within the five nuclear weapon countries has accelerated. The United States today has the highest number of nuclear warheads deployed all around the world (over 6,500 strategic nuclear warheads, 7,000 tactical warheads in Europe and 4,000 in the Pacific), posing risks of nuclear thefts, problems of command and control, etc. The Soviet Union has the second largest nuclear force. The French and the British come next with their seaborne missile capabilities, and thereafter China with its relatively vulnerable silo-based force. Here one could see that there is a correlation between the force levels in terms of nuclear weapons, their sophistication, their relative invulnerability to destruction by hostile action, the order in which these powers have emerged and their overall industrial capability. In other words, it is the earlier weapon powers who are the engines of proliferation in terms of their technological lead, the development of nuclear doctrines and the credibility with which they are in a position to pose nuclear threats. While the earlier nuclear powers developed doctrines in the offensive use of nuclear weapons and not merely in exercise of nuclear deterrence, the later nuclear powers tend to emphasize the concept of deterrence and do not stress the first use of nuclear weapons.

Most of the nuclear doctrines in regard to bipolar deterrence were evolved in a strategic milieu dominated by the cold war when the issue was considered purely as a two person game. There was also an underlying attitude of self-righteousness. When the ABM was developed, there was considerable debate about its destabilizing potentialities in regard to stable deterrence. Although it is easy to see that their destabilization potential is vastly greater, the submarine-based missile, the MIRV, the MARV and the airborne ICBM have not generated analogous debates on their destabilizing potentialities.

The Soviet Union has not crossed the Central European line agreed upon in Yalta in 1945. The Greek communist party was decimated without the Soviets raising a finger to assist them. The Soviets were made to evacuate northern Iran. Barring the two in-bloc instances—Hungary and Czechoslovakia—there has not been any other case of Soviet military intervention. Still, in all the nuclear debates in the West it is axiomatic that the Soviet Union had to be deterred with nuclear weapons. As against this, the United States has a long record of intervention operations in Cuba, the Dominican Republic, Vietnam, Cambodia and Laos, besides various covert operations in many countries of the world. While the Russian interventions were against its own bloc countries, the US intervention operations had extended to nonaligned countries of Indochina. Still it does not occur to those in the Western world that in the security calculations of a number of nonaligned countries, there may justifiably be a perception of a threat from the United States more deep-rooted than the one that exists in the minds of the Western world in regard to the Soviet Union. While the Western world has never been under colonial subjugation of the Soviet Union, most of the nonaligned countries have recent memories of colonial occupation by the Western powers. The United States, instead of dissociating itself from the imperialist powers has, in the past 27 years, voluntarily donned their mantle, continued to support the imperialist powers' efforts to prolong colonialism and in the case of Indochina continued the colonial war.

The full implications of the war in Vietnam and the way in which it was fought have not been widely understood. Chemicals and herbicides were used extensively against a poorly armed peasant population, in a situation where there could be very little retaliation. There was resort to a degree of intensity of bombing never before witnessed in history, again in a situation where there could be no retaliation. Seventeen megatons of explosives were used and, according to a congressional report, it will take more than 150 years for the country to recover from the ecological damage inflicted.

The experience in Vietnam, and the circumstances that led to the use of nuclear weapons on Japan when compared with experiences of confrontation in the central European line and the Sino–Soviet border, suggest that mass destruction agents like nuclear weapons, ecocidal agents, etc. tend to be used only when there is no fear of retaliation and when there is no sense of mutual deterrence.

These aspects and their psychological impact on the third world have not received much attention in strategic discussions in this part of the world. In the case of nonalignment, the Nonproliferation Treaty and the Arab–Israeli issue, while we in India have been candid in speaking out, the other nations of the third world have been more diplomatic—adjusting their policies and pronouncements to the exigencies of the situation. Similarly, in the case of perception of threats emanating from the nuclear powers, the third world countries have been generally reticent in giving expression to their views. It is not, however, very difficult to infer their perceptions from the proceedings of the nonaligned conferences.

Today in the United States, China has become popular. In 1954 Walter Robertson, assistant secretary of state, testifying before the Congress declared that it was the policy of the United States to keep up pressure on China until its government broke up from within [2]. In that administration, Richard Nixon was the vice president. The government which effectively controlled more than 750 million people was denied representation in the United Nations for 20 long years at a time when the majority of the UN members used to vote along with the United States. Explaining why the United States has decided to make up with China (it was the president of the United States who made his journey to Peking, not the Chinese leaders to Washington) Herbert Klein, the presidential aide, said that a nation of 800 million with nuclear weapons could not be ignored [3]. But this nation, according to President Eisenhower, was threatened with nuclear weapons in 1953 and in 1958 [4]. Some of the documents recently made public reveal that the threat to use nuclear weapons in 1958 at the time of the Quemoy and Matsu crisis was mostly due to the overenthusiasm of some Pentagon officials to have ready recourse to nuclear weapons and that it was not necessarily justified by the situation [5]. Such nuclear threats must have had inevitable impact on China in its decision to go nuclear. Similarly, some people in India feel that the international insensitivity to India's tremendous problems in 1971 and the correlation between Chinese nuclear capability and the change in US policy toward China may have led to the Indian decision to enlarge on their nuclear option. Very rarely in the strategic literature are these reasons for nuclear proliferation—namely, the behavior of the current nuclear weapon powers as the primary motive force behind proliferation—discussed. Instead, these reasons are obscured by such irrelevant and totally unrelated considerations as prestige, the desire to have regional dominance and so on. How can there be credibility in the behavior of nuclear weapon powers when at the drop of a hat they issue a nuclear alert, as happened on the 25th of October 1973. Some of the senior US officials of the State Department have confessed that they themselves could not understand the justification for that action [6]. Whether it was to put pressure on the Chinese in Korea or during the Quemoy—Matsu crisis, or on the Soviets during the Cuban confrontation or the Arab—Israeli war of 1973, or even when the Vietnamese peasants besieged the US garrison at Khe Shanh, there was advocacy of the use of nuclear weapons by the most powerful military power in the world. This in turn has its impact on the perceptions of the third world nations in regard to the credibility of the pledges of nuclear weapon nations that they will not use nuclear weapons against non-nuclear weapon states.

There is a curious pattern underlying the thinking of those who have been the most active contributors to the doctrine of ready resort to nuclear weapons and who at the same time vigorously advocate that in the interest of international peace other nations should not acquire nuclear weapons. This is somewhat analogous to the argument that the Vietnamese had no regard for human life because hundreds of thousands of them died fighting in their own homeland. By implication it was sought to be conveyed that those who inflicted those casualties had somehow very great regard for human life. Perhaps it does not occur to many academicians in the nuclear weapon countries that their advocacy of nonproliferation totally lacks credibility in countries like ours. It reminds us of a medieval drunkard emperor in our country who sought to introduce prohibition. Understandably enough, he did not succeed.

In the entire literature on nuclear war there is no credible analysis of how exactly a nuclear war will be fought, commanded and controlled. Will the authorization to use the weapons be given individual weapon-wise, category-wise, theatre-wise or target-wise? We have the UN secretary general's report [7] which says that the devastation caused in a war using tactical nuclear weapons will not be very different from what will be caused by the use of strategic nuclear strikes. There seems to be an implicit assumption that as soon as the first nuclear shot is fired there will be hot-line parleys with the other side and the war will be brought to an end. There is no literature which analyses the results of sustained war in which a few hundred tactical nuclear weapons are to be fired by both sides. Still the Dutch proposal to reduce the 7,000 tactical warheads in Europe was opposed on the ground it will undermine confidence of certain people in the credibility of nuclear deterrence [8]. It may be argued that these issues are dealt with in secret studies. Recently we had Dr. James Schlesinger telling the world that before he took over as defense secretary the US targeting doctrine did not permit the president any flexibility in regard to graduated escalation and now his retargeting doctrine provides that flexible option. But before Dr. Schlesinger came out with this announcement no nuclear strategist pointed that out or analyzed its implication. Even now one cannot be certain that a successor of Schlesinger may not tell the world that his doctrine is much too rigid and he has been able to improve upon it. To the third world a nuclear war between two well-matched nuclear powers does not appear to be a credible proposition. Most of the strategic debate about the throw weight, first-strike capability, etc. reminds us of some of the esoteric medieval theological debates. No one who deals extensively with the possibility of an adversary reaching a first-strike capability and therefore advocates increases in warheads and throw weights explains credibly why the second strike at sea will not thereafter be used to devastate his cities. The 7,000 tactical nuclear warheads in Western Europe, 4,000 tactical warheads in the Pacific theatre, the current 6,000-odd strategic warheads (rising to over 12,000 by 1980) and the corresponding Russian figures do not make any rational sense at all to us in the third world. It would appear that the military-industrial-intellectual complexes of the industrial nations are in the grip of an irrational cult which has engulfed them in a dangerous closed system of logic.

A recently reported exchange between the US Arms Control and Disarmament Agency and the Pentagon illustrates the above point. Dr. Fred Ikle, director of ACDA, disclosed that the explosion of nuclear weapons will result in depletion of the ozone layer which forms a protective filter against excessive ultraviolet radiation reaching the biosphere. He raised the issues whether such depletion will be only transient or long-lasting, whether it would merely increase the hazard of sunburn or destroy critical links in the intricate food chain of plants and animals and thus shatter the ecological structure that permits man to remain alive on this planet. He has initiated a study into this new potential danger from nuclear war [9].

The Pentagon concedes that the amount of ozone over the temperature regions could be reduced by 50 to 75 percent if the nuclear arsenals on both sides—but particularly the larger weapons of the Soviet Union—were unleashed. But Pentagon officials contend that such a reduction would have the effect of lowering the ozone content over the temperate regions to about the level that normally prevails over the tropical region. Since life goes on in the tropical region they see no reason to conclude that a substantial depletion of the ozone layer over the temperate region would have a serious adverse effect on living matter [10].

This reply of the Pentagon validates the charge of Dr. Ikle that we are not only unable to express the human meaning of nuclear war—the only meaning that matters—we are also unable to express the full range of physical effects of nuclear warfare, let alone to calculate these effects, because the damage from nuclear explosions to the fabric of nature and the sphere of living things cascades from one effect to another in ways too complex for the scientists—he should have added strategic analysts and militarymen—to predict. The above reply of the Pentagon prima facie lacks credibility. If the temperature of temperate zones goes up to the level of tropical zones the effect will not just be limited to the temperature. It will affect climate, the arctic ice cap and the entire ecology of the hemisphere.

The noncorrelation between the nuclear weapon stockpiles, strategic requirements and doctrines; the nonavailability of detailed analysis of command and control problems in regard to nuclear war; the casualness and negligence in deploying nuclear weapons and resorting to nuclear threats by the nuclear weapon powers; and the inadequacy of meaningful discussion of these issues in the West raise serious problems of credibility in regard to the entire range of strategic doctrines on the basis of which the current nuclear proliferation in the West is being sustained. Added to this problem of credibility is the absurdity of some of the arguments used against new countries acquiring nuclear weapons and the seriousness with which such arguments are propagated. It is asserted that while responsible powers like US, USSR, Britain, France and China can be safely entrusted with nuclear weapons, it is dangerous for Arabs and Israelis, Indians, Pakistanis and others to have such weapons. The unstated assumption is that these lesser breeds of human beings will not act as responsibly as the former and they have frequently resorted to wars. But may one ask which war resulted in more casualties; use of chemicals, herbicides and other ecocidal agents; 17 megatons of explosives—the four Arab–Israeli wars and four Indo–Pakistani wars all taken together or the war fought by the United States in Vietnam? What is the record of the five nuclear weapon powers? Two of them still, recently, were imperialist powers, two of them are self-anointed messianic powers and one is currently engaged in telling the world that if oil flow is stopped it will resort to the use of force. All five of these powers were involved in two world wars. In recent years the American legislature has discovered that it was misled in regard to the facts that led to the Tonkin Bay resolution, on the thousands of sorties on Cambodia and on the bombing that went on in Laos since 1964. Are we in a position to categorize that decisionmakers in certain countries are more responsible than in other countries? Did not Germans under Hitler consider themselves the most responsible people in the world, with a special mission to regulate the less responsible people in other countries? Is this attitude really qualitatively different from the one that underlay the myth of Aryan superiority? Is not this a twentieth century version of the doctrine of the white man's burden? If the Western nations believe sincerely that proliferation of nuclear weapons is bad, let them first practice it. This doctrine of nonproliferation has not been accepted by China, India, the USSR, the USA, Indonesia, Japan, Brazil, the UK, France, Pakistan, West Germany, Italy, Algeria, Argentina, Chile, Cuba, Egypt, South Africa, Israel, Spain and Taiwan among other states. Some of them may have gotten themselves formally licensed to proliferate under the so-called Nonproliferation Treaty, some may have signed the treaty but not ratified it, some others have verbally endorsed it but not signed it, and yet others honestly and openly disagree with it. The majority of mankind does not believe that the nuclear weapon powers are going to stop proliferating or that they are sincere in their advocacy of nonproliferation. It is easy to get states which in any case have no nuclear option or which are safely under somebody else's nuclear umbrella to sign and ratify a so-called nonproliferation treaty. But it contributes little to nonproliferation. Those who signed the treaty have not been able to get one warhead or one launcher reduced in spite of the preamble and Article VI of the treaty.

Neither SALT I nor the Vladivostok agreement have contributed to credibility in the doctrine underlying the so-called Nonproliferation Treaty. Since the Nonproliferation Treaty was signed the United States has increased its strategic warheads from 1,720 to 6,922 and the Soviet Union is estimated to have increased its from 1,580 to 2,337 [11]. The Strategic Arms Limitation Treaty only legitimized the existing stockpiles of launchers and further licensed the two weapon powers to continue with their proliferation of strategic weapons in the form of MIRVed warheads. While the Nonproliferation Treaty tried to declare that in a world of 138 equal sovereign states five states are more equal than others, SALT further refined that process and declared that among the five equal nuclear states, two are even more equal. The Vladivostok agreement makes it abundantly clear that the preamble and Article VI of the Nonproliferation Treaty are not likely to be taken seriously by the two superpowers. They are engaged in capping an arms race by the 1980s by increasing the weapons by a factor of four and including a new category of weapons and at the next round in the 1980s on the basis of their past performance the nations of the world will be asked to rejoice at an agreement which will cap it at a further multiple of four or eight and will include a further new category of weapons. We are even told that the fact that they negotiate at all constitutes a great victory for arms control. Which nation will be reluctant to conclude such arms control measures?

The problem of credibility in regard to Western postures on nonproliferation is further compounded by the recent disclosures regarding the casualness with which nuclear weapons are deployed and fissile materials are dealt with by the weapon-producing countries. Representative Clarence D. Long of the military construction subcommittee of the House Appropriations Committee filed a few months ago a written warning that the US nuclear weapons kept abroad were highly vulnerable to terroristic attempts to destroy or capture them. Among his findings are: (1) facilities to protect nuclear weapons against raids by terrorists or subversive elements were mostly 15 to 20 years old; (2) ill-protected weapons were exposed to the grave danger of destruction or seizure in case of an attack by a group of subversive elements with sufficient firepower; (3) a well-financed terrorist group could carry out such a raid anywhere in the world; (4) an abnormally large number of US nuclear weapons are widely scattered in various countries and thereby poses a grave problem of control (in one foreign country 700 warheads are deployed with 82 warheads in one airforce base alone); (5) in a country where more than 200 nuclear weapons were stockpiled there were slums where antigovernmental elements were known to have been hiding for years within a distance of just 80 meters from such weapons [12].

Another staff report of US Senate Foreign Relations Committee has revealed that 7,000 nuclear warheads in Europe are stored at over 100 sites and two-thirds of the sites contain weapons to be used by the European forces [13]. Though the weapons are technically under American control, they have been assigned to allied forces for use in the event of war. This procedure, and the availability of the weapons to allied forces, give rise to concern about their safety at a time of tension and crisis as in the case of the recent Turkish-Greek confrontation. Given the enormous wealth available to some of the Arab countries and the reports that Israel either has or is on the point of acquiring nuclear weapons, the chances of weapons deployed in Europe being diverted clandestinely are quite significant.

In January 1974 congressional testimony was released showing that a total of 3,647 persons with access to nuclear weapons were removed from their jobs within a single year because of drug abuse, mental illness, alcoholism or discipline problems. Another study found that 8 percent of the US army men surveyed in Germany used hashish daily and 53 percent said they had tried it at least once [14]

There have been fights between blacks and whites on US aircraft carriers [15]. One cannot rule out the possibility of a nuclear missile submarine being taken over by mutineers, since such fights could as easily occur among blacks and whites in a submarine. There has not been any discussion in the West about nuclear weapons falling into the hands of Black Panthers, dissident groups in the Soviet Union or the followers of a future Lin Piao in China and its implications for international security.

The recent disclosure [16] that there are considerable quantities of fissile material unaccounted for in US nuclear installations has not received the attention it deserves. There have been attempts to fob it off as possible bookkeeping errors, but this is not likely to carry much credibility in the rest of the world. The statistical approach which leads to the conclusion that the greater the nuclear weapon power the higher the probability of the use of such weapons should also have led to the conclusion the larger the number of nuclear facilities in a country and the greater the throughput of fissile materials through the installations of a country the higher the probability of theft, pilferage and diversion of fissile materials. If this logic had been adopted the US, with the largest number of power reactors, should have submitted itself to the most rigorous inspection and stringent safeguards system. Instead they exempted themselves from such inspection under the so-called Nonproliferation Treaty. The casualness with which they handle the fissile material is adequately dealt with in the Mason Willrich and Theodore Taylor study, *Nuclear Theft: Risks and Safeguards* [17]. The United States has the highest number of skilled and knowledgeable personnel who have the know-how to assemble a weapon. It also has some of the best-organized criminal organizations (like Cosa Nostra) in the world. The law of probability would indicate that the risk of either fully assembled weapons being stolen or parts pilfered and assembled into a weapon are the highest in the United States. Recently the Libyan leader Colonel Gaddafi talked of acquiring nuclear weapons [18]. Obviously they are not going to be produced in Libya. The strange silence of the US and Western governments and academicians (with exceptions such as Willrich and Taylor) on this issue raises very serious problems of credibility in regard to their approach to the question of proliferation. It cannot be accidental that during the sixties there were a plethora of studies on the issue of proliferation by way of new countries acquiring nuclear weapons but that there are very few studies on the issue of nongovernmental proliferation emanating from nuclear weapon countries. When the problem involved is likely to be embarrassing to one's own country it does not get studied. When such issues are raised in conferences, generally there is an uncomfortable silence and then the subject is changed. This procedure is not resorted to in this part of the world alone. The third world, while harboring a total lack of credibility about the western nuclear doctrine and approach to proliferation, has also been making its views known by a deafening silence in such discussions with the Western scholars and officials and expressing their apprehensions about US attitudes by discussions on imperialism at like-minded assemblies of the nonaligned.

The concept of nonalignment, the ideas that the Soviet Union is a responsible power, that the People's Republic of China was there to stay and had to be brought into international community, that Vietnam would prove a quagmire for the United States, and that there can be no peace in western Asia until restitution is granted the Palestinians took time to gain acceptance. It may be necessary to wait longer before it is realized that the Western nuclear doctrines and their approach to proliferation and arms control do not command much credibility elsewhere in the world. No one seriously believes that the industrial nations are going to fight a nuclear war among themselves. The doctrines of deterrence developed as a two person game are becoming somewhat obsolescent. It is obvious that the Western European powers do not seriously perceive any threat from the Soviet Union. Otherwise they would not be reducing their defense burden. The NATO—Warsaw Pact parleys today constitute bloc negotiations in which each side is trying to obtain the better of a bargain. The figures on warheads, launchers and throw weights bandied about have long ago crossed the threshold of irrationality. The fact to be noted here is that we are dealing with decisionmakers who are under the hypnosis of some irrational and highly esoteric dogmas. It is likely that there may be many among them who recognise the irrationality of their dogmas, but in such closed groups it must be difficult for one to come out openly in criticism of the dogma. As in religion, the practice of which finally depends on one's declared faith in a central dogma, one will not be a reputed strategic analyst, or reach influential positions in the defense department, unless one practices the faith. It is perhaps as difficult for a nonbeliever in the Western strategic doctrines to rise to influential positions in either the academic world or government as it would be for an avowed non—Marxist in the Soviet Union or non—Maoist in China. Some declare their faith in religion, some in Marxism—Leninism, some in Maoism and some others in the late twentieth century dogma of the nuclear strategic doctrines. Faith most often is a matter of one's birth and parentage and the society in which one has to function. Only a small minority dare to exercise independent judgment.

Today the real danger is not that of a nuclear war between nuclear powers but a nuclear threat to a non-nuclear power. In the Second World War Hitler used poison gas not against other armies which were in a position to retaliate but against the Jewish prisoners. Carpet antipopulation bombing, use of chemicals and herbicides, etc. were resorted to only in situations of asymmetry when there was no fear of retaliation. Nuclear threats have been held out against non-nuclear powers (Korea, 1953; Quemoy—Matsu, 1958; Khe Sanh, 1968). The superpower détente has contributed significantly to increasing the threat to the nonaligned nations. By mutually reassuring each other that neither superpower will resort to any action against the other or its allies or any area of vital interest the uncertainty about likely escalation in case of nuclear threat or strike against a nonaligned nation has been reduced. These developments and historical perceptions tend to increase the sense of insecurity of nonaligned nations in regard to future nuclear threats or strikes. These constitute more credible factors in the perceptions of nations than the dubious statistical approach relating the increase of threat to international security by an addition in the number of nuclear weapon powers.

It is possible to argue that the increase in the number of nuclear weapon powers will contribute to the reinforcement of general deterrence since with the number of actors increasing the uncertainty for each actor must increase. In a two person game it was argued that mutual assured destruction (the MAD doctrine) contributed more effectively to deterrence, but should this necessarily be true in a multiactor game? Someone may point out that this is the old Gallois thesis. It may be and what is wrong with it? On the other hand, one may also point out that the thesis that the greater the number of nuclear nations the greater the danger of a nuclear war breaking out is a hangover of the fifties when there was not adequate understanding about the nature of nuclear war. In banning production, stockpile and use of biological agents, our historical experience contradicts the dogmatic assertion that the greater the number of nuclear weapon powers the higher the probability of a war breaking out and the more difficult it is to reach an agreement on banning production, stockpiling and use of the nuclear weapons. This has also been the experience in the case of chemicals covered by the Geneva Protocol. As pointed out earlier it broke down in Vietnam only under conditions of extreme asymmetry.

There is no study on the comparative risks to international security caused by the increase in the nuclear power stations in the nuclear weapon powers and their material throughput, the increase in nuclear weapons in the hands of nuclear weapon powers, their increasingly scattered deployment, and the command and control problems caused thereby and the marginal increase if any by a new nation acquiring a small nuclear force. One would like to see it convincingly argued that the marginal increase of a nuclear weapon power will in fact increase the risks to international security. There are reasonable grounds to believe that such studies are not undertaken mainly because it will prove that the rate of increase of risk to international security is substantially higher with increase in nuclear stockpiles of the nuclear weapon powers than by the marginal addition of a new nuclear power.

It was earlier mentioned that the rate of proliferation in terms of new nuclear weapon states has significantly slowed down compared to the first and second decades of nuclear era. Those who advocated a domino theory on the advance of communism in Asia have also displayed a similar simplistic perception in regard to nuclear proliferation. The latter is as credible today as the former. India was not taken seriously when she argued that the first domino theory was absurd and she has similar problems of credibility with the same people in regard to the second domino thesis. The Bourbons of the twentieth century learn nor forget anything.

In order to obfuscate the real issue that nations are likely to go nuclear when they feel that they are likely to face threats from nuclear weapon powers, conventional wisdom tries to ascribe the idea that prestige is a major factor in proliferation. It was so in the British and French nuclear weapon programs, but the Soviet Union and China developed their weapons under threats to their security. The Nonproliferation Treaty instead of stripping the nuclear weapons of prestige enhanced prestige by conferring on the British, with their minimum deterrent, a status equal to that of the other nuclear weapon powers. Prestige weighs very much with the fading imperial powers, and the same motivation is ascribed to others. The decisionmakers who were responsible for a war fought 10,000 miles away and who spent over 50,000 lives and $190 billion on considerations of false prestige readily adopt this thesis. The danger of such self-fulfilling argument is that a nation like Japan which aspires to become a permanent member of the security council may be persuaded that it will have to pay the club entrance fee by developing a nuclear weapons capability.

Most of the nations which are mentioned as likely candidates to become the sixth, seventh, eighth nuclear weapon powers (India is not a weapon power and maintains that she does not intend to become one) are Israel, Japan and South Africa, all of which accept Western nuclear doctrines and strategic perceptions. They were not influenced by the Indian concept of nonalignment as a large number of other nations were. They all voted for Nuclear Nonproliferation Treaty, unlike India which opposed the treaty in its present form and abstained from it. Israel and Japan depend on the US for their security, and their intellectuals, soldiers and statesmen interact closely with those from the US. Therefore, if these nations turn and develop a nuclear weapons capability, they will be following the example of the five nuclear weapon powers and not India. There is widespread belief that Israel already has nuclear weapons. Though this has been mentioned in the US Congress since the middle of 1970, there has been no significant reaction in the Western world. Japan is not likely to go nuclear so long as its dependence on external fuel for its reactors continues. Japan says it does not have much use for peaceful nuclear explosions. Hence its turning nuclear can be only for weapon purposes. The continuation of fuel supply after Japan turns nuclear will mean the formal burial of the Nonproliferation Treaty by the nuclear weapon powers. South Africa does not face a threat from any nuclear power. In any case nuclear weapons are no answer to the increasing insurgency threat it is likely to face from the majority of its own population.

Brazil and Argentina are often mentioned in regard to their ambitions to try peaceful nuclear explosions. Brazil, indeed, joined India in its opposition to the Nonproliferation Treaty on the ground that the treaty did not differentiate between peaceful explosion and weapon tests. Both Brazil and Argentina have not signed the NPT. Their signatures to the Tlatelelco Treaty are subject to interpretations not acceptable to other parties to the treaty [20]. While Brazil has only a US–type boiling water reactor (which produces plutonium with maximum Pu-240 content), Argentina has a natural uranium reactor and a plutonium separation plant. Argentina is generally considered to be nearer to a peaceful nuclear explosion than Brazil if it chooses to develop this technology. If either of these two countries choose to exercise such a option one could perhaps trace some influence of India on their decision.

Pakistan used to be cited by the nonproliferation lobbyists as a certain follower of India if the latter were to go nuclear. Pakistan, while voting for the Nonproliferation Treaty, did not sign the treaty, ostensibly because India had not signed it. Following the Indian PNE, Pakistan initiated a campaign to obtain nuclear guarantees from other nuclear weapon powers and so far this attempt does not seem to have succeeded. The Pakistani prime minister has threatened from time to time [21] that his country would also turn nuclear and in the recent months has linked up this threat of going nuclear with continuation of US arms embargo in respect of his country. On receiving assurances during his recent visit to Washington that the arms embargo will be reviewed, he has declared that all Pakistani reactors will be put under international safeguards [22]. The Indian assessment is that Pakistan will take about eight years to be in a position to construct its own natural uranium reactor [23].

In these circumstances an accelerated proliferation triggered by the Indian PNE seems to be a far-fetched assumption. It is not unlikely that in the fourth and fifth decade of the nuclear era there may be additional nuclear powers (both weapon powers and nonweapon powers) but while such a slow gradual proliferation of nuclear powers seems to be undesirable, it is not preventible by the kind of approach to proliferation so far adopted by the Nonproliferation Treaty.

INTERNATIONAL PERCEPTIONS OF INDIAN NUCLEAR CAPABILITY

Having thus dealt with the issue of credibility of the conventional wisdom on the proliferation issue from the Indian point of view it is necessary to admit that the Indian government's stand has a similar problem of credibility for the rest of the world under the influence of the conventional wisdom. It is, however, possible to establish that the Indian government's current stand that it did not intend to go in for the manufacture of nuclear weapons is credible and can be substantiated. At present, India does not have plutonium producing facilities free of safeguards to sustain a reasonable weapons program. The US reactor at Tarapur and the Rajasthan reactor at Kotah built with Canadian assistance are both under IAEA safeguards. The Indian-designed reactors will be commissioned only from 1977–1978 onwards. Second, if India intended to go in for nuclear weapons in the near future, command and control problems will have to be tackled, steps for surveillance and early warning will have to be taken and the armed forces will have to develop doctrines. In an open society like India it is easy to see that these steps have not so far been initiated. Third, India has no credible delivery systems and its electronic industry is still far from meeting the requirements of a credible nuclear posture. Fourth, India does not need nuclear weapons against Pakistan as an adversary, while a few nuclear weapons are of no use either against China or to resist pressures of superpowers and this is fully understood in India. Therefore, there is nothing in the Indian government's declaration that India does not intend to go in for the manufacture of nuclear weapons that contradicts any of the actions and past pronouncements of the country. The PNE is a logical culmination of a policy advocated by India as far back as 1965 in the ENDC [24]. Since 1970 India had repeatedly made it clear in the Indian Parliament that the country would conduct a PNE as soon as investigations were completed [25].

Unlike the five nuclear weapon powers, India does not aspire to any international status. India turned down the offer of a permanent seat on the Security Council in place of China in the fifties and there is no strong lobby in India, as there is in Japan, for a permanent seat on the council. The Indian prime minister declared during her address to the Canadian Institute for International Relations that India did not want to be a power nuclear or otherwise [26]. India has no historical tradition of exercising its power beyond its own borders. The Indian influence outside India has always been cultural. India has consistently turned down all ideas of being a rival of China in Asia and of exercising leadership (an idea sponsored by President Kennedy and communicated in person by Vice President Johnson) in this part of the world [27]. The Indian behavior pattern has generally puzzled Americans and Westerners, and American observers have wondered whether India has a will to power at all [28].

While it is not status quo-oriented, India has no ideology to export and does not conceive for herself even the role of a regional power. The Indian armed forces have no strategic mobility. They are equipped to operate only for a few miles across India's land frontiers. There have been attempts in the West to portray India as a significant regional military power after 1971 and there was even an amusing study in the United States about India attempting to seize the West Asian oil fields after a decade or two [29]. These perceptions arise mainly because in the West the tendency is to ascribe to other nations one's own pattern of behavior and aspirations. In the last three decades India has been mostly concerned about its own problems of development and security. Jawaharlal Nehru played an international role because of his personality and certain objective conditions of the period. It is widely recognized that India has been playing a less salient role in international affairs in the sixties and seventies than it did earlier. In Southeast Asia the complaint is that India plays a low profile role in that region [30]. Contrary to the impression spread in certain sections of the Western press, India and Iran have developed very cordial relations and share certain common perspectives. It is true that in certain sections of the ruling elite in the countries around India, especially among those educated in the West and influenced by the West, there are references from time to time to India's size and power and the threat India may pose to their security. This is often used to involve links with the West, not as a countervailing factor but to strengthen themselves vis-à-vis their own masses, just as the Communist bogey has been used in certain countries by the elite to obtain Western support against the local population.

India is likely to be increasingly absorbed in its internal problems and these are likely to preoccupy India for the next two to three decades. In fact, the Indian concern is how to shield herself effectively from external interventionism during this difficult period. It is recognized that India's security problems have arisen by and large due to the interaction of great powers in the South Asian strategic environment [31]. The Indian diplomatic efforts in the last three decades have been directed toward reducing the great power presence around her. It is therefore hardly likely that India will do anything which will alarm her neighbors and make them invite big power countervailing presence in her neighborhood. Indian history of the eighteenth and nineteenth centuries leading to the subjugation of India is a sorry tale of small principalities in pursuit of their short term interests inviting external interventions. The present Indian policy of promoting a consensus in the Indian Ocean area to exclude the presence of external powers is derived from these historical memories. The idea of India taking upon herself the role of an international power is a figment of a fertile imagination conditioned by roles played by powers elsewhere in the world in the last two centuries.

India has been involved in five wars in the last 28 years since her independence, besides one decolonisation operation. All these wars were initiated by either China or Pakistan and were partly or wholly fought on Indian soil. The size and shape of the Indian armed forces and the Indian security policy are influenced largely by these developments. The Indian forces went to the Congo, Gaza, the Sinai, Indochina and Korea on UN missions. They were deployed in and around Sri Lanka in 1971 at the request of a democratically elected government of that country. India liberated Bangladesh after ten million refugees were pushed onto her soil and nine months of patience failed to produce any results. India had assisted Nepal to deal with an insurgency in 1950 and sent some assistance to Burma when the United Nations government was hard pressed by insurgents in the late forties.

India has a treaty of peace, friendship and cooperation with Bangladesh which requires India to enter into consultation if the security of Bangladesh is threatened to decide on such steps as may be considered necessary to counter such a threat. India has an open border with Nepal but has no legal or formal responsibility in regard to its security. While there is a treaty between Bhutan and India in regard to the former receiving advice on external affairs from the latter, India has assumed no formal responsibility for the security of Bhutan. A look at the map of the Indian subcontinent and its neighborhood will reveal that India's security is inextricably linked up with that of Pakistan, Nepal, Bhutan, Bangladesh, Burma, Sri Lanka and Maldives. India does not, however, believe in mutual defense pacts in order to ensure security of the region, preferring to deal with each situation on merits as it arises. India is extremely sensitive to the disproportionality of her own size and resources and those of her neighbors and has been trying to bend over backwards to avoid giving an impression of playing the role of big brother. This is one of the reasons why in regard to overseas Indian minorities India has adopted an attitude of no intervention on behalf of Indian nationals abroad.

The development of Indian nuclear technology has no relevance to India's relations with any of its neighbors and the Indian prime minister has already assured the Pakistani prime minister that the Indian nuclear technology will not be used against the interests of any other nation [32]. Since the five nuclear weapon powers carried out their nuclear tests for the purpose of development of weapons there is a presumption in most of the Western world and among those who are influenced by the strategic thought from the Western world that the Indian test should be in pursuit of the same purpose. It is likely to take some more time before the Western world accepts that the Indian atomic energy program is not just an imitation of Western programs but had an independent origin, motivation and strategy of development.

The Indian atomic energy program was started as early as March 12, 1944, 15 months before the Alamogordo explosion, when Dr. Homi Bhabha, in a letter written to the Sir Dorabji Tata Trust, proposed the setting up of an institute for fundamental research. In that letter he declared "[W]hen nuclear energy has been successfully applied for power production, in say, a couple of decades from now, India will not have to look abroad for its experts but will find them ready at hand." Let it be noted that even at that stage it was the peaceful and developmental use of atomic energy which impelled Dr. Bhabha to initiate the program.

India was among the first eight countries where an atomic energy commission was set up. At the first Atoms for Peace Conference in Geneva in 1955, Dr. Bhabha predicted the coming in of fusion power in the next two to three decades. In Asia, India set up its first research reactor, Apsara, in 1956, ahead of China and Japan. Similarly, the first plutonium separation facility outside the nuclear weapon states was also established in India entirely through indigenous efforts. India also formulated a long term three stage strategy for the development of nuclear power. The first stage will consist of a natural uranium heavy water reactor producing plutonium. In the second stage the plutonium will be used to fuel fast breeder reactors and the same breeder reactors will convert thorium, which India has in abundance, to U–233, which is a fissile material. In the third stage, U–233 will be the fuel for reactors [33]. These facts would show that the Indian atomic energy department had a long range and independent view of the nuclear development strategy for this country and did not follow the beaten track of other nuclear powers.

A similar approach is now applied to the technology of peaceful nuclear explosions. There is no expectation that the development of this technology will yield any short term gains. The Indian scientific community and the government are aware that it may take years for India—as it will for the USA and USSR—to master the technique of nuclear explosive technology. But following Dr. Bhabha's vision when he moved for the establishment of the Tata Institute of Fundamental Research, when mastery over nuclear explosive technology is established and peaceful explosions become normally applicable to constructive uses, Indians should not depend upon the Americans, the Russians, the Chinese, the French or the British to carry out such projects in their country. Knowing full well that this technology will take another 10 to 15 years to develop, India has decided to make a beginning now.

Considerable skepticism has been expressed in the West about India's ambitions to use peaceful nuclear explosions for constructive purposes. Part of the skepticism arises from the appreciation that since the US has not so far used PNEs constructively and has no immediate plans to do so, a developing country like India cannot be in a position to do so. One may recall how in the fifties and early sixties the oil lobbies used to decry the economics of nuclear power. While the application of the PNE may not have made much progress in the US it seems to have been more widely and purposefully employed in the Soviet Union. The *New Scientist* reported in December 1972 that there was evidence of Soviet silver having been mined using PNE [34]. According to a report of the Lawrence Livermore Laboratory the Soviet program of nonmilitary detonations appears to cover a wider range of field experiments than does the comparable Plowshare program in the US. While the US program has concentrated in recent years on development of technologies to recover natural resources such as gas, the Russians appear to have pressed forward with a more diversified effort, including canal excavation tests, the sealing of runaway gas wells, the stimulation of oil field flows and the creation of underground storage reservoirs for oil and gas. A tabulation compiled by the Californian scientists is reported to have shown about 35 presumed nuclear detonations in the Soviet Union over the last seven years in areas away from sites normally used for weapons tests. In the IAEA the Soviet Union is reported to have described 14 experimental and industrial nuclear explosions [35].

The editor of the scientific journal *Nature,* Dr. David Davies, reported in an article recently that the Soviet Union has exploded more than 30 nuclear devices since 1967 to build canals, increase the flow from oil fields, store water and put out oil well gushes [36]. In the *New Times* of Moscow in February 1975, Igor Dmitriyev has confirmed that with a PNE the Soviet Union had blasted a 50,000 cubic meter underground reservoir to store gas condensate. The Soviet experts claim that this new nuclear method cuts construction costs to one-third. The same article describes another interesting experiment concerning the proposed 112.5 kilometer canal along the Pechora–Colva divide. The purpose of this canal is to divert the water of the north-flowing river into the Caspian Sea. The Russians have apparently carried out three nuclear blasts on the route of the proposed canal and now estimate that the use of nuclear blasting will cut construction costs by two-thirds or more and complete the job faster [37]. It is perhaps relevant to note that the Soviet Union has accepted the Indian stand that the nuclear test of May 18, 1974 was in pursuit of peaceful nuclear explosive technology.

The Indian national goals, to quote Dr. Vikram Sarabhai, involve leap-frogging from a state of economic backwardness and social disabilities—attempting to achieve in a few decades a change which has historically taken centuries in other lands. Even as the country is being industrialized with setting up of steel mills, machine-building plants, fertilizer plants, oil refineries, ship-building, etc., it has been found necessary to catch up with more sophisticated industrialization in electronics, aerospace, nuclear technology computerization and automation. Deep down in the Indian consciousness there is the historical memory that India was colonized mainly because it fell technologically behind the West and, further, that during the colonial period the country was deliberately denied industrialization. Today, there is considerable fear that the technological gap between the superpowers and other nations may widen and measures like the Nonproliferation Treaty may be used as convenient instruments to establish the technological hegemony of a few nations. This fear is shared by others as well. Mrs. Alva Myrdal, a fervent protagonist of the Nonproliferation Treaty, has expressed her disillusionment in the following words: "This split into two discontinuous categories of 'super powers' and 'other nations' has not only become more apparent to us during the disarmament negotiations, it has been made even more bluntly manifest by a conscious design on their part. The best example of this is, of course, the NPT. . . . What we are witnessing today it seems to me, is the emergence of a duopoly of the two super powers in regard to modern technology, giving them a more and more dominating hegemony over the World Affairs" [38]. Again she has written in regard to PNE, "The nuclear weapons states are the sole possessors of the know how and the means necessary to use nuclear devices for such potentially very lucrative activities as digging canals and extracting oil, gas and minerals. Thus, the mighty have a virtual control over the application or nonapplication of these techniques, a fact that cannot but cast ominous shadows over the future, if an international licensing system cannot be introduced" [39]. These are among the factors that motivate India in going ahead with its development of nuclear technology, including peaceful nuclear explosions.

After the PNE there has been some concern in the Western world whether India would start exporting the nuclear explosive technology. Some tendentious comments were made on an Indian agreement with Argentina on nuclear cooperation following the Indian nuclear test. But the agreement specifically excludes information on research of a secret character [40]. India has no more interest in exporting nuclear explosive technology than the other five nuclear weapon powers. Even China, an ardent advocate of breaking the nuclear monopoly of superpowers, has not chosen to transfer that technology to other nations.

On present indications India does not seem to be looking for any specific payoffs from her nuclear status. The idea that nuclear powers constitute a privileged group is incorporated in the Nonproliferation Treaty and the Western attitude toward it. India has consistently disagreed with this approach. Over the last three decades a mystique has been developed around nuclear weapon capability and the Nonproliferation Treaty seeks to legitimize this by according the nuclear weapon powers special privileges (such as exempting them from inspection procedures, giving them the sole right to conduct PNEs, not requiring them to cut back on their armaments and legitimizing those armaments, etc.). In fact, by carrying out the PNE and not claiming any special privileges, India is attempting to denigrate this special mystique developed around the nuclear weapons. India has exercised some of the privileges reserved for nuclear weapon powers and at the same time declared that she will not go in for weapons. By this step India is trying to disassociate the possession of nuclear weapons from other special privileges.

The Soviet Union decried nuclear weapons at a time when it was in a vulnerable position vis-à-vis the United States. It no longer does so. The Chinese called nuclear weapons paper tigers, but once they acquired them their behavior has conformed to the conventional wisdom on the subject. India alone, after demonstrating its capability to produce the weapons, has tried to denigrate the weapon by proclaiming that it does not consider it necessary to go in for its manufacture. It is not unlikely that this strategy may have some deterrent value (not of the kind implicit in the MAD doctrine) while at the same time contributing to strip nuclear weapons of the prestige invested in them by the so-called Nonproliferation Treaty. Just as nonalignment gave considerable maneuverability to the countries practicing it, the policy of developing nuclear capability without ostensibly converting it into a weapon capability has similar advantages. During the days of nonalignment the two power blocs, though they opposed each other, were tacitly agreed that there was no third way and hence both were opposed to nonalignment—though China and the Soviet Union displayed greater flexibility of approach much earlier than the US and the Western alliance. Similarly, at present the nuclear weapon powers, as well as countries who cannot conceivably expect to go nuclear on their own, may conclude that there is no third position possible between nuclear weapon status and non-nuclear status. The current Indian posture is one of nuclear nonalignment—neither with nuclear powers, nor with those who have renounced acquisition of nuclear capability, including PNE capability—for all time to come.

Nonalignment constituted a significant challenge to the international system as the two power blocs sought to develop it in the years immediately after the war. The entire Western hemisphere, Western Europe and the few sovereign states outside Eastern Europe with a few odd exceptions formed a system of interlocking alliances. The Soviet Union, China and Eastern Europe formed the other system of interlocking alliances. Chairman Mao Tse Tung felt that a country had to lean to one side or the other and there was no third road available. (He seems to have changed his mind since then.) The Indian position then was incredible as it is today. It was interpreted by both sides in a hostile manner. Scholars of the West who could not grasp the significance of the concept of nonalignment called it neutrality and proceeded to berate it in many learned volumes. Today 90 countries have adopted that policy. It is not unlikely that part of the opposition to the Indian nuclear posture may arise out of the realization that this poses a challenge to the current structure of international power, which equates the nuclear weapon status with the big power status and seeks to freeze the Yalta settlement arrived at before the developing world was decolonized. India consciously does not pose this challenge as she did not even when she formulated the nonalignment policy based on the perception of her national interests and conditioned by her colonial history. But when one-sixth of mankind adopts a policy, it is bound to have a major impact on the international system.

While the Indian motivation to go in for the PNE and the current stand on manufacture of nuclear weapons can be adequately explained, the Indian declaration about not exercising the weapon option has to be interpreted as being subject to certain external and subjective conditions. As a parliamentary democracy, the declarations of this government are binding only on this parliament and they may continue to influence the future governments from the same party. This declaration is based on the perception that India sees in the near future no nuclear threat to its security [40]. If nuclear weapon nations resort to nuclear threats or use of nuclear weapons elsewhere in the world such action is bound to have its impact on future Indian perception. Interventionism by nuclear powers under the cover of their nuclear might is likely to have an adverse impact on the Indian decision not to go in for the manufacture of nuclear weapons [41]. Large scale buildup of external navies in the Indian Ocean, Chinese missile tests in the Indian Ocean, and development of strategic and tactical doctrines which emphasize the use of nuclear weapons will also influence the Indian decision on going nuclear. Therefore, discussion on whether or not India will exercise its nuclear weapon option will be meaningful only if the external strategic environment is defined. Most of the discussions on a possible Indian decision have skirted around the issue of the behavior of the existing nuclear weapon powers and have tried to portray the Indian decision as though it is unrelated to and uninfluenced by external factors. Consequently, most of those scenarios have been unrealistic.

Some ask the Indians to define the doctrine of use of Indian nuclear force and its composition and size. Historically all countries developed the weapons first and the doctrines later. As pointed out in the earlier section of this chapter the doctrines developed on the use of nuclear weapons from time to time in the West have not been noted for their sustained credibility. Even now after 30 years since the first nuclear weapon was developed it is difficult to identify a doctrine for the use of nuclear weapons which is generally accepted among the nuclear weapon powers. The whole issue of nuclear war is wrapped in terrible and gigantic uncertainties. If India decides to develop nuclear weapons it will be to deter nuclear threats and the use of nuclear weapons against her. It will not be based on considerations such as equalizing conventional superiority with the use of nuclear weapons. It will be based on the doctrine that when irrationality dominates decisionmakers in some powerful countries, they will have to be deterred by factors which are logical within their irrational framework.

In considering India's future nuclear intentions attention is bound to be focused on the Pakistani proposal for a South Asian nuclear weapon–free zone. India has taken the stand that the initiative for the creation of such a zone should come from the states concerned. In the Latin American nuclear weapon–free zone such an initiative came from the states concerned and led to a considerable amount of fruitful discussion among the Latin American states before it came up before the UN. None of the Latin American states which are signatories to the Tlatelelco Treaty has a security problem with a nuclear state. Cuba has, and it is not a party to the treaty. The geographical configuration of Asia and the Indian Ocean is such that it is difficult to separate a South Asian nuclear weapon–free zone from an Indian Ocean nuclear weapon–free zone. Lastly, whether Tlatelelco Treaty has really achieved the purpose of a nuclear–free zone is open to doubt. Argentina and Chile have not ratified the treaty. Cuba and Guyana have not signed it. The Soviet Union has not acceded to the Protocol II to the treaty, and without one of the major nuclear powers recognizing the area as a nuclear-free zone it cannot be deemed to be one. The treaty provides that no political entity should be admitted, part or all of whose territory is the subject of a dispute or claim between an extracontinental country and one or more Latin American states so long as the dispute has not been settled by peaceful means. In South Asia there are disputes between China and India, Pakistan and India, Afghanistan and Pakistan. Neither Pakistan nor China has agreed to sign a declaration that they will not use force to alter a border. India feels that without these basic issues being discussed among the countries concerned, bringing up the issue before the UN will only complicate the issue. Apart from that it is doubtful whether this planet, which a satellite circumnavigates within 90 minutes, can really be apportioned realistically into compartmentalized nuclear-free zones and whether such acts do not lend themselves to increasingly legitimize a nuclear arms race among a few selected nations.

The Indian stand on the Nonproliferation Treaty and the current Indian PNE have served to strengthen the hands of those nations which, having signed the NPT, are still attempting to fight the battle of arms control and disarmament. The Indian PNE is the only event since the ratification of the NPT which is likely to put some pressure on the sponsoring nuclear weapon powers to have a second look at the treaty and the way in which they have failed to fulfill their obligations under Article VI of the treaty. Otherwise the NPT would in all probability have been declared a successful arms control measure which needed no revision at all. In retrospect it is clear that many of the nations which are genuinely interested in arms control had surrendered too easily their nuclear option under the NPT without using it as a bargaining leverage to bring about genuine arms control. It has now been left to countries outside the NPT to exert pressure. History may well record that those who stayed out of NPT contributed more effectively to arms control than those who yielded to pressure and signed away their nuclear option without obtaining anything in return.

NOTES TO CHAPTER SIX

1. Dr. Kissinger's address to the Indian Council of World Affairs, October 28, 1974, reprinted in *Strategic Digest* (New Delhi), November 1974.
2. Quoted in *India in World Affairs, 1954-56* (New Delhi: Asia Publishing House), p. 272.
3. *Times of India,* July 18, 1971.
4. Eisenhower, Dwight David. *Mandate for Change.* The White House Years. [1st ed.] 2 vols. Garden City, NY: Doubleday, 1963-65.
5. *Statesman,* December 25, 1974.
6. *International Herald Tribune,* December 9, 1974.
7. EFFECT OF POSSIBLE USE OF NUCLEAR WEAPONS AND SECURITY AND ECONOMIC IMPLICATIONS FOR STATES OF THE ACQUISITION AND FURTHER DEVELOPMENT OF THESE WEAPONS. A/6858. New York: United Nations, 1968.
8. *The Guardian,* December 12 1974.
9. *Times of India,* September 10, 1974.
10. *International Herald Tribune,* October 22, 1974.
11. *US News and World Report,* November 4, 1974, p. 31.
12. *International Herald Tribune,* September 28-29, 1974.
13. *International Herald Tribune,* September 23, 1974.
14. *International Herald Tribune,* December 18, 1974.
15. *US News and World Report,* November 27, 1974, p. 27.
16. Thousands of pounds of Plutonium were unaccounted for at 15 commercial plants regulated by the USAEC. See the *New York Times,* December 30, 1974.
17. Mason Willrich and Theodore B. Taylor. *Nuclear Theft: Risks and Safeguards* (Cambridge, Mass.: Ballinger Publishing Company, 1974).
18. *Patriot,* January 21, 1975.
19. *New York Times,* July 17, 1970.
20. Stockholm International Peace Research Institute, World Armaments and Disarmament: See *SIPRI Yearbook, 1974,* Stockholm: Almquist and Wiksell, 1974, notes, pp. 491-492.
21. *The Hindu,* October 15, 1974, and *Times of India,* December 2, 1974.

22. *The Hindu,* February 7, 1975.

23. The assessment was made by Dr. H. N. Sethna, chairman of the Indian Atomic Energy Commission in a recent statement before the Parliament Consultative Committee attached to the Department of Atomic Energy. See *The Hindu,* February 13, 1975.

24. For example see V. C. Trivedi's statement in the First Committee of the UN, October 31, 1966.

25. For example, Defence Minister Jagjivan Ram told the Lok Sabha on May 2, 1972 that "in the field of atomic energy we are already amongst the advanced countries of the world" and that the Atomic Energy Commission was studying the technology to conduct underground nuclear explosions for peaceful purposes.

26. Prime Minister Shrimati Gandhi's address to a luncheon meeting of the Empire Club and Canadian Institute of International Affairs, Toronto. For text see *Sainik Samachar* Annual 1973 (August 12, 1973), pp. 7–9.

27. John Kenneth Galbraith, *Ambassador's Journal* (London: Hamish Hamilton, 1969), p. 122.

28. William P. Bundy's statement before the US House Foreign Affairs Subcommittee, July 20, 1971. For text see USIS release dated July 21, 1971.

29. *Indian Express,* January 10, 1975.

30. This is the general complaint of Lee Kuan Yew, the Singapore prime minister, and came up for discussion in the Conference on the Security of Southeast Asia held in Singapore under the joint auspices of IISS, London and ISEAS, Singapore, May 31–June 3, 1974.

31. India Ministry of Defence. *Annual Report* 1971–72. New Delhi Government of India Press, 1972.

32. Prime Minister Shrimathi Gandhi's letter to Prime Minister Z. A. Bhutto of Pakistan. For text see *Patriot,* June 8, 1974. Also Shrimathi Gandhi's interview to American Broadcasting Corporation wherein she reaffirmed India's policy not to use nuclear energy for military purposes or threat to anybody. *Financial Express,* June 17, 1974.

33. Dr. Raja Ramanna, "Development of Nuclear Energy in India, 1947–1973," *Nuclear India,* September 1974.

34. *Mainichi Daily News* (Tokyo), December 29, 1972.

35. *Dawn* (Karachi), July 9, 1974, and *Times of India* October 23, 1974.

36. *Patriot,* January 11, 1975.

37. *Indian Express* and *Tribune,* January 19, 1975.

38. Alva Myrdal. "The Game of Disarmament," *Impact of Science on Society* 22, no. 3 (July–September 1972): 229–30.

39. *Ibid.,* p. 224–25.

40. *Financial Times* (London), May 31, 1974.

41. Deputy Defence Minister J. B. Patnaik's statement in the Rajya Sabha on August 29, 1974. See *The Hindu,* August 30, 1974.

Discussion Essay: Nuclear Safety, Weapons and War

RISK ASSESSMENT

Several types of risks are associated with the proliferation of nuclear energy plants and nuclear explosive capabilities. One is the risk of unintentional catastrophic events occuring during the nuclear energy fuel cycle (Hohenemser). Another is the possibility of (intentional) nuclear theft by terrorists and the intendent risks to public order (Taylor). And a third, also intentional, is the increased risk of nuclear war posed by the existence of more nuclear states (Subrahmanyam).

The three chapters illustrate the complexity of risk assessment. Judgments about nuclear risks involve evaluating the physical characteristics of the fuel cycle, the influence of technological developments upon a nation state's (or terrorist's) capacity to produce explosive materials independently and efficiently, the distribution of nuclear materials and the nature of the international strategic system. Focusing upon reactor safety, Hohenemser concludes that risk assessment is at a very rudimentary stage, although the risks are probably sufficiently clear to warrant using truncated fuel cycle operations at as many nuclear plants as possible. Taylor, by contrast, assumes that the new nuclear states will not accept trancated fuel cycle operations, and that safety and safeguards risks must be met with other preventive procedures, either bilateral or through the International Atomic Energy Agency (IAEA). Subrahmanyam argues that the risks of intentional nuclear catastrophe, or war, are not greater and, in fact, potentially fewer if the two major superpowers lose their nuclear monopoly.

In each area—safety, safeguards and war—risk assessment is a very new, primitive and speculative enterprise. One of the major difficulties is in deciding which events to assess. Assumptions regarding the field of relevant events tend to become polarized into ideological positions (Kates). One of these is that the events which are being assessed are only a very small portion of the potentially catastrophic events (the tip-of-the-iceberg approach). Conversely, another common position is that too many highly improbable events are assessed. Still a third possibility is that the wrong events are monitored. Risks must compete for attention and the competition is not necessarily related to scientific merit. Assessing risks to human safety and implementing precautionary measures always take place in the context of other human values (McClintock). Nuclear energy issues, like others, are influenced by value priorities.

SAFETY

In addition to these caveats about the general limitations of risk assessment, several specific nuclear safety issues arise. For example, Hohenemser's suggestion that truncated fuel cycles should be used in as many reactors as possible was based upon his assessment of the number of potential risks occuring at different stages of the fuel cycle. But the truncated fuel cycle produces reactor wastes which include not only actinides but also plutonium, with its far longer half-life (Wilbur). If these wastes are released into the environment through inadequate disposal methods, the truncated fuel cycle may increase health risks, albeit perhaps spread over a longer time span, to higher levels than if the plutonium were recycled into the reactor. Unfortunately, the additional environmental hazards presented by waste plutonium over waste actinides also are uncertain (Hohenemser).

Ideally, nuclear energy production risks would be evaluated against risks from other energy sources. Coal and oil also pose unique risks (Taylor). One dramatic prediction is that the electrostatic precipitators used to reduce coal burning pollution will induce an ice age by the reflection of the sun on the small particles above the earth's surface. Another ice age scenario is based upon changes in the coefficient of friction between the ocean's surface and the wind, caused by oil spills. Conversely, the higher levels of carbon dioxide released into the atmosphere from burning fossil fuels have a countervailing warming effect. Similar disagreements arise over thermal imbalances created in the production of solar energy. At minimum, regardless of the hazards foreseen, it is "flat out wrong to say that the only catastrophies that may be incredible are associated with nuclear energy" (Taylor).

WEAPONS

Basic political assumptions are interwoven with nuclear risk assessment. The truncated fuel cycle is less acceptable than a complete cycle to nation states which are intent upon developing nuclear explosives. As preferences for nuclear explosives multiply, so too will the number of complete fuel cycle reactors. Catastrophic risk assessment, then, includes intentional catastrophes.

The increased risks of war from proliferation usually are attributed to the greater number of persons in the position of making decisions to use nuclear weapons. More specifically, the risk is of an irresponsible decisionmaker gaining control over weapons—either a government leader or a terrorist (Schoettle).

A tip-of-the-iceberg approach to assessing intentional risks is (1) that any nuclear detonation would entail the Armageddon of a wider nuclear war; (2) that leaders of second order governments are more likely to be irresponsible; and (3) that the number of nation states possessing nuclear weapons is more relevant to risk assessment than is the type or number of nuclear weapons held by the superpowers.

All these assumptions were challenged. Contrary to the escalation premise, a limited nuclear war could remain limited (Kemp). Second, nuclear weapons could have the same stigma attached to their use by a second order state as they have for the superpower (Subrahamanyam). Third, proliferation could have a restraining effect upon the superpowers against further deployment of nuclear weapons, and contribute toward deterring against their use in any situation.

SAFEGUARDS

Few assessments of intentional risks are accepted both by the superpowers and by the new or potential nuclear states. Such vague risk of war considerations as are presented will not themselves constrain a state from producing nuclear explosives. Safeguarding a nuclear reactor system against diversion by the host government is a limited approach to reducing hazards. Safeguards against the theft of nuclear materials are more promising (Taylor; see Chapter Five).

Nuclear theft is highly probable if the history of thefts of conventional weapons is an accurate guide (Kemp). The major risk of theft must be confronted when nuclear materials are missing or not in sight; the threat alone is not a sufficient warning of catastrophic risk (Taylor). Enough information about nuclear explosives is available so that an educated layman can make a threat which will be credible if materials are missing. Also, as the number of nuclear engineers increases, the number of potential real threats from possible counter-elites also increases.

The risks of nuclear theft increase with increased numbers of reactors, nuclear engineers and widely available information. Concern about theft has increased also and numerous proposals for safeguarding against theft have been made. Among the approaches to nuclear safeguards are:

1. Creating nuclear parks with both reactors and any reprocessing or enrichment facilities within the area (Wilbur). The nuclear park concept would both minimize theft risks by centralizing facilities for security purposes and be responsive to some safety problems by lessening the amount of materials transporting necessary.
2. Maintaining comprehensive plutonium inventory systems (Wilbur).
3. Maintaining records of persons with nuclear engineering skills (Leitenberg).

4. Maintaining security forces of moderate size (Taylor).
5. Creating international technical assistance programs which emphasize safeguards procedures. New or potential nuclear states have as much self-interest in preventing theft as do the superpowers, but many do not have comprehensive safeguards systems (Taylor). The United States also could benefit from such advice. Canada has been testing safety and safeguards procedures for the Candu reactor which could be applied in other settings (Taylor).
6. Instructing the International Atomic Energy Agency to develop recommendations on safeguards. The agency does not have statutory authority but it can promulgate recommendations. It has made significant contributions to what is known about procedures for safe handling of radioactive materials (Taylor).

The last proposal for international safeguards procedures raises the issue of nuclear technological competition. Significant differences exist among the nuclear states in terms of technological capacity. For example, the United States is several years behind France in the development of breeder reactors, partly because of US public hesitancy over nuclear power (Wilbur). The fear is that industrial espionage will take place during inspection. As a result, in the past, inspection has excluded nuclear reactors (Lawrence; see Chapter Three).

Safeguard and safety procedures are subject to both political and economic constraints. All the policies proposed above are costly. National sensitivities and the "garrison state" image of the securely safeguarded nuclear plant are major political concerns (McClintock, Kemp, Leitenberg).

The economic costs of a sufficient safeguards system could be established within the upper limit of 2 percent of total production costs (Taylor). With full recyling, they could be as low as 0.9 percent. These figures can change over time, but the current price of safeguards is relatively low (Taylor).

Whether or not one agrees that the financial costs of safeguards would be willingly borne by the nuclear industry (Rosen did not), other long-run economic costs can be anticipated. For example, in an economy in which a high proportion of the total energy used is porduced by nuclear fuel, the costs of safety and safeguards procedures will be higher in terms of energy lost from facility shutdowns (Baker).

The possible public fear engendered by strong safeguards and safety procedures may threaten the nuclear industry with stagnation and provide disincentives for adopting those procedures (Bray). Conversely, if the safety and safeguards measures are widely regarded as sufficient, the procedures might have positive economic effects (Taylor).

Safety and safeguards systems probably would be most reliable if the cost and hazard decisions were not made by the same group of persons. Although the nuclear industry is under private control in only a few countries (the United States, Japan and, to some degree, France), publicly owned industry can also be subject to the influence of economic considerations (Taylor).

Both nuclear risk assessment and the formulation of policies designed to prevent nuclear catastrophes warrant sustained attention in the future. Research is likely to be most rewarding in the area of safety and safeguards against terrorism. Moreover each new nuclear technological innovation potentially raises new issues for safety and safeguards procedures and makes necessary a continued dialogue between the scientific community and policymakers.

Part Three

Chapter Seven

Nuclearization and Stability in the Middle East

Steven J. Rosen

Is it possible that we stand at the edge of a transformation of the Arab–Israel conflict to a nuclear stage? Observers with different opinions on the substance and process of the conflict are coming to agree that nuclearization could happen very suddenly, if indeed it has not already happened. For example, at a recent conference of 25 scholars on proliferation, a straw poll was taken on the likelihood that Israel and/or one or more members of the Arab coalition would come into possession of atomic weapons. Twenty of 25 agreed that Israel either already has an atomic explosive or would have one within two years, while almost as many believed that at least one Arab state would have atomic weapons under its own control within ten years.

We may soon need a conceptual apparatus to comprehend a regional "balance of terror." Yet academic analysis of the nuclear issue in the Middle East has been restricted until now largely to what might be called prenuclear policy questions: Does Israel have the capacity to develop the bomb? Why should or shouldn't she go ahead? Can the Arabs also get it? What are the benefits and costs of a secret option? At what point should covert development stop? What can be done to halt the spread of nuclear weapons and to create incentives for a nuclear-free zone?

The purpose of the present analysis is to explore the consequences of the open introduction of nuclear weapons in the Middle East and particularly to examine the possibility of a regional balance of terror as a path to stabilization [1]. The central hypothesis is that apocalyptic visions and doomsday images of what an openly nuclear Middle East would look like have been accepted too readily and out of proportion to the arguments that are given. A stable system of mutual deterrence may be entirely possible, and such a system may make a positive contribution to the political deescalation of the Arab–Israel conflict.

LOSS OF FAITH IN THE CONVENTIONAL BALANCE OF POWER

The balance of terror is one of three conceptually distinct approaches to stability in the Middle East, and interest in this approach can be related to a loss of faith in the other two. The first and preferred method is political settlement, which in this case would take the form of a territorial compromise—"a piece of territory for a piece of peace"—and the relaxation of mutual fears and hostilities. But given the intensity of this conflict between two nationalist movements over the past 50 years, a negotiated settlement of all outstanding issues as a method to achieve stability seems to be a rather remote possibility, however desirable a "just and lasting peace" may be in principle.

The second and assertedly more "realistic" approach assumes the permanence of conflict and seeks stability not in reconciliation but in a stable balance of power. If potentially aggressive states are held at bay by the strength of the potential victim, they will eventually moderate their demands and accept a compromise. *Si vis pacem, para bellum.* The containment policy looks for its example to Central Europe, where the peace that has prevailed for 30 years is based not on a perception in either camp that the postwar lines of division are equitable or optimal but simply on a universal belief that no change in the status quo can be achieved by force of arms at an acceptable cost. Similarly tense but stable equilibria exist between China and Taiwan, North and South Korea, India and Pakistan, Iraq and Iran, and other hostile dyads that hold each other off by the venerable method of a balance of power.

Pacification of the Middle East by a regional balance has been the goal of American policy for almost 30 years. In practice, "balance" has meant an Israeli margin of superiority to dissuade an effort to eliminate the Jewish state. Abba Eban stated the objective succinctly in 1965, two wars ago: "We want to create doubt, and eventually resignation and despair [in the Arab mind] about the dream of eliminating Israel from the world's map" [2]. But Israel has now decisively won four wars, the most recent under worst-case conditions, yet the Arabs are far from "resignation and despair" and indeed some seem ready to contemplate a fifth, sixth, Nth round rolling indefinitely into the future [3]. Many observers are beginning to ask of the second approach, *Can a stable military equilibrium be created between Israel and the Arab states?*

Why has the prediction of the balance theory failed, and why have at least some Arab leaders not given up on military solutions? Some Israelis explain it as a kind of Arab madness, an endless *jihad* or holy war in which there is no concept of ends-means rationality nor any calculation of the relationship between actions and their consequences, costs and benefits [4]. But three "rational actor" explanations deserve attention before resort is made to socio- or psychopathological analysis. These refer to (1) the reasons given for past defeats; (2) reasonable expectations of future improvements in the military balance; and

(3) subtle understanding of the conversion process between military defeats and diplomatic victories. These reasons for the failure of the conventional balance approach must be understood to comprehend the nuclear alternative.

Past Defeats

Unless one assumes a racial theory of Arab inferiority, tiny Israel's past victories against the populous Arab world must be regarded as somewhat "unnatural" events to be explained by special circumstances. For example, a crude comparison of the quantities of combat aircraft and battle tanks in the inventories of Egypt, Syria, Iraq and Jordan with those of Israel on the eve of the three more recent wars and today would lead to the appearance that the Arabs have consistently been stronger. Ignoring differences in the performance characteristics of the various types of equipment, the arms ratios have favored the Arab side by margins of two and three to one (see Tables 7-1 and 7-2) [5]. The gross figures for Egypt alone exceed those of Israel in all four years.

For those who seek them, explanations are available for the "unnatural" Israeli victories in each of the wars. For 1956 the "special condition" was the intervention of two major powers in support of Israel, as well as the inexperience of the Egyptians in the use of modern weapons. In 1967 the "special condition" was the uniquely successful Israeli surprise attack on the Egyptian airfields, as well as the poor leadership and coordination of the Arab armies and perhaps also the failings of Nasser as supreme commander. In 1973 the Arabs did better, but mistakes included a weakly defended expanse of the canal line between the

Table 7-1. Combat Aircraft

	1956	1967	1973	1974
Four Arab States	299	677	1222	1136
Israel	110	290	488	466
Ratio	2.7:1	2.3:1	2.5:1	2.4:1

Source: Derived from figures in the International Institute of Strategic Studies, *Military Balance* (London, IISS), various issues.

Table 7-2. Tanks

	1956	1967	1973	1974
Four Arab States	870	2395	4570	5550
Israel	400	800	1700	1900
Ratio	2.2:1	3:1	2.7:1	2.9:1

Source: Derived from figures in the International Institute of Strategic Studies, *Military Balance* (London, IISS), various issues.

Egyptian Second and Third Armies; the failure of the Egyptians to press their advance into the Sinai and of the Syrians to press theirs into the northern plains of Israel in the first days of the war; and most notably the failure to activate the Jordanian third front. Moreover, throughout these wars Israel has had certain technological advantages that have been critical, especially the superiority of her air force. If Soviet research and development close the technological gap with the American equipment and the Arabs are supplied with more advanced models, the Arabs, if they can reduce their own technological gap with Israel, will have the means to erase these Israeli advantages. Thus, it is possible to argue that the past Israeli victories are not necessarily a guide to the future.

Future Trends

Moreover, almost every development since October 1973 favors the Arab side of the equation with regard to the future military balance. While the Arab coalition enjoys unprecedented wealth and influence, Israel is increasingly isolated and under tremendous financial strain. The additional oil revenues of the Arab members of OPEC exceeded $60 billion in 1974, while Israel's entire GNP was approximately $8 billion. Israeli military expenditures as a percentage of GNP increased from 17 percent in the first year after the June War (i.e., 1968) to approximately 40 percent in the first year after the October War (1974). Israel does not have the resources to continue the present rate of acceleration in the arms race, while the Arabs can continue to escalate. Israeli Ambassador to Washington Simcha Dinitz [6] and Defense Minister Shimon Peres [7] have both begun to speak of a one to three quantitative arms ratio as the "minimum defense posture" Israel must have to maintain the military balance, and even these requests will have rough going in Washington. The United States is beginning to weigh Israel's unilateral dependence against its own interests in the Arab world and in Europe. The Soviet navy is increasing its fleets and operating schedules in the Mediterranean and Indian oceans. The Arabs are gradually closing the industrial and technological gap that has been Israel's main advantage. The Arab coalition has finally achieved and seems likely to sustain a measure of unity in fact that previously existed only in theory. Against such a long list of new advantages, why not assume that time is on the side of the Arab cause? There is a revolution of rising expectations. In the words of President Sadat, Israel is "in a corner. . . . Right is on our side. We have the military power, the strength of Arab solidarity and its various means. We have the upper hand" [8].

I do not mean to convey an impression of gloom from Israel's point of view, for the truth is much more complex than these simple images. But it would not be unreasonable for an Arab strategist to conclude, from a certain ordering of the facts, that the day may come when the Arabs can defeat Israel in the field. Indeed, *some* Israelis and their friends abroad have reached exactly this conclusion.

Military Defeats Equal Diplomatic Victories

The final reason that the Arabs have not been dissuaded from a policy of military challenge is a sophisticated strategic notion based on observation of the relationship between what happens on the battlefield and what happens at the conference table. The Arab–Israel conflict is not autonomous but is in many ways a dependent subsystem of the world balance of power. The local wars are permitted to continue as long as the Arabs are on the offensive. Once they begin to lose badly, the superpowers intervene to impose a cease fire, as neither the United States nor the Soviet Union believes it can tolerate the possible consequences of a full humiliation of the Arabs. Then, at the conference table, the pressure is on Israel rather than the Arab states to make concessions. The purpose of active warfare, in other words, is not necessarily to achieve movement in combat, but in diplomacy. Like the mouse that roared, the Arabs can win by losing.

For all these reasons, it appears that the conventional balance of power cannot alone bring stability to the Middle East, even if Israel can sustain the margin of superiority she has enjoyed through great exertions in the past. Secular changes in the external environment, uneven rates of economic growth in the Arab camp and in Israel, and the special nature of perceptions in the Middle East make it unlikely that the European or Korean models of military stalemate can be achieved in this region.

THE NUCLEAR DETERRENCE APPROACH

If the search for a stable conventional balance of power produces nothing but an unending series of wars, and the conditions for a comprehensive political settlement don't exist, it seems reasonable to conclude that Israel may eventually turn to a policy of nuclear deterrence to convince the Arab populations and their governments of the futility of continuing their confrontation with the Jewish state. Ironically, the case for this Israeli approach has been put forth most eloquently by an Arab scholar, Fuad Jabber.

> Where conventional power has failed, weapons of mass destruction would be expected to succeed in convincing Arab populations first and their governments second of the futility of continuing their confrontation with Israel. With the realization that Israel cannot be militarily defeated, the rationale behind the permanent state of war, the economic blockade, and the policy of non-acceptance and non-recognition might be expected to break down. Moreover, whatever tendencies towards recognition and negotiations may already exist in the Arab world would be enormously strengthened. Hitherto, governments willing to negotiate had not dared to act because their position at home would have become untenable. In a nuclear context, the survival imperative might provide enough justification to make such approaches possible. . . . The psychologically erosive effects

of the nuclear logic would be at work on the Arab will, gradually producing that pervasive feeling of "doubt—and eventually resignation and despair—about the dream of eliminating Israel from the world's map" that Israel's doctrine of deterrence had always sought to create [9].

Curiously, other Arab writers have shared this "bullish" appraisal of the utility of nuclear weapons in enforcing stability in the Middle East and guaranteeing the ultimate security of Israel, while Israeli writers have been much less confident [10]. Taking the Arabs first, here is a selection from the very scant analytical literature on the subject [11].

Ahmad Samih Khalidi, in "An Appraisal of the Arab–Israel Military Balance," writes that, when Israel gets the atomic bomb there will be a "stalemate." Egypt will also get atomic weapons "and then neither country will dare attack the other" and "the sands will have run out for the Arabs" [12].

Colonel Hussan Mustafa wrote in an Iraqi daily that "the atomic weapons would serve as a permanent deterrent to the Arabs, would compel them to avoid provocations." Moreover, "Israel's acquisition of atomic weapons will decrease the value of the ordinary weapons in the hands of the Arab armies and the importance of their numerical superiority" [13].

Finally, the most extensive statement on the subject, by Muhammad Hassanein Heikal, confidant of Nasser and Sadat and for many years the editor of Cairo's *Al-Ahram,* gave three reasons why the atomic bomb is likely to be essential for Israel:

> *Mentality* ... Israel is aware that she stands alone and that she may one day find herself in circumstances where she will no longer be able to rely on the foreign forces which have helped her so far. ...
>
> *Security* ... Israel is aware of a rising and strengthening pressure of opposition. ... Egypt would soon reach a military condition where she could stand up to Israel [as] a dimension of quality [is added to] the quantitative superiority of the Arabs. ...
>
> *The ruling cliques* ... The top level of the Israeli establishment feel that time is not working in their favor, both on account of the changes occurring in the world at large and because of the revolution in the Arab world and the shifts in the traditional balance of power [14].

Serious Israeli analysis of the possibility of nuclear deterrence in the Middle East is also scarce, and as Simha Flapan observes, "The Israeli public has shown until now a total lack of interest in the implications of nuclear power" [15]. But most of the literature that does exist is far less confident of the benevolent effects of atomic weapons for Israel than the Arab statements cited above. Yigal Allon, now foreign minister, has long been known for his doubts on nuclear deterrence. "Israel would be exposed to a fresh danger of the utmost gravity if any Arab country got hold of nuclear bombs, irrespective of whether or not

Israel herself possessed similar bombs for retaliatory purposes." The regimes of the Arab countries may at any time become "militant, unstable, and irresponsible" and there would be little consolation for Israelis "huddled together in anti-radiation shelters" to know that "they could engage in nuclear retaliation" [16].

A similar view is given by most Israelis writing as private citizens, including Avigdor Haselkorn—"If nuclear weapons were to become part of Middle Eastern arsenals Israel would be faced then and only then with a *real* threat to its physical existence" [17]; Yair Evron—"It is most unlikely that an Israeli nuclear option would change either the basic pattern of conflict or the basic strategies employed by the two sides" [18]; Simha Flapan—"The 'balance of terror' is valid least of all in the Middle East" [19]; and Nahum Goldmann, president of the World Zionist Organization—"I see no advantage whatsoever nor any moral or security justification for the manufacture of atomic bombs" [20]. It is a striking fact that almost the entire literature by Israelis on the subject of nuclear weapons is broadly hostile to the strategic and political consequences of the open introduction of nuclear deterrence capabilities by one or both sides in the Middle East [21].

The premier objection given by all pro–Israeli analysts who fear the introduction of atomic weapons in the Middle East is the territorial issue. The population of Israel is highly concentrated and vulnerable to massive damage from even a few nuclear weapons [22]. All major cities are within easy striking range of bombers flying out of Egyptian or Syrian airfields. If these could not penetrate Israel's interceptors and surface-to-air defenses, the Scud surface-to-surface missile with a range of 160 to 180 miles could reach any targets from Syria, Jordan or the northern Sinai (though only the southernmost part of Israel if fired from the western side of the Suez Canal). While a substantial portion of the US population might survive a very small scale Soviet counterforce attack [23], it is difficult to imagine any "postattack environment" in which the tiny state of Israel would be habitable.

However, all this argument shows is that it would be even more unacceptable for Israel to fight a nuclear war than for a larger country. The main purpose of atomic weapons is deterrence, not fighting wars. The relevant question for Israel is whether she can threaten her likely adversaries with levels of damage that they too will find unacceptable [24], and here the answer is certainly affirmative. The population of Egypt is concentrated in the upper third of the country, including Cairo, Alexandria and the towns along the Nile delta, and a third of the Syrians live in Damascus, Aleppo, Homs, Hama and Latakia. The Israeli air force has enjoyed for 20 years the freedom to strike in depth at these Arab targets virtually without limit. Should this freedom be lost in the future to advances in surface-to-air missile defenses, Israel is reported to have a Jericho offensive missile of its own manufacture with a striking range of 280 to 300 miles [25]. Cairo, Alexandria, Suez, Damascus, Amman and the Euphrates Dam could be hit from Jericho launching sites behind the pre-1967 borders of Israel, though Aswan would require a base at Sharm el Sheik and Baghdad would be out of range of Israeli land-based missiles but not of airplanes. The Lance missile now being supplied by the US gas a shorter range (45 to 75 miles depending on the weight of the warhead) but could reach Damascus, Amman, Beirut and possibly the Mitla pass. By any combination of delivery modes, an Israeli strike force armed with even a few atomic bombs of modest yield could threaten damage adequate to deter any rational Arab adversary. No city of the Arab world would be immune to destruction if Israel were attacked with nuclear weapons or if overwhelming massed conventional forces penetrated across the green line to threaten the heartland of the Jewish state.

Some analysts have even suggested that the major cities of south-central Russia, including Odessa and Tbilisi and the Baku industrial region could be possible targets of Israeli Phantoms on one way suicide missions [26]. If so, this would give the Soviet state a direct interest in the moderation of the Mideast conflict and in preventing confrontations which could give rise to an Israeli mood of desperation. However, this arresting strategic idea seems to ignore the fact that the USSR had in 1974 over 2,900 strategic defense interceptor aircraft and 1,200 counterair tactical fighters [27], as well as the highest density surface antiaircraft defenses in the world. Also, an Israeli attack wing would have to overfly Iran, Greece or Turkey, or naval craft would have to move undetected through the Bosporus and Dardanelles into the Black Sea [28]. Finally, as Bell notes, Israel would have to consider the possibility of Soviet reprisals against the hostage population of three million Russian Jews. Yet the possibility cannot be excluded that a desperate Israel could evade Russian radars and interceptors by overflying weakly defended areas at treetop levels, successfully executing a strike on southern Georgian cities. Even a small possibility that unrestricted escalation of a Middle Eastern crisis could result in atomic destruction of a Soviet city could be expected to act as a moderating influence on Russian policy in this region.

Thus, the problem of limited territory does not rule out a stable deterrence system of "mutual assured destruction" if we assume the existence of two opposed parties armed with nuclear weapons [29] and the availability of delivery systems able to penetrate enemy lines and inflict unacceptable levels of damage. Two other objections to stable deterrence in the Middle East require more extended analysis: (1) the ability to protect retaliatory, second-strike forces against disarming first strikes; and (2) the problem of rationality and moderation in command and control.

Second-strike Capabilities

The protection of a second-strike retaliatory force from destruction by an incoming counterforce surprise attack is a classic design problem for stable deterrence. Atomic stability depends on a belief in the mind of the potential aggressor that he will suffer retaliation at an unacceptable cost, and this means above all that a considerable portion of the defending state's strike force must survive the initial assault. The two superpowers guarantee their second-strike capabilities by burying land-based ICBMs in reinforced concrete silos that can withstand all but near hits and by deploying sea-based missile forces in nuclear submarines whose location is concealed from enemy intelligence by the benign properties of sea water (which impedes the passage of detection impulses over long distances). Several writers have argued that it will be impossible to reproduce these favorable conditions for second-strike deterrence in the Middle East, and that therefore a stable "balance of terror" cannot be established [30].

Ciro E. Zoppo gives the most systematic case against second-strike deterrence. The most likely delivery system in the Middle East is not missiles, he argues, since the fitting of warheads "would require very sophisticated technology and testing to miniaturize them." Instead, Middle Eastern atomic powers will rely on available high performance bombers. These could not survive an incoming attack on the ground, given the "difficulty of 'hardening' airfields." The result is a necessity to get them airborne in "extremely short warning time," which would result in a "hair-trigger situation with strong urges to pre-empt." Moreover, given the small territories and high population densities of the Middle East, "neither Egypt nor Israel could absorb a minor nuclear attack and retain the measure of national viability needed to retaliate." Therefore, the kind of deterrence situation that would exist in the Middle East is "one of the most unstable one can imagine" [31].

Every term of this analysis is open to question. The argument that hardened airfields on the ground could not survive an incoming first-strike ignores the possibility of protective dispersion of aircraft in underground caverns in remote areas adjacent to highways suitable for takeoff. Such installations already exist in Sweden to protect the Viggens and in Switzerland where they are regarded as "underground aircraft carriers." The hills of Eilat or Sinai or the northern Galala Plateau, for example, afford many possibilities for dispersion and concealment.

Moreover, the use of missile delivery systems certainly cannot be excluded. Zoppo's problem of fitting miniature warheads to medium range missiles exists only for Israel, if we assume that Egypt or Syria were supplied by France, China or Russia (see below). And the problems of small warhead construction may be, according to weapons specialists, well within the design capacity of Israel's science and technology. *Jane's All the World's Aircraft 1972–73* reports without comment expectations that Israel would have suitable nuclear warheads of its own design available by 1970 to fit its MD–660 (Jericho) missile [32]. The shorter range but highly mobile Lance, now being supplied by the United States, also is expected to present no insuperable obstacles to Israeli atomic warhead development [33]. It has been argued that Israel's interest in these missiles makes sense only if atomic warheads are assumed, since they are not cost-effective (at $100,000–plus) with conventional explosives against tactical targets [34].

Assuming that medium range ballistic missiles fitted with miniaturized nuclear warheads come into the possession of Middle Eastern governments, could they be protected from destruction in a hostile environment? The method of preference to secure a land-based retaliatory force is "hardening" the missile launchers in permanent underground reinforced concrete silos. The intercontinental attack missiles of the superpowers have not, up to the present time, achieved the accuracies that would be required to destroy superhardened silos. However, the distances between adversaries in the Middle East are much shorter, and the technical problems of accurate targeting are substantially reduced. *How secure would hardened sites be against counterforce attacks in the Mideast?*

Two effects of nuclear explosions require shielding: (1) the shockwave of tremendous overpressures created by the sudden expansion of the air around the point of explosion; and (2) the generation of intense electromagnetic pulses that can destroy the complex and delicate electronic guidance circuits of the missiles inside silos even if the silos remain intact [35]. Countermeasures against the electromagnetic pulse, including electronic shielding, are probably well within the design capabilities of available engineering. Therefore, the "kill probability" of an attack on silos will be a function primarily of three overpressure variables: (1) the *yield* of the blast generating the shockwave; (2) the *accuracy* of the strike vehicle measured in circular error probability (CEP), which will determine the distance between the hard-point target and ground zero; and (3) the *hardness* of the silo expressed in resistance to peak overpressures. Table 7-3 relates accuracy and overpressure to kill probability for 20 and 200 kiloton warheads, using three unclassified methods of estimation.

This means, for example, that an Israeli or Egyptian offensive missile with a circular error probability of 200 feet and a 20 kiloton warhead would have a probability above 97 percent of destroying an enemy missile site hardened to 100, 300 or 500 pounds per square inch, but a lower kill probability for a silo hardened to 1000 psi. An attack missile with a CEP of 1,200 feet armed with a 20 kiloton warhead would have a kill probability of less than 50 percent against even the weakest silo. A missile with a CEP of 5,000 feet would pose little danger to a hardened deterrent force.

What values of yield, hardening, CEP and kill probability are realistic for an analysis of the Middle East? The warhead size of 20 kilotons is the yield most often estimated for Israeli weapons, on the assumption of five to eight kilograms of plutonium and one to one and one-half bombs per year production. Also, the first successful test conducted by each of the past six entrants into the nuclear club was in this range. The alternate yield of 200 kilotons is suggested privately by one well-informed source as within the capacity of Israel and is given here for comparison.

The choice of silo hardness depends on available technology, configuration of surrounding terrain and above all expense. Tsipis estimates that half of the American silos are hardened to 1,000 psi and half to 300, while all of the Soviet

Table 7-3. Survivability of Missile Silos

Yield and Peak Incident Overpressure (hardening)	Single Shot Kill Probability (percent)	Circular Error Probability (feet)		
		Tsipis Equation[a] (feet)	RAND Computer[b] (ground burst) (feet)	Brode Equation[c] (feet)
20 Kiloton				
100 psi	50	(827)	900	970
	75	(583)	640	
	97	369	400	
300 psi	50	550	590	645
	75	388	410	
	97	246	270	
500 psi	50	458	480	535
	75	323	340	
	97	203	220	
1000 psi	50	352	350	420
	75	251	250	
	97	159	170	
200 Kiloton				
100 psi	50	(1783)	2000	2100
	75	(1258)	1400	
	97	796	860	
300 psi	50	1184	1250	1400
	75	837	875	
	97	530	560	
500 psi	50	986	1000	1150
	75	696	730	
	97	437	450	
1000 psi	50	759	900	900
	75	541	550	
	97	342	340	

a. Kosta Tsipis' equation:

$$K = \frac{Y^{2/3}}{(CEP)^2}$$

where K is the lethality of a re-entry vehicle to a silo, Y is the yield of the warhead in megatons and CEP is the circular error probability in nautical miles. Tsipis relates K to silo hardness in his Tables 5B and 6. He does not estimate K for 100 psi silos at 50 percent and 75 percent kill probabilities; these are extrapolated for the present study. Tsipis, pp. 16, 23-24. See also his "Physics and Calculus of Countercity and Counterforce Nuclear Attacks," *Science*, vol. 187, February 7, 1975, pp. 393-397.

Table 7—3. *(Notes continued)*

b. The Rand Corporation, *Bomb Damage Effect Computer,* AFP 136–1–3 (Santa Monica, California, 1964), based on US Atomic Energy Commission, *The Effects of Nuclear Weapons* (Washington, D.C., 1962). A later version corrects for certain sources of error, but yields vulnerability codes for silos rather than usable data. See D. C. Kephart, *Damage Probability Computer for Point Targets with P and Q Vulnerability Numbers,* R–1380–PR (Santa Monica, Cal.: The Rand Corporation, 1974), and Defense Intelligence Agency, *Physical Vulnerability Handbook: Nuclear Weapons,* AP–550–1–2–69–INT (Washington, D.C., 1969). The margins of error in the earlier computer are tolerable for present purposes.

c. H. Brodes equation:

$$P = \frac{3300W}{r^3} + \frac{192W^{1/2}}{r^{3/2}} \text{ psi}$$

where P is the peak overpressure, W is the yield in megatons and r is the range in thousands of feet. H. Brode, "Review of Nuclear Weapons Effects," *Annual Review of Nuclear Sciences, 1968,* vol. 18, p. 180. While the Tsipis and Rand methods relate CEP and yield to kill probability, the Brode equation relates these to peak overpressures. But, since half the strike vehicles are expected to fall closer to the target than the estimated CEP, thus subjecting the target to higher overpressures, and half will land further away with the opposite result, the Brode figures can be related to silo vulnerability at 50 percent kill probability.

silos are less than 500 and most are only 100. (Incident overpressures of only 5 psi will demolish an ordinary brick house.) A security-conscious state may go for the higher psi.

The level of kill probability that will be acceptable for a counterforce attacker is a political choice. Given the large amount of damage that would be inflicted on an aggressor by even minimum retaliation in the Middle East, it is reasonable to assume that a rational attack would be undertaken only with a very high degree of confidence in the antisilo kill probabilities.

The final term, accuracy, is the most sensitive assumption of our analysis. In Tsipis' equation, accuracy is as important as the cube of yield [36]. A relatively small quantitative change of a few hundred feet may correspond to a fundamental qualitative change in the survivability of a second-strike force. According to Table 7–3, a missile with an accuracy of 150 feet would be effective against even the hardest target with the smaller warhead, while a CEP of 2,200 feet would be ineffective against even the weakest target with the larger warhead.

While information is available on the guidance systems of some of the known missiles in the Middle East, only fragmentary reports, rumors and intelligent guesses exist on the CEPs. The Jericho is "said to have guidance problems" [37] and has been rated as accurate within one kilometer (3,280 feet) [38]. The standard Lance also is no sharpshooter. It depends on a simplified inertial guidance system limited to general directional control with meteorological compensation for wind speed and direction, the major sources of error [39]. The Arabs' Scud–B is reported in one unpublished study as having an error probability of 400 meters (1,312 feet) [40], and the unguided Frog–7 is known to be inferior to the Scud.

If we estimate the CEPs of the four missile systems to fall within the range, 1,200–3,500 feet, the following conclusions are indicated: (1) If armed with 20 kiloton warheads, none of the attack missiles would be effective as silo-killers. (2) If armed with 200 kiloton warheads, none of the attack missiles would be effective against silos hardened above 100 psi, and even the most accurate missile would have less than 75 percent reliability against the 100 psi silo. (3) Therefore, none of the four offensive missile systems presently available in the Middle East possesses destabilizing counterforce first-strike capability, and *the technical conditions for reliable hardening of second-strike missile forces in underground silos do appear to exist in 1975* [41]. This makes it unnecessary to rely on vulnerable piloted delivery vehicles, solves the problem of short warning without resort to "hair-trigger" readiness and assures both sides that their strike forces could absorb a surprise attack and retain enough firepower to retaliate at levels of damage that will be unacceptable to a rational adversary [42]. In short, we have the technical preconditions for a minisystem of mutual assured destruction.

This is reasonable conclusion in 1975, but as noted above it is sensitive to the assumed levels of missile accuracy. If the Middle Eastern MRBMs were improved to achieve the error-to-distance ratios of existing US and Soviet ICBMs, the assumed CEPs could be reduced to a range of five to 125 feet, making them reliable even with the smallest warheads against the hardest targets [43]. Moreover, recent advances in microcircuitry have made possible an entirely new set of miniaturized devices and sensors that in turn make possible jam-proof guidance of cruise missiles with accuracies within 100 feet regardless of distance [44]. The implications of this revolution in precision-guided munitions (PGMs) are only dimly understood [45], but one consequence for both the global balance of terror and any regional minibalance may well be the depreciation of "hardening" as a method to secure second-strike forces. Already, a terminally guided submissile warhead is under development by the LTV corporation for the Lance [46], comparable to an arrowhead which separates from the main shaft in midflight and guides itself to the bullseye. Israel is among the many countries that have the technological infrastructure in electronics, airframes and jet engines to develop advanced missiles, and they may become a "normal" part of the burgeoning international arms trade by 1985 [47].

The introduction in the Middle East of very accurate nuclear-tipped missiles that are invulnerable to electronic countermeasures and ABMs is a worst-case. But if hardened silos are in fact rendered obsolete by improvements in accuracy, *do alternative means exist to protect a second-strike force?* Three major alternatives can be identified: (1) the use of naval platforms, especially submarines and missiles boats; (2) the use of mobile land-based launchers; and (3) the stationing of permanent missile sites in politically secure environments.

Naval launch could depend on missile-firing boats, such as the Reshef and Saar (Israel) and the Komar and Osa (Egypt and Syria). The offensive potential of the Saar was demonstrated during the Yom Kippur War, when after disabling at least seven of the Arab missile boats, the Israeli craft patrolled the coastline firing at coastal installations, radar stations, military complexes, and SAM sites [48]. Alternatively, Israel will soon have three advanced submarines [49], and Egypt has 12 undersea vessels of the Soviet W and R classes [50]. It is not impossible that these could be adapted to nuclear use, though the technical problems are formidable and navies far more sophisticated than those of the Middle Eastern powers are in danger of becoming vulnerable to relatively inexpensive terminally guided attack missiles [51].

The alternative of mobile land-based forces is more promising. Overland transportability is presently available for the Jericho and the Lance, as well as the Scud–B and Frog–7, all of which are fired from self-propelled launching ramps. The Lance platform is a tracked vehicle that is air-transportable and air-droppable, amphibious, and self-contained. It can go virtually anywhere under its own power or by helicopter transport. The launcher also carries the entire crew of six men needed to move the missile into position, aim and fire it [52]. One can imagine an agile land-based missile force that could evade hostile detection and possibly survive a first-strike. No surprise attack could be relied upon to destroy *all* the nuclear weapons available to the defending side, and since even minimal retaliation would cause unacceptable damage to the attacker, the uncertainty factor would "render a first strike a suicidal proposition in any and all cases" [53].

Mobile systems may be deployed in one of two patterns, namely, (1) continuous circulation or (2) remote positioning. Continuous circulation limits the target acquisition probabilities for the adversary, since a moving objective must be located in real time. Existing reconnaisance methods for constant surveillance depend on satellites such as the United States' Big Bird, an 11 ton camera which ejects film cartridges to be caught by waiting air force planes. The next generation of satellites will permit electronic relay of pictures to transmit the information instantaneously. Unless mobile launchers can be camouflaged entirely, it may be assumed that their movements will be known to the superpowers. But will the great powers feed this information to their Middle Eastern clients, to aid and abet a first strike? One may doubt the willingness of the United States to encourage an Israeli nuclear pre-emption, though the policy of the Soviet Union may or may not be as restrained. Without satellite data, surveillance of continuously circulating mobile launchers will depend on aerial reconnaisance by advanced jets. This seems to be a much less reliable method, as surveillance is intermittant and incomplete in its coverage, and reconnaisance aircraft themselves are vulnerable to counter–air operations. In summary, a continuously circulating mobile force may achieve a margin of security by evasion and concealment.

The strategy of remote positioning of a mobile strike force would have asymmetric advantages for the Arab side. Such a force would be stationed in a rear area sanctuary and moved to launch sites only when needed. While all of Israel is within missile range from the Arab states, extensive sections of the barren Egyptian southwest, the northern salient of Syria, and most of Iraq and Saudi Arabia are beyond the range of Jerichos or Lances launched from Israel. Israeli attacks against the remote sanctuaries would depend on penetration by manned aircraft over great distances against sophisticated air defenses. Moreover, the ample warning time of an approaching air attack would signal the Israeli intention to the Arab side. Thus, an Israeli attack on remote sites could not be undertaken without a substantial risk of unleashing an atomic response against Israel.

The third alternative, if hardened silos are rendered obsolete, is the stationing of permanent missile sites in politically secure environments. If the previous concept of remotely stationed mobile forces confers an asymmetric advantage for the Arab side, the idea of politically secure environments applies uniquely to the special circumstances of Israel. The Jewish state now exercises effective military administration over Arab population centers in the West Bank and the Gaza Strip comprising over one million Arab inhabitants, One can conceive of Israeli missile sites placed above ground or in silos in the heart of such cities as Nablus (population 50,000), Hebron (40,000), Bethlehem (40,000), Gaza (40,000) or even the Old City of Jerusalem (Arab population over 70,000). Barring the most extreme provocations, an atomic surprise attack on these cities by an Arab state would be highly improbable. (Arab evacuation of the cities would signal a possible attack.) This "collateral damage" strategy assumes that Israel will continue to control at least some of the Arab centers [54] and that it will be willing to bear the onus of a policy that makes military use of these populations and of Jerusalem as sanctuaries in which to base an atomic strike force. Such a desperate and repugnant idea would be accepted only in an extreme situation. But the notion of using civilian populations as hostages could gain credibility if all other options seemed to point to the eventual destruction and strangulation of Israel in an endless cycle of wars.

In summary, the following conclusions on second-strike deterrence are supported: (1) The technical conditions for secure hardening of second-strike forces in underground reinforced silos do exist at the present time. (2) Emerging technologies of terminal guidance for attack missiles could, if introduced into the Middle East, reduce the margin of security of hardened forces in the future. (3) If silos become vulnerable in the future—the worst-case—as yet untested alternatives exist for both sides in continuously circulating mobile forces, for the Arab side in remote positioning, and for Israel in politically invulnerable sites. Overall, the technical problems of a stable system of mutual assured destruction appear to be solvable.

Rationality and Moderation in Command and Control

If the purely mechanical problems of deterrence can be solved, we are left with the political problems of assuring responsible and moderate behavior and avoiding miscalculation. The Arab–Israeli conflict is characterized at times by intense and uncompromising emotions and fears, expressed in extreme utterances and spectacular images. The practical issues are overlayed by elements of religious war and moral imperatives that may justify the sacrifice of millions. The Arab side in particular is widely believed to have a tendency to emotionalism and loss of self-control [55]. This view is maintained, for example, by Ciro Zoppo:

> Should Egypt and Israel travel the nuclear road to military options, the slender reed of rational calculation may not sustain, in a nuclear confrontation ... the weight of the Arab quest for justice. Arab policies have in the past, on occasion, appeared suicidal [56].

Can Arab regimes be relied upon to moderate their behavior in a nuclear Middle East? [57] No strictly scientific and objective judgment is possible on this speculative question, but there are some reasons to doubt the most alarming views. Many Arabists find a substantial trend to moderation in the transition from Nasser to Sadat in Egypt and in the evolution of the Ba'ath regime in Syria. Moreover, past perceptions of the Arab states may have been clouded by the climate of intense hostility and by the lack of normal intercourse. Comparison with the case of China suggests that an adversary whose behavior seems irrational from a distance may look quite restrained after relations are improved and more objective and detailed information becomes available. Future studies of Arab policy may well find, retrospectively, a consistent and rational pattern of behavior that was not perceived earlier. If so, casual statements about "crazy states" may come to be seen as simplistic and even racist as applied to the Arabs, and we may come to appreciate the care with which the leading Arab regimes assess the possible costs and benefits of alternative policies. Moreover, nuclear weapons themselves may induce a respect for moderation and reduce the credibility of supermilitants within the Arab coalitions, while at the same time creating new incentives for the great powers to restrain their Middle Eastern allies.

If the intentions of the top leadership are moderated, the problem of command and control is reduced to the familiar unintentional nuclear hazards, including: (1) miscalculations leading to uncontrolled escalation; (2) unauthorized use of nuclear weapons by subordinates who take matters into their own hands; (3) accidental detonation of nuclear explosives; (4) seizure of weapons by terrorist groups and zealots; and (5) future transfer of power to immoderate successor regimes. In general, these raise political, technical and administrative issues essentially similar to those in other regions and they have been addressed extensively in the literature [58].

Consequences of a MAD System

Let us assume for the following discussion that the major requirements of a minisystem of mutual assured destruction can be fulfilled in the Arab–Israel conflict, including reliable and secure second-strike delivery systems capable of inflicting unacceptable levels of damage, and also moderation and rationality in command and control [59]. Many consequences will follow in the transformation of the conflict to a nuclear system.

Subnuclear Wars. Critics of regional nuclear deterrence argue that it still will be possible to have fairly large scale subnuclear military engagements. Indeed, the more stable the relationship of atomic deterrence, the wider the scope for subnuclear provocations without fear of uncontrolled escalation. If a substantial part of the captured territories are not returned, it is even conceivable that the Arab states might mount an October-style assault to regain the Sinai and the Golan Heights, and possibly also the West Bank, in defiance of the Israeli deterrent. Will Israel resort to nuclear weapons if the fighting is restricted to the occupied areas? It is difficult to imagine Israel committing national suicide to hold on to Abu Rudeis or Hebron or Mount Hermon. The atomic capability is only a last-resort deterrent to guarantee the ultimate survival of the state. Therefore, the threat of MAD will insure the inviolability of the Israeli heartland, but will not in itself necessarily halt the endless round of border wars in the peripheral areas.

The Territories. Large scale subnuclear wars in the border territories will raise incalculable dangers of uncontrolled and disastrous escalation. If, for example, an Egyptian force should break through the Sinai passes and move rapidly toward the pre-1967 borders, Israel would not know whether this unrestricted advance of Arab armies flushed with victory would stop at the green line or penetrate into the Jewish state itself. Similarly, an Israeli advance into the Damascus plains past the Saasa lava ridges or across the Suez Canal to within striking distance of Cairo would force the Arab states to consider the nuclear option. Any scenario involving full mobilization or conventional war must assume that there would be a tense nuclear alert, with unprecedented dangers of catastrophic miscalculation. The world has not yet seen a full scale conventional war involving directly opposed armies of two nuclear-armed parties.

Therefore, a stable regional system of mutual assured destruction would have as an almost unavoidable political corollary the assumption of a return of the captured territories [60], combined with agreements for the limitation of forces in the areas surrendered by Israel. Indeed, *the major virtue of nuclearization of the Middle East is that, while it will make a territorial settlement necessary, it will also render it possible* [61]. The stumbling block of settlement until now has been the inability of Arab moderates to sell a policy of full détente with Israel to their own coalition partners and constituents, especially as new oil wealth and improved military performance have generated a revolution of rising diplomatic expectations. Nuclearization will induce both moderation and a revolution of declining expectations, as the end-of-the-world character of nuclear war is understood by the least educated, most isolated Egyptian villager. The so-called "rejection front" of regimes and factions refusing to have anything to do with Israel and compromise will lose all credibility. The "Zionist entity" will at last be a permanent fact that must be accepted, and the preconditions will exist for the trade of "a piece of territory for a piece of peace" that is so elusive today.

The Palestinian Commandos. However, it is entirely possible that a general political settlement between the established states of the region will leave the Palestine Liberation Organization out in the cold. The Palestine Covenant, Arafat's first address before the United Nations Assembly and many official utterances of the PLO reinforce the principle that the final and uncompromisable goal of the Palestinians is the return to the lands that now compromise Israel. Many of their sympathizers in Western countries believe that the Palestinians will in fact soon settle for less, possibly only a truncated state in what is now the occupied West Bank. At the present time these predictions are only speculation. If the PLO will not settle with Israel, or if Israel is unable due to popular antipathy to the terrorists to settle with the PLO, the hypothetical territorial resolution postulated here could leave the Palestinian fighting organizations unsatisfied.

If the Palestine question remains unresolved, it is reasonable to expect that Fedayeen raids into Israel will continue from bases in Arab countries. Israel's past pattern of response has been to retaliate in force against the centers behind Arab lines that are assumed to be staging areas for the commandos. These reprisals in depth have never been considered tolerable by the Arab governments, but they have lacked the means to respond convincingly. Lobbing missiles tipped with conventional high explosive charges into Israeli villages and installations might be a deterrent threat, but this has been impossible because of the enormously greater Israeli threat to Arab cities. The introduction of atomic deterrent capabilities on both sides will actually limit the Israeli freedom of action in the future because of the danger of escalation [62]. Therefore, the guerrillas will paradoxically have a freer hand in a nuclear Middle East, and Israel will be more constrained in its policy of reprisals. Scenarios can be constructed in which the nuisance attacks by the Fedayeen escalate to the point that Israel is forced to take the risk of full confrontation with the sanctuary state(s) even up to the nuclear brink. This could be an element of continuing volatility in a MAD system.

Preventive War. The transitional period, during which atomic weapons are being introduced in the Middle East and before the new relationships settle into routine patterns, could threaten other kinds of instabilities. One is the danger of preventive war by one side to arrest the atomic development of the adversary. Israeli Major General Ariel Sharon, a leading hawk, warned in 1975 that, in his opinion, "if we discover that the Egyptians are working on a nuclear weapon of their own ... we'll have no choice but to wipe them out" [63]. This echoes a similar warning to Israel by President Nasser 15 years earlier, that if it were certain that Israel was constructing a bomb, "it would mean the beginning of war between us and Israel, because we cannot permit Israel to manufacture an atomic bomb. It is inevitable that we should attack the base of aggression...." [64] Others have discussed the possibility of a Soviet pre-emptive attack [65]. But these prophesies of preventive doom have a long history. Harrison Salisbury and others adumbrated Soviet strikes against the Chinese atomic facilities in Lanchow, Paotow and Lop Nor [66], while the *Women's Home Companion* of March 1948 found it necessary to warn readers against the simplistic idea of "using the atomic bomb against the Russians before they crack the formula and use it against us" [67]. The idea of "surgical strikes" to excise an adversary's nuclear cancer has always been more appealing to scenario writers than to commanders in chief.

The degree of instability caused by the transition to nuclear weapons will depend partly on the circumstances of their introduction and disclosure. If disclosure is incremental and takes the form of a series of rumors, followed by reliable reports, followed by official pronouncements or demonstrations, with adequate time periods between stages, there may be a process of "weaning" or assimilation during which the parties will have time to adjust to the new idea without undue alarm. Conversely, if atomic weapons possession is suddenly disclosed at the height of a crisis, as when nuclear tips for the Scuds were rumored in Alexandria while Israeli forces threatened to destroy the trapped Egyptian Third Army in the closing days of fighting of the Yom Kippur War, the possibilities of precipitous action and miscalculation are heightened. But while the disclosure process ideally will proceed in phases, the actual introduction of the weapons themselves into the arsenal of a state should precede even the first rumors, to minimize the possibility of a preventive strike when the adversary suffers the initial "trauma" of the disclosure. A preventive strike undertaken after the physical introduction of nuclear weapons would have the character of a counterforce first strike and would run the risk of a nuclear exchange, while a preventive strike against a facility which has barely initiated the production process leading to atomic weapons might appear to someone to involve a more tolerable element of risk.

In the case of Israel, more than a few observers have concluded that the atomic threshold has already been crossed after years of rumors [68], and no disarming attack has occurred. Therefore, the above "advice" on disclosure applies especially to the Arab side, which might do well to emulate the Israeli policy of "ambiguity."

The Numerical Balance. The conversion to a MAD system may introduce new problems in the numerical balance of both atomic and conventional arsenals. If the Arabs are supplied from abroad and Israel is self-reliant, Israel will face asymmetrical limitations in the availability of fissionable materials. The United States, which is now virtually the only reliable outside source for Israel's conventional arms, will be constrained by law and a profound antipathy to proliferation from transferring atomic weapons or fissionable materials. Unless she can find another source of supply (possibly South Africa), Israel will be limited to the production of less than two 20 kiloton warheads per year based on the present output of Dimona. However, an Israeli minimum deterrent would require no more than a dozen bombs of Hiroshima yield to destroy the principal cities of Egypt, Syria and Jordan as well as the Aswan and Euphrates Dams, and a countervalue strike by the Arabs could be equally modest in size. The quantitative balance would therefore be relevant only for counterforce strikes, and this threat could be controlled by the means outlined earlier irrespective of the size of the attacking force.

The open introduction of atomic weapons could also have an affect on the willingness of the United States, and also possibly the Soviet Union, to supply their clients with conventional arms. If Washington alone restricted the flow, either on grounds of Israeli self-sufficiency or as a punitive sanction against the Israeli atomic policy, the effect might be to upset the conventional arms balance. However, the United States might be inclined to adopt the opposite posture, and to keep Israel well supplied with conventional weapons to provide "flexible response" and to reduce the probability that she will feel compelled to resort to nuclear retaliation to a conventional attack. Perhaps in this relationship also there will be a period of trauma after which behavior will return to the normal pattern. In practice, this would mean a temporary interruption of military aid. However, it is clear from past behavior that any restriction in the flow of American arms of however brief duration will be regarded as a serious handicap in Jerusalem [69], and this will be a major disincentive to develop nuclear weapons or to disclose them if they exist [70].

IS A NUCLEAR MIDDLE EAST INEVITABLE?

Even if the introduction of atomic weapons in the Middle East is rejected by each party as an optimizing policy choice, proliferation could occur as the result of a process of foreclosing alternatives. The Arabs or Israelis may be forced to arm, not for perceived advantage, but because they believe the other side to be doing so.

Israeli leaders have repeatedly stated that "Israel will not be the first to introduce atomic weapons in the Middle East" [71], but nobody seems to believe them. The disclaimer leaves three semantic ambiguities that may be seen as loopholes:

1. "... *the first* ...": Nuclear weapons have already been introduced into the Middle East on board ships of the American Sixth Fleet and on Soviet ships

in the Mediterranean [72]; also, nuclear-tipped Scud missiles were widely rumored to be in Alexandria during the October war [73].

2. "*...to introduce...*": This could be interpreted to allow advanced development without final testing and deployment.
3. "*...atomic weapons...*": Israeli acquisition of fissile material and fabrication of devices could stop just short of final assembly of deliverable explosives.

Yigal Allon added to the uncertainty with his caveat that, while Israel would not be the first, "we will not be the second either" [74], seeming to imply a "bomb in the basement" which could be disclosed at will. President Ephraim Katzir did nothing to dispel the verbal confusion when he announced in December 1974 that Israel "possesses the potential to produce atomic weapons" and will do so "if we need it" [75].

These subtleties of wording have given rise to a rich literature specifically on the functions and dysfunctions of the Israeli policy of calculated ambiguity, including two strikingly parallel recent articles by Israelis in American publications [76]. But almost everyone outside Israel seems to believe unambiguously that she either already has the bomb or is going ahead full steam to get it. Egyptian President Anwar el-Sadat told Iranian publisher Farhad Massoudi (December 1974) that he believes Israel already has nuclear weapons [77]. Arafat said that "our own sources inside Israel" report the existence of three to five atomic bombs [78], and the PLO leader asserted before the General Assembly in November 1974 that a fifth Middle East war would "forebode nuclear destruction and cataclysmic annihilation" [79]. A leading book on the Yom Kippur War reports flatly that "Israel's stock of atomic bombs is, by superpower standards, small. Kissinger has said privately that Washington believes Israel to have three nuclear devices. Israeli sources mention up to six" [80]. Similar though less categorical reports are given by SIPRI [81], the London *Daily Telegraph* [82], *Jane's All the World's Aircraft* [83], *Der Spiegel* [84], the *New York Times* [85] and the Soviet newspaper *Moskovski Komsomoletz* [86]. The one book-length study of Israel and nuclear weapons, sponsored by the International Institute of Strategic Studies (London), asserts that, "Declaratory policy aside, Israel has gone steadily ahead in the development of the capacity to build atomic bombs...." [87] Former CIA director Richard Helms told a closed-door hearing of the Senate Foreign Relations Committee in 1970 that Israel had the capacity to produce atomic weapons [88]. Intuition finds the idea of an atomic Israel credible. Jewish science is resourceful; Israel lives under a constant state of seige; how could a responsible Jewish government *not* take at least the preliminary steps to a "last resort" deterrent, given the history of this people and the widespread belief that the Arabs are bound to get stronger? [89]

With so many rumors and so much smoke, most observers conclude that there must be fire. *But what if everybody is wrong?* An Israeli government faced with strong domestic and foreign criticism and secure in its conventional military advantage might be inclined to arrest nuclear development at a point substantially short of the fabrication of weapons. At least one major link in the production chain leading to nuclear status is reported to be missing in Israel—a chemical separation plant. Jabber, in the most thorough and exhaustive analysis based on a combining of public sources of data, concludes that the Israeli program has traveled all the stages along the road except this last one [90].

Nonetheless, there is a widespread belief that Israel has or soon will have atomic weapons, and this seems to have spurred a decision on the part of Egypt and other Arab states to get the bomb also. As early as 1965, Nasser approached the Soviet Union to purchase atomic weapons, but was rebuffed [91]. A close associate of the Egyptian president told a British journalist in the same year that "We would go to the Devil, if necessary, in order to get a bomb" [92]. Colonel Muammar el Qaddafi of Libya is said to have offered over $1 billion in 1970 to China or France for fissile materials or weapons. H. Heikal reported after the October War that the Arabs had made several attempts to buy atomic weapons, especially from China, and having failed, had turned to the possibility of developing their own capability on a consortium basis [93]. Sadat himself said that "We shall also [like Israel] find a way of having atomic bombs" [94].

Production of weapons grade plutonium by Egypt or other Arab states is a fairly remote possibility at the present time. The Inchass reactor in Egypt is rated at only two megawatts (Dimona, 24) and is known to be inefficient. Iraq has a small Soviet-built reactor south of Baghdad, but this produces very small amounts of plutonium and is covered by an IAEA safeguards agreement. The proposals discussed between Egypt and the US and France during 1974 would mean the building of new facilities in the 500—plus megawatt range, but it is not known whether these transfers will materialize nor what inspection and controls would be established if they did. Finally, it might be some time before Egyptian technologists working alone could fabricate weapons even if a reliable source of plutonium was established [95].

For these reasons, speculation on an Arab breakthrough centers on foreign transfers of arms or weapons grade material. The Soviet Union has given assurances to Egypt, Syria, Iraq and Algeria that it will provide them with nuclear devices if it is proved that Israel possesses atomic weapons, according to Arab diplomatic sources in Beirut. Each country already has a Soviet-trained nuclear warfare section in its armed forces, and Egyptian and Syrian specialists have received training in arming Soviet-made missiles with nuclear warheads [96]. But the idea of the Soviet Union giving autonomous warheads to the messianic and at times anti-Soviet Arab movement strains credulity in Western capitals. Even the idea of Soviet forces carrying nuclear arms on Egyptian territory is doubted because of the risk of seizure by Egypt or even of capture by Israel.

Perhaps more likely would be the sale of a limited number of warheads for an exhorbitant price—$1, $2, $5 billion?—by China or France or India. If Israel is perceived to be rattling a nuclear sabre at the Arabs, and if any of the three potential suppliers thinks that one of the others will open the door anyway, the political conditions may exist for a sale that could have very large financial and political rewards. Jean–Jacques Servan–Schreiber, publisher of *L'Express* and head of France's Radical party, fears that the day is fast approaching when Paris will sell atomic arms [97]. Here is one scenario: a desperate India accepts a Saudi [98] offer to buy up and cancel part or all of her foreign debt burden in exchange for a small quantity of fissile material. India is sympathetic to the Arab cause, sees the Saudi government as responsible and moderate, and believes that it is only redressing the imbalance created by an already nuclear Israel. The Saudis take the bill of sale to Paris, to say, look, the genie is already out of the bottle, and thus to salve the conscience of a more important supplier. They can also threaten to go to Peking.

Whatever the plausibility of these lines of speculation to disinterested third parties in the outside world, they are bound to be very powerful images in Israeli think tanks. Therefore, *even if nuclear deterrence is rejected as the strategy of choice, Israel may be forced to develop atomic weapons.* Conversely, driven by a belief that Israel has crossed or will cross the atomic threshold, one or more Arab states will seek an independent atomic strike force. The prevention of a two-sided atomic relationship may no longer be possible.

NOTES TO CHAPTER SEVEN

1. This study was supported in part by a travel grant from the Abram Sachar International Fellowship Program. It is part of a larger work in progress on the search for long term stability in the Arab–Israel military balance. Of the many people who contributed valuable comments and ideas, I am especially indebted to Kosta Tsipis, Michael Nacht, Enid Schoettle, Anne Cahn, Walter Clemens, Dale Tahtinen, Stanley Roth, Thomas Friedman and Robert Harkavy.

2. *Jewish Observer and Middle East Review,* July 2, 1965. Quoted in Fuad Jabber, *Israel and Nuclear Weapons,* an International Institute of Strategic Studies monograph (London: Chatto and Windus, 1971), p. 104.

3. E.g., Lt. Col. Haytham Al-Ayubi, head of the military research department of the Palestine Research Center (Beirut): "The fifth round will not be the last round. There will be a sixth, seventh, eighth ... until both sides finally realize that the only solution lies in setting up a [secular] democratic state...." "Is War Inevitable?" translated from the Beirut weekly *Monday Morning* in *Atlas,* January 1975, p. 36.

4. Images of Arab "irrationality" are questioned below.

5. Derived from figures in the International Institute of Strategic Studies, *Military Balance* (London, IISS), various issues. Inventories of equipment increase over this period by 300 to 500 percent, significantly, the proportions are fairly stable over time.

6. Wolf Blitzer, "Dinitz: Israel Requests Tied to Huge Arab Arms Supplies," *Jerusalem Post Weekly*, March 4, 1975, p. 4.

7. *Time*, March 3, 1975, p. 33.

8. *New York Times*, March 24, 1975, p. 14.

9. Jabber, pp. 146–147; also, p. 133.

10. A partial explanation may be differences in the prevailing climates of opinion among Israeli and Arab intellectuals and academicians. Israeli intellectuals tend disproportionately to be associated with the smaller and more dovish parties of the left, from the left wing within the ruling Labor alignment to Moked, while Arab intellectuals, particularly Palestinians, tend to be fervently nationalist and relatively hard-liners.

11. The three selections that follow are dated, but the author knows of no counterliterature to suggest that they are unrepresentative of much current thinking. However, it is not possible on the fragmentary evidence that is cited to conclude that "the Arabs" think this or "the Israelis" think that. One can only summarize tendencies in the available published literature.

12. Ahmad Samih Khalidi, "An Appraisal of the Arab–Israeli Military Balance," *Middle East Forum* XLII (3) 1966, 55, 63.

13. From a series of articles in January 1961 in the Baghdad daily *Al-Ahali* written by a retired colonel. Translated in Avraham Ben-Tzur, "The Arabs and the Israeli Reactor," *New Outlook, Middle East Monthly* (henceforth *New Outlook*), March–April 1961, p. 21.

14. H. Heikal in *Al-Ahram* (Cairo), translated in *New Outlook*, September 1965, pp. 54–57, as "Atomic Danger on the Middle East Horizon".

15. Simha Flapan, "Nuclear Power in the Middle East (Part I)," *New Outlook*, July 1974, p. 46.

16. Yigal Allon, *The Making of Israel's Army* (London: Sphere Books, Ltd., 1970) p. 78, 112.

17. Emphasis in original. Avigdor Haselkorn, "Israel: From an Option to a Bomb in the Basement?" in Robert M. Lawrence and Joel Larus eds., *Nuclear Proliferation: Phase II* (Lawrence: University Press of Kansas, 1974), p. 165.

18. Evron weighs very marginal gains against substantial risks and costs. Yair Evron, "Israel and the Atom: The Uses and Misuses of Ambiguity, 1957–1967," *Orbis* XVII (4), Winter 1974, 1342; see also pp. 1332, 1333.

19. Flapan, p. 54.

20. Statement by Dr. Nahum Goldmann, endorsed by Ygal Yadin, one-time chief of staff of the army, and other prominent panelists in *Ma'ariv* interview. Translated in "The Atom Bomb in Israel: A Symposium," *New Outlook*, March–April 1961, p. 15.

21. More has been written by Israelis on this sensitive subject in English than in Hebrew. There have, however, been some Hebrew articles by writers associated with the harder line elements in the present Labor alignment giving a more positive view of an Israeli deterrent. Evron (pp. 1330–1331) cites the following: Avraham Schweitzer, *Ha'aretz*, August 14, 1962; Schmuel Segev and Avraham Schweitzer, "Symposium on Nuclear Policy," *Hotam*, August 1962; Rimalt, *Tmurot*, September 1962.

22. Given, e.g., by William Bader, *The U.S. and the Spread of Nuclear Weapons* (New York: Pegasus, 1968), p. 90.

23. See, for example, Secretary of Defense James R. Schlesinger, *Briefing on Counterforce Attacks,* hearings before a subcommittee of the Committee on Foreign Relations, US Senate, January 1975 release.

24. The classic expression of this view is Pierre M. Gallois, *The Balance of Terror* (Boston: Houghton–Mifflin, 1961), pp. 1–14.

25. One such report is in Riad Ashkar and Ahmed Khalidi, *Weapons and Equipment of the Israeli Armed Forces* (Beirut: Institute of Palestine Studies, 1971), p. 77. Some reports question the existence of this missile, while others give it as operational or "in production."

26. Haselkorn, p. 165 ff.; J. Bowyer Bell, "Israel's Nuclear Option," *Middle East Journal,* Autumn 1972, pp. 386–387; Lawrence Freedman, "A Nuclear Middle East?" *Present Tense,* Winter 1975, p. 22.

27. William D. White, *U.S. Tactical Air Power: Missions, Forces, and Costs* (Washington, D.C.: Brookings Institution, 1974), p. 115.

28. An attack on the Russian fleet would face similarly formidable obstacles.

29. The probability that nuclear arms are or will come to be in the possession of Israel and one or more Arab states is discussed briefly below. With regard to formal constraints on nuclearization, it may be noted that Israel has neither signed nor ratified the Nonproliferation Treaty; Egypt signed July 1, 1968 but never ratified; Syria signed July 1, 1968 and ratified September 24, 1969; Iraq signed July 1, 1968 and ratified October 29, 1969; Jordan and Lebanon have both signed and ratified; Libya has signed but not ratified; and Saudi Arabia and Algeria have neither signed nor ratified. A state which has completed the formalities of adherence is required to give three months prior notice if it decides to withdraw because "extraordinary events . . . have jeopardized the interests of its country."

30. "The 'balance of terror' is valid least of all in the Middle East." Flapan, p. 54. See Haselkorn, p. 150; E. Rutherford, "Israel and the Bomb," *Cambridge Review,* December 2, 1967, pp. 157–160; and Y. Nimrod and A. Korczyn, "Suggested Patterns for Israeli–Egyptian Agreement to Avoid Nuclear Proliferation," *New Outlook,* January 1967, p. 9.

31. Ciro E. Zoppo, "The Nuclear Genie in the Middle East," *New Outlook,* February 1975, p. 24.

32. John W. R. Taylor, ed., *Jane's All the World's Aircraft 1972–73* (London: Jane's Yearbooks, 1973), p. 565; also, Jabber, pp. 96–97. But the London Times "Insight" Team reports in *The Yom Kippur War* (Garden City, New York: Doubleday, 1974), p. 283, that "while Israel might have miniaturized its nuclear warheads sufficiently to fit the Jericho, the absence of test facilities makes this unlikely." Military sources "close to the Israelis" say that "the bombs are rather large and unwieldy and that a couple of transport aircraft have been converted to take them."

33. "According to weapons experts, it would not be too difficult for Israel to develop an atomic warhead to fit into the relatively small Lance missile." John W. Finney, "Israel to Receive 200 U.S. Missiles," *New York Times,* Janu-

ary 24, 1975, p. 6. The Lance carries a warhead up to 1,000 pounds, while the Jericho is said to have a 1,200 pound payload.

34. Funds for the conventional warhead for the Lance deployments by the US army were cut by Congress because of questions over the cost-effectiveness of the missile with such a warhead. Clarence A. Robinson, Jr., "Lance Delivery to Israel Expected Soon," *Aviation Week and Space Technology*, February 17, 1975, p. 46. But one Israeli need could be served by the Lance—suppression of Arab SAM sites at a reduced cost in IAF fighter-bombers.

35. Kosta Tsipis, *Offensive Missiles*, Stockholm Papers number 5, (Stockholm: SIPRI, 1974), p. 13.

36. Robert M. Leggett, "Two Legs do not a Centipede Make—Congressman Debunks Soviet Strategic Superiority," *Armed Forces Journal International*, February 1975, p. 30.

37. London Times "Insight" Team, p. 283.

38. *Le Monde*, April 25, 1968; Jabber, p. 96; Freedman, pp. 20–21.

39. "Funds for Advanced Lance Sought by Army," *Aviation Week and Space Technology*, February 24, 1975, p. 20.

40. Robert E. Harkavy, "The Strategic and Diplomatic Implications of the Israeli Nuclear Weapons Program" (Paper, Kalamazoo College, Michigan, 1974), p. 39.

41. I am also assuming adequate shielding against electromagnetic pulse.

42. Compare Jabber, p. 142.

43. The Minuteman III, with an error-to-distance ratio of 0.002 percent, is 100 times more accurate than the Scud, which has a ratio of 0.2 percent.

44. Kosta Tsipis, "Long Range Cruise Missiles," *Bulletin of the Atomic Scientists*, April 1975.

45. A reflective essay on the subject is James F. Digby, "Precision Guided Munitions: Capabilities and Consequences," Rand Paper P–5257 (Santa Monica, Cal.: Rand Corporation, 1974).

46. "Lance Delivery to Israel expected Soon," *Aviation Week and Space Technology*, February 17, 1975, p. 46.

47. Tsipis, "Long Range Cruise Missiles."

48. London Times "Insight" Team, pp. 212–213.

49. *SIPRI Yearbook, 1974*, p. 270.

50. IISS, *Military Balance 1974–75*, p. 33.

51. See "Navy Faces Grave Cruise Missile Threat," *Aviation Week and Space Technology*, January 27, 1975, pp. 101–105.

52. *Jane's All The World's Aircraft 1972–73*, p. 581; "Funds for Advanced Lance Sought by Army," p. 21.

53. Jabber, p. 142. On the question of low- versus high-confidence security for a second-strike force, see Haselkorn, p. 170 and his note 89.

54. Present Israeli policy seems to leave open the possible return of the West Bank but not East Jerusalem or Gaza. In the event of major territorial remission, the collateral damage strategy would make increased use of Gaza and Jerusalem, as well as sites in the Golan Heights and on the Jordanian border that would create a fallout hazard to Syrian and Jordanian populations.

55. One study of "the Arab mind," for example, finds "particularly interesting . . . the frequency with which self-control gives way to temper, the ease with which the flood of anger, violence, or other intense emotion sweeps over the dam of self-control and in an astonishingly short time transforms the entire personality." Raphael Pitai, *The Arab Mind* (New York: Charles Scribner's Sons, 1973), p. 160. See also Sania Hamady, *Temperament and Character of the Arabs* (New York: Twayne, 1960), pp. 205, 211, 222.

56. Zoppo, p. 25.

57. The leading study of Arab perceptions of Israel is not entirely comforting on the capacity for moderation. Y. Harkabi, *Arab Attitudes to Israel* (Jerusalem: Israel Universities Press, 1972), *passim* and esp. p. 68, 79, 81.

58. For example, Herman Kahn, *Thinking About the Unthinkable* (New York: Horizon, 1962), pp. 39–58.

59. We must also assume that a kernel of the command and control apparatus will survive a first-strike to execute the retaliation.

60. Jabber (pp. 141–144) sees Israeli nuclearization having precisely the opposite effect: freezing Israel's hold on the occupied territories.

61. This will reduce the efficacy of the collateral damage strategy; cf. note 52.

62. Cf. Jabber, p. 138.

63. Interviewed in the *New York Times,* February 8, 1975, p. 4.

64. Gamal Abdel Nasser, quoted in *Jewish Observer and Middle East Review* December 23, 1960, p. 3–4; cited in Bader, p. 96.

65. E.g., Haselkorn, pp. 168–169.

66. E.g., Harrison Salisbury, *War Between Russia and China* (New York: Bantam Books, 1970), pp. 126–27.

67. Excerpted in Peter G. Filene, *American Views of Soviet Russia, 1917–65* (Homewood, Ill.: Dorsey Press, 1968), p. 235.

68. Discussed below.

69. On the relationship between Israel's nuclear option and US arms transfers, see Fuad Jabber, "Not by War Alone: Curbing the Arab–Israeli Arms Race," *Middle East Journal,* Summer 1974, p. 240; also, Aubrey Hodes, "Implications of Israel's Nuclear Capability," *Wiener Library Bulletin,* vol. XXII, no. 4.

70. George Quester, "Israel and the NPT," *Bulletin of the Atomic Scientists,* June 1969, p. 44. Quester asks, Why should Israel change the game when she has been winning under the old rules?

71. E.g., Levi Eshkol in the Knesset, May 18, 1966. Dayan told *Le Figaro* that neither short-term nor long-term Israeli military planning is based on nuclear weapons. Quoted in Leonard Beaton, "Why Israel Does *Not* Need the Bomb," *New Middle East,* April 1969, p. 11.

72. Quester, p. 44.

73. E.g., London Times "Insight" Team, p. 411.

74. *Jewish Observer and Middle East Review,* December 24, 1965. Allon repeated the statement several times in December 1974 while foreign minister.

75. *New York Times,* December 5, 1974.

76. Evron, "Israel and the Atom: The Uses and Misuses of Ambiguity,

1957–1967"; and Haselkorn, "Israel: From and Option to a Bomb in the Basement?"

77. "The Middle East Ready to Explode, Sadat Says," *New York Times*, December 17, 1974.

78. *New York Times*, April 4, 1975, p. 3.

79. *New York Times*, November 14, 1974, p. 22.

80. London Times "Insight" Team, p. 282.

81. *SIPRI Yearbook 1972*, p. 312.

82. Quoted in Freedman, p. 20.

83. *Jane's All the World's Aircraft 1972–73*, p. 565.

84. *Der Spiegel*, May 5, 1969; cited in Dale R. Tahtinen, *The Arab–Israel Military Balance Today* (Washington, D.C.: American Enterprice Institute for Public Policy Research, 1973), p. 34.

85. Hedrick Smith, "U.S. Assumes the Israelis have the A–Bomb or Its Parts," *New York Times*, July 18, 1970; see also William Beecher, "Israel Believed Producing Missile of Atom Capability," *New York Times*, October 5, 1971.

86. Quoted in Freedman, p. 22.

87. Jabber, *Israel and Nuclear Weapons*, p. 123.

88. Smith, "U.S. Assumes the Israelis have the A–Bomb."

89. This view is given eloquently in Bell, pp. 383–384.

90. Jabber, *Israel and Nuclear Weapons*, p. 77.

91. *New York Times*, February 4 and 21, 1966; cited in Jabber, *Israel and Nuclear Weapons*, p. 141.

92. Quoted in Simha Flapan, "Nuclear Power in the Middle East: Part II, The Critical Years," *New Outlook*, October 1974, p. 39.

93. *Philadelphia Enquirer*, November 24, 1973, p. 3A; cited in Harkavy, p. 2.

94. "Mideast Ready to Explode . . . ," *New York Times*, December 17, 1974.

95. Cf. Yair Evron, "The Arab Position in the Nuclear Field: A Study of Policies up to 1967," *Cooperation and Conflict* X (1), 1973, 19–31.

96. "Soviets Said to Give Arabs Atom Pledge," *New York Times*, December 5, 1974.

97. Quoted in *Time*, March 3, 1975, p. 44.

98. The Saudi king could be motivated by considerations of the balance of power in the Persian Gulf (vis-à-vis Iran or the danger of US intervention in the oil fields); by fears of a nuclear Israel; or by hopes of a psychological break in a political impasse on the 1967 territories and Jerusalem.

Chapter Eight

Determinants of the Nuclear Option: The Case of Iran

Anne Hessing Cahn

Since that awesome early morning in the summer of 1945 when mankind first witnessed the glare of the fireball and the ghostly violet glow of the sky, one of the most fateful, fundamental and far-reaching decisions nations have faced has been whether or not to pursue a nuclear option.* In contrast to the myriad of governmental decisions which are incremental in nature, dilatory or temporizing in their intent, or ambivalent or discretionary in their interpretation or implementation, the decision to develop a nuclear weapons capability possesses none of these attributes. The magnitude of the resources required in scientific expertise, technical and engineering capabilities, skilled labor, capital investment and nuclear materials ensure that the decision to become a nuclear weapons state is determined by a calculus which in some way expresses the essence of that nation: its past history, present capabilities, future hopes and aspirations, its self-image and its perceived role in local, regional and global arenas. Within the next few years nuclear power reactors are scheduled to spread widely and rapidly throughout much of the world. What does this bode for the proliferation of nuclear weapons? What are the implications for the political system of which nations choose a nuclear weapons option, why, when and how they do so? Will the paths chosen by the first few additional nations set examples for others to follow?

To unravel such interrelated questions it may be useful to examine in detail one nation which has proclaimed its intention to acquire nuclear energy rapidly and in a major way—Iran.

This chapter examines the major determinants of the nuclear option in Iran, looking at the present and future capabilities of Iran to acquire nuclear weapons

---

*By nuclear option is meant the determination to acquire nuclear weapons, not to acquire nuclear reactors for power generation or research purposes alone. While nuclear weapons may in the future be obtained by theft or by purchase, this study will deal with the national determination to produce nuclear weapons indigenously.

and then turning to a search for indicators of intent—what appear to be the incentives and disincentives for Iran to pursue a nuclear option in the future.

GENERAL BACKGROUND

Iran, with a population of 32 million and a total area of 628,000 square miles, is about one-fifth the size of the United States. Its history, stretching back 2,500 years, has witnessed the empires of Cyrus and Darius, Shah Abbas and Nadir Shah, in successive periods of Persian hegemony over territory extending from the Peleponnesus to the Indus and from the Volga to the Nile. Equally vivid in the national consciousness of contemporary Iranians are the conquests by the Greeks, the Arabs and the Mongols and the more recent humiliations imposed by the British and the Russians. This interplay between pride in a glorious past and shame at frequent subjugations explains in part the sense of political fragility expressed so often by Iran scholars and by many Iranians themselves.

In considering Iran's nuclear options it is important to bear in mind that Iran signed the Nonproliferation Treaty (NPT) on the day it was opened for signatures, 1 July 1968, ratified the treaty in 1970 and signed safeguards agreements with the International Atomic Energy Agency (IAEA) in 1973. Iran has thus pledged to forswear the acquisition of nuclear weapons under any circumstances and has obligated itself to place all its nuclear facilities under international safeguards.

The unitary basis of Iran's wealth—oil—makes it extraordinarily dependent upon perennially high petroleum prices both for continued prosperity in general and for short term economic gains in particular. Furthermore, the present affluence is of extremely recent origin. In 1962 per capita income was $197, in 1963 only one million Iranians were employed in industry, construction of the country's first steel mill began in 1968 and as recently as 1970 the World Bank granted Iran a loan of $60 million to expand the electric network around Tehran.

Yet by March 1975 Iran had become a nation which could commit itself to spend $15 billion on American goods and services over the next five years. Included in the US—Iranian communique, issued at the conclusion of the latest meeting between members of the United States—Iran Joint Commission, is an agreement in principle for Iran to acquire in the next decade as many as eight large nuclear reactors [1]. This is in addition to the four nuclear power plants Iran has already agreed to buy from France and West Germany. Iran has thus firmly announced its intention of becoming a major consumer of nuclear energy.

How realistic is such an announced intention? What is the probable timetable for its fulfillment? To what extent does this major commitment to nuclear power generation affect Iran's nuclear options?

PRESENT AND FUTURE CAPABILITIES

Access to Materials

The capability of a nation to pursue a nuclear weapons policy depends on two basic elements: access to nuclear materials and access to the technology of nuclear explosions. Nuclear weapons can be made either from highly enriched uranium or from plutonium. These fissionable materials can be obtained from nuclear reactor power plants in a variety of ways.

The nuclear fuel cycle, whereby nuclear fission is used to generate electric power, can be separated into three processes: enrichment, fabrication and reprocessing. Reactors are designed to use either natural uranium, or low enriched uranium (U) (in which the concentration of fissionable U–235 has been increased from 0.7 percent—the concentration in nature—to 2 to 5 percent), or uranium of high enrichment (in some cases to over 90 percent U–235). While the low enriched and natural uranium are not directly suitable for weapons use, the more highly enriched uranium is. On a commercial scale, the enrichment process—by far the most difficult technical aspect of the nuclear fuel cycle—has been performed only by the gaseous diffusion technique. A typical plant using gaseous diffusion covers 90 acres, uses 400 million gallons of recirculating cooling water per day and requires 1,300 MWe. A multinational European enrichment plant now being built in France is estimated to cost close to $3 billion by the time it is completed in 1980. At the present time the United States and the Soviet Union are the world's dominant suppliers of enriched nuclear fuel, with the enrichment capacity of the United States now fully committed.

Another enrichment technology using the gas centrifuge method is currently being developed in the US and abroad and would, if successful, require much smaller plants and about one-tenth the electricity as the diffusion method. The smaller size and low electric power requirements would permit the secret construction of centrifuge enrichment plants in remote sites if desired.

Additional enrichment techniques are areodynamic processes such as the Becker nozzle process and laser-based processes. Each is being investigated intensively by scientists in several countries including Germany, South Africa, Israel and the United States. The nozzle technique has the advantage of being technologically simpler than either gaseous diffusion or the gas centrifuge. However at present it consumes twice as much electric power as a gaseous diffusion plant and even with predicted improvements, the power consumption is still expected to be as high as that of gaseous diffusion. If successfully developed, the laser would require on the order of one-tenth the electric power of the centrifuge process. In addition it could probably be used efficiently on a very small scale, say in a laboratory. Since the same enrichment process can also be used for making weapons grade uranium, development of any of the alternative enrichment processes would have substantial implications for nuclear weapons proliferation.

The fabrication process for reactor fuel elements consists of taking uranium, enriched or natural, fabricating it into small pellets, encasing those pellets in long tubes and fashioning containers for the tubes. Because of the hostile environment in which the rods will be in the reactor vessel and because of restructuring and swelling that takes place, this process requires the application of sophisticated techniques of metallurgy and ceramics. At the present time, in addition to the six nuclear powers, Belgium, Brazil, Canada, West Germany, Italy, Japan, Netherlands and Sweden have commercial fabrication plants and Argentina has a pilot plant.

The type of reactor selected for nuclear power is directly relevant to weapons options. All of the United States commercial power reactors are light water reactors (LWR) and use low enriched uranium as fuel and ordinary water as coolant and moderator. LWRs may also be fueled by natural uranium mixed with an appropriate quantity of recycled plutonium. A country using the LWR is therefore dependent upon obtaining the requisite supply of enriched uranium or plutonium.

Furthermore, because the reactors are very large and operate at very high pressures (e.g., 2,000 p.s.i.), the fuel rods are usually changed only once a year and the reactor has to be shut down for the operation. Interrupting the reactor's operation and taking the head off the pressure vessel is a major undertaking and easy to verify; therefore inspectors need be present only at that time to detect clandestine diversion of material. LWRs have two implications for weapons proliferation: (1) because of the relatively long period of time the fuel rods are in the reactor, a higher proportion of plutonium (Pu) produced is in the form of Pu-240, rather than Pu-239 best suited for weapons use; and (2) the only time plutonium could be diverted for weapons use is when the reactor is shut down and the vessel is opened.

The other types of reactors currently in commercial use are the heavy water reactor (HWR), fueled with metallic natural uranium and moderated and cooled by heavy water; and the steam generating heavy water reactor (SGHWR) which uses slightly enriched uranium fuel. The relevant aspect of the HWR is that it enables a country to bypass the enrichment process and thus eliminates one dependency. Since the HWRs use natural, not enriched, uranium, there is no economic motivation to reprocess the spent fuel and thus reprocessing, if it is done, can be assumed to be for weapons purposes, unless the country also has LWRs which could use recycled plutonium. Neither of the heavy water reactors normally operates under high pressure, and the fuel rods, which are pressurized individually, are designed so that they can be changed more easily and frequently without shutting down the reactor. This means that the plutonium produced, while present at a lower concentration, is also less contaminated with Pu-240. It also complicates greatly the inspection problem because foolproof detection would seem to require that inspectors need to be present continuously.

Two other types of reactors are being developed: the high temperature gas cooled reactor (HTGR) which uses highly enriched uranium as fuel, and the breeder reactor which produces more fissionable material than it consumes.

At the end of the nuclear fuel cycle comes the reprocessing or reclaiming of that part of uranium which is not consumed. It is by this chemical separation process, in which the unfissioned fuel is separated from the fission products, that plutonium suitable for weapons use is extracted. The separation of two elements such as plutonium and uranium can be done by relatively easy and cheap chemical processes. To separate isotopes of the same element, such as U-235 from U-238, one can use only differences of nuclear properties such as nuclear mass differences or nuclear magnetic differences. This makes the process of isotope separation technically much more complicated and difficult. It is for this reason that virtually all nuclear weapons have been made from plutonium rather than from enriched uranium.

If a nation wishes to pursue a nuclear option it is clear that it will seek self-sufficiency with respect to all steps of the nuclear fuel cycle. For, as William Van Cleave points out, "As long as a country is dependent upon a foreign source to keep its nuclear program in operation (whether it is a weapons program or an important civil program), it must face the possibility of having this assistance cut off and the probability of being coerced or influenced by such a threat" [2].

In light of the above discussion let us look at Iran and analyze the nuclear decisions it has made to date. In December 1972 Iran announced its intention to acquire nuclear power plants within the next ten years. A few days later the Ministry of Water and Power began a comprehensive study on establishing a nuclear power reactor in southern Iran [3].

Almost two years elapsed before the Iranian Nuclear Energy Organization was created in April 1974, headed by a Swiss-educated nuclear physicist, Dr. Akbar Etemad. Since that time there have been fequent reports of numerous agreements Iran has signed to acquire nuclear reactors.

As of this writing, however, only two firm contracts have been negotiated. In November 1974 Iran signed agreements for the construction of two reactors of 900 MWe each to be built by Framatone, a French affiliate of the Creusot–Loire group, and two reactors of 1,200 MWe each to be built by the West German Kraftwerke Union (KWU). Both contracts are for light water reactors; the French are Westinghouse reactors built under license in France, the German ones are independent of US license. Both the Framatone and the KWU orders include the initial loading of enriched uranium and ten years of reloads. The German reactors are to be installed at Bushapor and the French plants are to be located at Bandar Abbas [4].

These four reactors, when operational, will produce 4,200 MWe. Assuming that about five to ten kilograms of plutonium (Pu-239) are enough to make a bomb, these four reactors could produce enough plutonium to fabricate about a bomb per day, *if* Iran possessed or had access to reprocessing facilities and *if* Iran abrogated or violated her NPT obligations—two big assumptions to be discussed later.

To ensure a supply of enriched uranium to fuel these reactors Iran agreed to lend the French Atomic Authority $1 billion for 15 years for a 10 percent stake in the multinational enrichment plant being built at Tricastin. According to Iranian press accounts Iran's share in the Tricastin plant is later to rise to 15 percent [5].

In addition Iran has two firm contracts from the United States Atomic Energy Commission to fuel two additional 1,200 MWe light water reactors and there is provisional agreement to provide additional fuel for up to six more reactors for a total of 8,000 MWe. Both the firm and the provisional agreement must be incorporated within an effective agreement of cooperation between the governments of the United States and Iran and approved by the president. The agreement must then lie before Congress for 60 days and Congress can, if it chooses, veto the agreement by a concurrent resolution.

The provisional agreement is also contingent upon a generic finding by the Nuclear Regulatory Commission (NRC) that the utilization of recycled plutonium is permissible in light water reactors. That is, the NRC must rule that plutonium can be recycled and used as fuel for power reactors with adequate safeguards against terrorism or theft and that the impact on the environment is acceptable. Early in May 1975 the NRC announced it would postpone a final decision on plutonium recycling for at least three years [6].

As far as its initial power reactor purchases are concerned, Iran has chosen those which are less conducive to pursuing a nuclear weapons option. However Iran has announced plans to install up to 34,000 megawatts of nuclear power by 1995 [7]. If this goal is even approximately reached, Iran must acquire more than 25 nuclear power reactors. Perhaps with some of these future reactors in mind Iran has called on the French Atomic Authority to help it prospect for uranium in Iran [8] and has expressed interest in obtaining regular supplies of uranium from Australia [9].

This indicates that Iran is presently keeping its options open—obtaining sources of natural uranium for the day it either might want to buy heavy water reactors or it might be able to enrich fuel itself. Evidence that Iran is thinking far ahead in terms of attaining nuclear independence is provided by Iranian press reports that "depleted uranium from reactors would be stockpiled . . . to be used as fuel by fast breeders which Iran planned to buy after they were fully evolved in about two decades" [10].

An additional indication that Iran is anxious to acquire nuclear independence comes from recent accounts that it is pressing for reprocessing facilities to be located within Iran [11]. Dr. Dixy Lee Ray, then chairperson of the US Atomic Energy Commission, floated the idea of regional enrichment and reprocessing facilities for the entire Middle East on her trip to Iran in May 1974. But because reprocessing was not a reality yet for any country in the Middle East* (the only nuclear reactors operational are research reactors), the subject was not considered a high priority item. Now apparently the Iranians are pressing for autonomy and location of reprocessing facilities on Iranian soil. This matter is presently still a subject of negotiation between the US and Iranian governments.

It is clear that Iran is in a hurry to acquire nuclear power generating plants, but it is equally clear (to this author at least) that the announced timetables are highly unrealistic. The German-built facilities are scheduled to begin operation in 1980 and 1981 and the French ones in 1982 and 1983. That is a reactor construction time from 60 to 72 months for the German order and from 84 to 96 months for the French plants.

A recently published study of the construction of nuclear reactors in the US from 1964 to 1975 details the steady increase in reactor construction time from an average of 85 months for reactors ordered in 1965 to an average of 115 months for 1969–ordered reactors [12]. The study also found that in comparing earlier estimates of construction time with more recent estimates, misestimation of construction time appears to have been increasing. The report claimed that the slippage was caused by a variety of factors including regulatory delays, environmental concerns, and shortages of architectural engineers and building materials. Not all of these necessarily will be applicable to Iran. Yet the Iranian schedules do not appear to be realistic, particularly when one considers the location of the Iranian reactor plants.

Both the French and German plants are to be built in the southern part of Iran, a section that is very arid and has little population, few settlements, no railroads and poor roads. These nuclear power plants are to be "super turnkey" projects. This means that the contractor is responsible not only for designing, manufacturing and installing the reactors and then "turning over the key" to the Iranians to operate the facility, but in addition, he will have to provide all the ancillary facilities. These will range from enlarging the port facilities to enable the ship with the reactor vessel to enter, to reinforcing highways with substructure strength, to building housing for the construction workers. The infrastructure for such an undertaking is absent at the present time. Therefore, regardless of how high a priority Iran places upon acquisition of nuclear power, her options to acquire nuclear weapons by this means, if she should choose to do so, would seem to be a decade away.

Additional delays may well be caused by declining oil exports. Oil production in mid–1975 was 11 percent less than a year ago [13]. Abdul Majid Majidi, minister of state and director of Iran's Plan Organization, announced a slowdown in Iran's current five year development plan in June 1975. Majidi said that due to a 1.3 million barrel a day reduction in daily oil exports, Iran would halt

---

*With the probable exception of Israel.

secondary development projects and impose "priorities" to insure the continuation of more important projects [14].

Nuclear materials suitable for fashioning bombs can also be obtained from research reactors, which often utilize highly enriched weapons grade uranium. The Tehran University Nuclear Centre was established in 1959 in cooperation with the US. The centre's five MW pool-type reactor went into operation in November 1967.

Research reactors usually do not run continuously. They are often shut down as new experiments are being conducted or analyzed. But even with this in mind, a small five MW reactor still produces enough plutonium to make a bomb every five years or so.

Since Iran is a signatory to the NPT, this reactor, like all her nuclear facilities, falls under IAEA safeguards. Some of the means of verification under IAEA procedures are examination of the design of the facility, monitoring of records, examination of reports and physical inspection of the facility. The present IAEA budget for safeguards is $5 million per year. As of December 31, 1974 there were 462 nuclear power plants of 30 MWe and over operable, under construction or on order. If the IAEA safeguard budget is not increased this will average to less than $11,000 per reactor. Whether this will permit rapid enough detection of diversion of nuclear material to be meaningful is open to speculation.

Access to Technology

In order to be able to exercise a nuclear weapon option, a nation needs to have access to both nuclear materials and nuclear technology. The greatest obstacle to Iran's desire to become a great world power within the next 10 to 15 years is her acute lack of skilled and technically trained manpower. In Iran the number of university students in the social sciences and humanities has traditionally exceeded those studying natural sciences and engineering. This continues to be the case. In the fifth Plan (1973–1978) 32.2 percent of university places are allocated to students studying humanities and social sciences; 38.5 percent for teachers training, fine arts, law, medicine and agriculture; and only 29.3 percent for natural sciences, mathematics and engineering students [15].

According to George Quester, the Iranian Atomic Energy Commission at present has a staff of some 150 people at various levels of training in nuclear physics [16]. Quester reckons that by the time the reactors are all operational, Iran will need some 15,000 trained persons, an expansion of two orders of magnitude [17]. If Iran does not possess this requisite technology indigenously, it will have to acquire it from elsewhere.

Technology of any sort can be transferred from one country to another in a variety of ways. It can be imported, it can be transferred through cooperative arrangements with countries possessing the technology, it can be acquired abroad by one's own nationals who then return home or it can be transferred by on-the-job training. In the field of nuclear technology, as in many other areas, Iran is following all four paths simultaneously.

Importation. The Prime Minister has announced plans to import 49,000 skilled technicians, 42,000 engineers, 44,000 doctors and 770,000 workers. These are to come predominately from third world countries: South Korea, Pakistan, Phillipines [18].

In May 1974 it was reported that Rear Admiral Oscar Armando Quihillat, a former president of the Argentine Atomic Energy Commission, was in Tehran advising on atomic energy [19]. By now Argentinian scientists comprise over half of the foreign staff of the Iranian AEC [20]. When asked about this, a high Iranian official said that advisors on nuclear energy were also being imported from the US and Great Britain. He added that India had "lots of knowledge and experience in nuclear matters" as well as "a surplus of specialists" and that "there was no limit to Iran's importation of foreign technologists" [21].

Cooperation. Together with direct importation Iran is pursuing a policy of extensive cooperation with a variety of countries. The official communique issued at the conclusion of the visit of Indira Gandhi, prime minister of India, to Tehran in May 1974 stated that "contacts will also be made between the atomic energy organizations in the two countries in order to establish a basis for cooperation in this field" [22].

Iran, Pakistan and Turkey are members of the Regional Cooperation for Development (RCD), and the Atomic Energy Commission chairmen of the RCD countries met in January 1975 to discuss ground rules for a joint atomic energy organization [23].

The United States and Iran have had cooperative agreements in nuclear energy since 1959, and recently France agreed to establish a nuclear research laboratory in Iran [24].

Education Abroad. Another strategy Iran is following to acquire nuclear technology is sending young scientists overseas for specialized training. This year 100 BS students are being sent abroad for advanced training in the nuclear field. Three hundred students will be sent next year to the United States, Great Britain, France and West Germany. The Massachusetts Institute of Technology announced plans to increase the size of its nuclear engineering department to accommodate 23 to 27 Iranian students expected to enroll in fall 1975, with a similar number the following year [25].

On-the-Job Training. When Iran's nuclear power reactors become operational, they will transfer nuclear technology to Iranian scientists and technicians by on-the-job training. All the steps of the fuel cycle—from learning how to work with highly radioactive materials by remote control manipulation to having to account for materials to within 1 percent at all stages of the fuel cycle—will embody the transfer of technology. Furthermore, this knowledge will be applicable to nuclear weapons, if Iran should at some time choose that option.

The basic question, which cannot be answered in this chapter, is whether Iran's imported high technology eventually can be maintained by local energies. Will there be sufficient scientific and engineering skills, national cohesion, and long term capital support for the option of becoming a nuclear weapon state to be viable?

MILITARY-STRATEGIC CONSIDERATIONS

Iran's military buildup in recent years has received considerable attention [26] and only a few major points will be highlighted here. First, the buildup has been extremely rapid and proceeded from a very modest base. In 1966 Iran's military budget was $225 million; in 1975 it will be $10 billion. Seven years ago Iran had 25 helicopters; today it has more than 700 in its inventory or on order. Its navy has grown from coast guard dimensions to being the dominant local power in the Persian Gulf. Table 8-1 presents data illustrating the rapid increase in Iran's conventional armed might.

Second, as Dale Tahtinen has made clear, Iran today is the dominant military power in the Persian Gulf, on land, on sea and in the air [27]. The numerical advantage Iraq possesses in combat aircraft is offset by the fact that "Iran can deliver over twice the ordnance tonnage of Iraq and Saudi Arabia combined" [28]. Likewise, Tahtinen discounts Iraq's slight edge in tanks (1,065, to 920 for Iran) by considering both the armor capability and anti-armor weapons and here, too, Iran holds a significant edge [29]. In all other categories one finds Iran's military capabilities superior to its neighbors and the military balance of power in the Persian Gulf which leans heavily in Iran's favor appears to be widening.

Third, Iran's military capabilities are beginning to warrant comparison with countries other than her immediate neighbors. She already has the biggest hovercraft navy in the world. At Chah Bahar, Iran is building a $1 billion army, navy and air force base, the largest of its kind in the Indian Ocean. NATO runways are typically 8,000 feet long. At least one of the three planned for Chah Bahar will extend 12,000 feet. By the end of the seventies, Iran's air force will have more fighter bombers than any NATO country except the United States. Iran's F-4 Phantoms, with their tanker aircraft, have a refueled operational radius of up to 1,400 miles from their bases. This perimeter extends to Cairo, Socotra, Bombay and the Georgian Republic. When the 760 British Chieftain battle tanks and the 250 lighter Scorpion tanks are delivered, this armored force will be larger than that allocated to Britain's army in the Rhine. Unless other nations engage in similar military buildups, when the shah's placed orders of F-14s, attack helicopters, antisubmarine reconnaissance aircraft, tanker aircraft, destroyers, TOWs, Mavericks, Hawks, Tigercats, Rapiers and Sidewinders are in the inventory, Iran will be one of the ranking conventional military powers in the world.

Proliferation Disincentives

How do these military capabilities impinge on Iran's nuclear options? A strong argument can be made that this heavy investment in conventional arms will militate against pursuing nuclear weapons. As long as there are no nuclear arms in the area, Iran will be the strongest military power not only of the Persian Gulf, but of most of the Middle East and the Indian Ocean as well.

In a mini-unipolar world, Iran benefits from the flexibility afforded by conventional weapons. This was illustrated by the relatively quick and easy landing of Iranian troops on the islands of Abu Musa and the two Tumbs, at the mouth of the Persian Gulf, on November 30, 1971, one day before the British protection of the Gulf was terminated. To fulfill its perceived role as guardian of the Gulf, Iran does not need to threaten or to pose as a nuclear power. It needs (and has) a capability which contains the elements of surprise, mobility and speed. Such a capability is provided by conventional, not nuclear, weapons.

Another factor mitigating against Iran pursuing a nuclear option concerns its relations with the Soviet Union. There are at least two elements that must be considered. The first is that the only foreseeable nuclear threat to Iran in the near future comes from the Soviet Union. Protection against this threat can only be and is being provided by the protection of the nuclear umbrella of the United States, formalized in a bilateral US—Iranian defense treaty signed in 1959. If Iran were to pierce that umbrella by obtaining a few nuclear weapons of its own, that umbrella might fold.

Second, it is highly unlikely that the Soviet Union would view with equanimity Iran's acquisition of nuclear weapons. The two countries share a border 1,400 miles long. Although relations between them are cordial at the moment, with economic and technical cooperation agreements, exchanges of visits by the heads of state, and expanding trade, underlying tensions remain. The Iranians have a strong distaste for and distrust of Soviet intentions and the Soviets have equal misgivings about Iran's close ties to the West, its participation in the Central Treaty Organization (CENTO) and its massive arms buildup. Possessing a small and vulnerable nuclear force might invite a pre-emptive attack and hence result in diminished, not increased, security for Iran.

Finally, if Iran pursues a nuclear weapons option, nuclear weapons might rapidly proliferate in its immediate vicinity. India might well abandon all talk of nuclear energy for peaceful uses only, Pakistan has already muttered misgivings about not having the bomb, Iraq is unlikely to want to remain non-nuclear in a proliferated world and the Saudis would likely feel that they can more easily match Iran with nuclear weapons than they can with conventional arms.

Scenarios such as these may not seem so outlandish if one is willing to entertain the thought that in a few years there will be enough plutonium in the world so that fissile material will be available on a nuclear black market. A "nuclear weapons state" may then simply mean a nation that purchased or stole one or a few "primitive" nuclear devices and has one or more airplanes capable of transporting these bombs to their destinations.

Nuclear proliferation would have a great equalizing effect. Iran's quantitative and qualitative superiority would be diminished. From Iran's point of view it

Table 8-1. Iran's Military Capabilities, 1966–1975

	Years					
	1966–67	1968–69	1970–71	1972–73	1974–75	1975–76 (est.)
Population (millions)	25	26	28.4	30.5	32.2	33.1
Gross National Product ($ billions)	$6.4	$7	$8.9	$12.1	$22.5	$36
Defense Expenditure ($ millions)	$255*	$495*	$779*	$915*	$3200*	$10,000
Defense Expenditure as percent of GNP	4.0	7.1	8.7	7.6	14.2	27.8
Manpower:						
Army	164,000	200,000	135,000	160,000	175,000	
Navy	6,000	6,000	9,000	9,000	13,000	
Air Force	10,000	15,000	17,000	22,000	50,000	
Total	180,000	221,000	161,000	191,000	238,000	
Gendarmerie	25,000	25,000	40,000	40,000	70,000	
Combat aircraft	—	200	175	160	216	
Naval vessels	37	45	45	51	41	

*Does not include U.S. grant aid, grant training or credits which from FY 1966–1974 totaled $951 million.

Sources: The International Institute for Strategic Studies, London, *The Military Balance 1966–67*, pp. 28–29; *1968–69*, pp. 32–33; *1970–71*, pp. 39–40, 110; *1972–73*, pp. 31, 70; *1974–75*, pp. 33, 78.
Dale R. Tahtinen, *Arms In the Persian Gulf* (Washington, D.C.: American Enterprise Institute for Public Research, 1974) pp. 4, 16.
"Senate Budget Report," Senate of Iran, January 1975.

would not make strategic sense if it lost its hegemony because of nuclear proliferation.

Proliferation Incentives

There are, of course, incentives as well as disincentives for acquiring nuclear weapons. The predominant role of nuclear weapons since Hiroshima and Nagasaki has been for deterrence. Just as the nuclear arsenals of the superpowers are viewed as effective deterrents against actions hostile to the vital interests of each, so too might Iran think that one or a few nuclear bombs might deter intolerable interference if the Iraqui problem, presently healing, festers or boils or if some traditional Gulf sultanates are overthrown by outside forces or if Pakistan is threatened by India.

It is also possible that sometime in the future Iran will consider the acquisition of nuclear weapons as basic to its perceived military and security interests. Much of the area surrounding Iran has been unstable and volatile in the past as indicated by the breakdown of Pakistan, the left-wing military coup in Afghanistan, the continuing friction between Pakistan and India and between Pakistan and Afghanistan concerning Baluchis (who also live within Iran), and the Kurdish rebellion in Iraq.

If India increases her nuclear capability and remains economically poor but militarily capable, this might also become a source of anxiety to Iran (whose oil could "solve" India's chronic financial crises). A nuclear option in that case might appear to Iranians to be in their best national interest.

The Iranians might also argue—*à la force de frappé*—that having nuclear weapons at their disposal increases their leverage on their allies. According to this line of reasoning, the United States might take a nuclear Iran more seriously than a non-nuclear Iran, so that the nuclear protective umbrella of the United States might in fact be strengthened rather than punctured or folded by Iran "going nuclear."

The argument can be also made that Iran's increased conventional capabilities will result in greatly enlarged horizons and aspirations. That is to say, the perceived sphere of countries or geographical areas whose security Iran considers vital to its own national interests will become far-reaching.

The seeds of this are already sown. Iran's navy today patrols its 1,175 mile southern coastline, its islands in the Gulf, southward around Oman and Yemen, and extends protection to laden tankers heading for Europe as far south as the tenth parallel, just north of Malagasy, and it has recently acquired basing rights in Mauritius.

The shah on a visit to the Imperial Navy base in 1972 said: "Events in the world have taught us that the sea contiguous to the Gulf of Oman and I mean the Indian Ocean, recognizes no frontier." He continued, "We are thinking of Iran's security perimeter and I am not speaking in terms of a few kilometres . . .

the navy would be able to assume its new responsibilities only by increasing its ability to strike against any enemy and win in a wide radius" [30].

Shahram Chubin has documented Iran's increasing concern for Pakistan's territorial integrity since the 1971 Indo–Pakistan War [31]. The shah has stated his opposition to any effort to fragmentize Pakistan and declared that "if Pakistan were attacked, Iran would hasten to its aid, even with military hardware" [32].

On its southern borders, Iran has not only furnished the sultan of Oman with troops estimated to number 2,000, but there are now regular landings by C–130 cargo planes, a 50-bed hospital has been established at the main base of the Iranian unit [33], and most recently Iran "guaranteed" Oman's airspace against intruding foreign aircraft. An Iranian ground-to-air defense system is now operational in Oman [34].

Enlarged spheres of influence of course need not necessitate pursuit of a nuclear option. But history lacks many examples of mighty nations which have forgone acquiring for themselves the deadliest and the most technically advanced weapons available. The oft-repeated phrases of Iran's reemergence as "one of the world's five most powerful nations" and being "guardian of the world's oil lifeline" may become a powerful and driving psychological force to acquire this century's prestige symbol—nuclear weapons.

DOMESTIC CONSIDERATIONS

In analyzing the political factors of the nuclear options calculus one needs to separate the domestic and international elements. Domestically, the first question is: Who will make the decision? Who will be the relevant actors—in the bureaucracy, in the legislature, in the media and in the public? In Iran the answer to these questions is easy. The decision whether Iran develops a nuclear capability or not will be made by the shah.

Historically, all shahs have always been the principle decisionmaker, particularly in foreign and military affairs. As Prime Minister Amir Abbas Hoveyda phrases it, "In Western countries you talk too much about things, take them from committee to committee. Here we go to the Shah and then act" [35].

The recent decree that Iran is to become a one party state for "at least the next two years" [36] accentuates the absence of any meaningful political role for the legislature or the political parties. The now defunct majority and opposition parties, Iran–Novin and Mardom, had never been more than the façade of a two party system. The local press is tightly controlled and no scintilla of public criticism or opposition to the regime is allowed. On a question of such national importance as the nuclear option, it is clear that there will be no participatory decisionmaking.

If the shah alone will cast the die of Iran's nuclear option, on what basis will the decision be made? A helpful clue might be to examine the declaratory policies of the shah and his officials. The earliest statements by the shah on nuclear weapons were clear and unequivocal. "You know," he said at a graduation ceremony at the Command and Staff College on 30 September 1970, "that we are one of the first ones to sign the agreement, or decision to make the world nuclear free. We continue to insist on this, on the destruction of the world's nuclear stockpile, and on the peaceful use of atomic energy" [37].

More recent statements no longer mention Iran's obligations under the NPT but instead center on the actions and intentions of her neighbors. In an interview with a Kuwaiti newspaper in 1974, the shah mentioned the fact that "both Pakistan and India were talking about nuclear strength" [38]. By February 1975, in *Der Spiegal*, the shah declared that "Iran had no intention of acquiring nuclear weapons but if small states began building them, then Iran might have to reconsider its policy" [39].

It may well be that, in the future, pronouncements concerning Iran's nuclear options will become purposely vague and contradictory, playing the same "maybe we do and maybe we don't" game as the Israelis. Evidence for this comes from the shah's reported answer in June 1974 to a French interviewer who asked whether Iran would have its own nuclear arms: "Without a doubt and sooner than one would think" [40]. The statement was immediately denied by the Iranian embassy in Paris and the shah later said, "All I told the four journalists . . . was that any small country which plans to acquire nuclear weapons must think of this measure's grave consequences as well. . . . I reaffirm our stand that not only Iran, but also other nations in the region should refrain from planning to gain atomic arsenals" [41].

The Shah's declarations on Iran's nuclear intentions, of course, may turn out to be weak reeds indeed on which to pin predictions. Chubin and Zabih, in their penetrating analysis of Iran's foreign policy decisionmaking, point out that the shah "can make decisions on the spot, conclude agreements that in other states might take months, and reverse himself overnight if he so chooses" [42].

Chubin and Zabih also explicate the fuzziness and imprecision which surround Iranian proposals for a nuclear-free zone in the Middle East and the Indian Ocean as a zone of peace. According to the authors, since pronouncements on such policies "are personal and often improvised to accommodate hosts or to time with visits, they tend to be based on little study" [43].

Economic factors also may play a decisive role. All the options open to Iran, if it should choose to become a nuclear weapons state—the wherewithal to buy reactors, to import technicians, to send students abroad, etc.—are predicated on the continued high price of oil. Even if the price of oil stays at its present level, the director of Iran's Plan and Budget Organization recently stated that the current pace of development could turn Iran into a debtor nation within three years [44].

If this should become so, and if credit to finance Iran's ambitious development plans is not easily or readily available, major economic readjustments might become necessary. A serious Iranian economic slowdown could effect the nuclear question either way.

It might seem prudent in such a period to retrench and to cancel some of the more costly projects such as nuclear reactors and hence forgo or at least delay the option of acquiring nuclear weapons. Alternatively, an equally plausible Iranian reaction might be to see nuclear weapons as a relatively cheap and quick way to buy entry into major world status and use the prestige associated with nuclear weapons to allay social discontent stemming from unfulfilled domestic expectations.

One also cannot discount the fact that as the conventional weapons acquisition results in the buildup of a military bureaucracy, the shah may become more and more dependent on this bureaucracy for his power and be more and more inclined to give them what they want.

In addition, conventional arms tend to get national executives accustomed to thinking of national security in terms of weapons and military solutions. One could argue that this is especially so in Iran, where the shah looks upon the armed forces as a unifying force in the country. The services have been used to help reduce the illiteracy rate (about 60 percent at the present time), improve rural health care and build up the general infrastructure of the society.

Lastly, the decision to pursue a nuclear weapon option may be made not by this shah but by his successor. The internal politics of any post–Shah Reza Pahlevi government may well dominate the issue of "going nuclear." Given the great uncertainties concerning any Iranian successor government, the same "retrench" or "go nuclear" alternatives discussed above seem equally plausible.

INTERNATIONAL CONSIDERATIONS

On the international side the major disincentive for Iran to acquire nuclear weapons is that it would have to renunciate or violate its adherence to the NPT. The treaty makes withdrawal rather simple: Article X states that "Each Party shall ... have the right to withdraw from the Treaty if it decides that extraordinary events ... have jeopardized the supreme interests of its country. It shall give Notice of such withdrawal ... three months in advance."

But once a treaty is in force, a variety of organizational, political and military factors militate against abrogation. These include the consensual process which enabled the agreement to be reached, the commitment of decisionmakers to the treaty and inertia. However, these factors are less applicable in a highly centralized, authoritarian state such as Iran than in more open, decentralized and democratic nations.

Genuine disenchantment with the treaty itself could overcome these impediments and lead to abrogation. Such disenchantment on Iran's part at present is only articulated sotto voce. It includes the following elements: (1) the NPT regime is inherently unfair in its differential treatment of nuclear and non-nuclear weapons states; (2) the nuclear weapons nations have not fulfilled their NPT obligations; and (3) international prestige accrues to nuclear weapons states.

In the lengthy United Nations General Assembly debates on the NPT prior to its adoption, the delegate of Iran linked comprehensive arms limitations and an international security regime with minimizing the proliferation of nuclear weapons [45]. Article VI of the treaty obligates the nuclear nations "to pursue negotiations in good faith on effective measures relating to the cessation of the nuclear arms race at an early date and to nuclear disarmament, including a treaty on general and complete disarmament. . . .'

Yet in the seven years since the NPT was signed, the number of deliverable warheads in strategic missiles alone in the inventories of the US and the USSR has increased fourfold, from 2,600 to 8,000. These do not include the additional thousands of tactical nuclear weapons and those delivered by bombers.

The Iranian delegation at the NPT Review Conference held in May 1975 took note of this and commented that the "Vladivostok accord and the Threshhold Test Ban fell far short of an Article VI commitment" [46]. The Iranians further pointed out that the lack of nuclear disarmament makes it difficult for non-nuclear weapon states to sign the NPT [47].

Prominent Iranians privately voice the view that treaties such as the NPT and the more recent Strategic Arms Limitation Talks (SALT) had nothing to do with the security needs of countries like Iran. The criteria for decisions taken at those negotiations pertained to the economic, bureaucratic, political and security interests of the United States and the Soviet Union. Non-nuclear weapon states, such as Iran, are looking for assurances, guarantees or treaties which take account of *their* security needs as *they* perceive them.

Following negotiation of the NPT, the United States, the United Kingdom and the Soviet Union agreed in a Security Council resolution that "they would provide or support immediate assistance, in accordance with the UN Charter, to any non-nuclear weapon state party to the NPT that is a victim of an act or threat of aggression in which nuclear weapons are used." In the eyes of the Iranian delegation at the review conference, this commitment is too vague and is subject to the Security Council veto. They argued for "no use" pledges by nuclear weapons states against non-nuclear weapons countries party to the treaty.

Another major incentive Iran may have to withdraw from NPT stems from the enormous prestige which accrues from possessing nuclear weapons. To many in the area it was no coincidence that five months after India exploded her "peaceful" nuclear device, Secretary of State Henry Kissinger journeyed to New Delhi. A product of the Kissinger visit was the establishment of a joint Indian–American Commission for Economic, Commercial, Scientific, Technological, Educational and Cultural Cooperation. Many watchful analysts and policy-makers are now waiting to see what concrete benefits, if any, India acquires as a result of her detonation. Regardless of objective factors, internal Indian reaction was one of great pride and Indians see their nuclear explosion as an important step toward attaining major world status.

In part the psychological pressures to equate international influence and nuclear power stem from the fact that all permanent members of the Security Council are nuclear weapon states. This kind of political legitimation could be diluted by assigning permanent status to one or two non-nuclear weapon NPT signatory nations such as Iran.

To downgrade the prestige attached to nuclear weapons the nuclear weapon states have to change their own defense posture, they have to lessen their reliance on nuclear weapons, they need to be more responsive to the aspirations of non-nuclear weapon states, such as Iran. Only to the extent that one can be optimistic and hopeful about these conditions being met, can one be sanguine about Iran not choosing a nuclear option.

In the meantime, the psychological push for Iran to keep its options open, to gain as much nuclear independence and self-sufficiency as quickly as it can, is strong. National pride, historic remembrances of grandeur, its self-perceived role as a regional power, all point in one direction: Barring great unforseen changes in the international system Iran will choose a nuclear option before the end of the century.

NOTES TO CHAPTER EIGHT

1. "Iran Will Spend $15 Billion in U.S. Over Five Years," *New York Times*, March 5, 1975.
2. William Van Cleave, "Nuclear Technology and Weapons," in Robert M. Lawrence and Joel Larus eds, *Nuclear Proliferation Phase II* (Lawrence: University Press of Kansas, 1974), p. 34.
3. "Nuclear Plant Study Started," *Kayhan International*, December 19, 1972.
4. "Iran into Uranium in Big Way—Etemad," *Kayhan International*, November 30, 1974.
5. Ali Reza Jahan-Shahi, "Iran and France to Jointly Build Uranium Plant," *Tehran Journal*, January 5, 1975.
6. David Burnham, "U.S. Panel Delays for 3 Years Decision on Using Plutonium as Fuel for Reactors," *New York Times*, May 9, 1975.
7. Clyde H. Farnsworth, "France Gives Iran Stake in Uranium," *New York Times*, January 4, 1975.
8. *Ibid.*
9. "Iran is Interested in Uranium—Whitlam," *Kayhan International*, October 5, 1974.
10. Raji Samghabadi, "U.S. Uranium Service Will Save $700 M. Oil Per Year," *Kayhan International*, February 7, 1974.
11. Leslie H. Gelb, "U.S. Nuclear Deal With Iran Delayed," *New York Times*, March 8, 1975.

12. Irvin C. Bupp et. al. "The Economics of Nuclear Power," *Technology Review* 77 (4), February 1975, 14–25.
13. Eric Pace, "Iran is Exhorted to Develop Thrift," *New York Times*, May 28, 1975.
14. "Iran to Reduce Spending," *New York Times*, June 17, 1975.
15. "Report on Educational Development in Iran, 1971–73" (Presented to 34th Session of International Conference on Education, Geneva, September 1973).
16. George Quester, "The Shah and the Bomb" (Unpublished paper, 1975).
17. *Ibid.*
18. *Tehran Journal*, February 8, 1975.
19. James F. Clarity, "Iran Negotiates for Nuclear Energy Aid," *New York Times*, May 27, 1974.
20. Quester, "The Shah and the Bomb."
21. Private Interview, February 1975.
22. "Full Text of Iran-India Joint Communique," *Iran Almanac*, (Tehran: The Echo of Iran, 1974), p. 176.
23. Irfan Parviz, "Regional Atom Chiefs in Talks on Three-Country Organization," *Tehran Journal*, January 27, 1975.
24. Farnsworth, "France Gives Iran Stake in Uranium."
25. "Iran in MIT's Future," *The Tech*, May 16, 1975.
26. See for example, "Iran's Military Strength," *The New Middle East*, May 5, 1971, p. 39; Arnaud De Borchgrave, "Colossus of the Oil Lanes," *Newsweek*, May 21, 1973; R. D. M. Furlong, "Iran—A Power to be Reckoned With," *International Defense Review*, June 1973; Dale R. Tahtinen, *Arms in the Persian Gulf* (Washington, D.C.: American Enterprise Institute for Public Policy Research, 1974); "Iran Beefs up its Navy in the Indian Ocean," *Christian Science Monitor*, September 26, 1974; "The Master Builder of Iran," *Newsweek*, October 14, 1974.
27. Tahtinen, p. 2.
28. *Ibid.*
29. *Ibid*, p. 14.
30. *Kayhan International*, September 11, 1972.
31. Shahram Chubin, "Iran: Between the Arab West and the Asian East," *Survival* 16 (4), July–August 1974, 172–182.
32. *Iran Almanac*, 1974, p. 166.
33. "Army's Morale is at its Highest," *Kayhan International*, January 29, 1975.
34. Eric Pace, "Iran is said to Guarantee Sky of Oman," *International Herald Tribune*, February 5, 1975.
35. Louis Kraar, "The Shah Drives to Build a New Empire," *Fortune*, October 1974.
36. "Shah Decrees Iran a One-Party Nation," *New York Times*, March 3, 1975.
37. "Shah's Speech Emphasizes Military Self-Reliance," *Foreign Broadcast Information Service*, Department of Commerce, October 1970, vol. 1.
38. *Tehran Journal*, January 6, 1975.

39. *Der Spiegal,* February 8, 1975.

40. John K. Cooley, "More Fingers on Nuclear Trigger?" *Christian Science Monitor,* June 25, 1974.

41. "The Shah Meets the Press," *Kayhan International,* June 29, 1974.

42. Shahram Chubin and Sepehr Zabih, *The Foreign Relations of Iran* (Berkeley: University of California Press, 1974), p. 301.

43. *Ibid,* p. 303.

44. "Oil, Grandeur and a Challenge to the West," *Time,* November 4, 1974.

45. *Official Records of the General Assembly,* 22nd sess. 1st comm.—vol. II, 1562nd meeting, May 7, 1968, p. 6.

46. NPT Review Conference, General Debate, SR 4, p. 5.

47. *Ibid.*

Chapter Nine

South Africa's Foreign Policy Alternatives and Deterrence Needs

Edouard Bustin

South Africa's international position has been and continues to be unique, and many of the analytical tools commonly applied to the study of foreign policy and international relations have to be recalibrated when dealing with any situation in which the government in Pretoria is involved. South Africa's enforced isolation from many forms of multinational interaction, culminating with last year's partly successful attempt to remove it from effective participation in the United Nations, typifies that country's dismal image in the eyes of an overwhelming majority of national actors and clearly represents a decisive (though possibly not immutable) parameter in the conceptualization and formulation of its foreign policy goals and strategies. At the same time, however, South Africa's deviance from prevalent norms of international acceptability is largely a function of its domestic political system, and particularly of its racial policies, rather than from any specific forms of aggressive or noncooperative behavior. Although one may legitimately regard South Africa's commitment to white supremacy or its continued occupation of Namibia in defiance of United Nations rulings as forms of permanent aggression, the fact remains that Pretoria's goals have been to protect its own system of exploitative racialism rather than to promote its diffusion throughout the continent. Thus, in terms of broad categories, South Africa's foreign policy has been and continues to be defensive rather than offensive or even revisionist [1]. In this perspective, South Africa's policy has been strongly motivated by the desire to preserve a stable regional environment, and its related strategies accordingly relied on such conceptual referents as containment, collective security and buffer zones. Because of the fundamentally unacceptable nature of the status quo it was trying to maintain, however, South Africa has been severely constrained in its choice of allies and of tactical alternatives. Offers of mutual security guarantees emphasizing the notion that South Africa was not seeking for itself "one square inch of additional territory" [2] failed to impress the black African states, who correctly perceived the irrelevance of the territorial issue and had, in any case, no significant interest in matching such meaningless guarantees with a tacit endorsement of white supremacy. Unilateral offers of alignment with the Western powers, which have been made in one form or another over the past 30 years, have fallen on more receptive ears, particularly since the closing of the Suez Canal and in the strategic context of the Indian Ocean, but such tentative linkage remains predicated on South Africa's ability to demonstrate that its potential contributions to Western security would outweigh the additional demands and liabilities which it would place upon the system—an issue that will be subsequently reexamined in this chapter. Even in its relations with France or Israel, South Africa has not been able to develop anything more than a narrowly defined, ad hoc mutuality of interests predicated on short term tactical gains for its partners rather than on any fundamental convergence of foreign policy goals.

South Africa's material leverage on the viability of its black neighbors (particularly the enclaved or semienclaved territories of Lesotho, Botswana and Swaziland) generated another type of relationship or, more properly, clientship, in which the republic clearly had the upper hand and which, for that reason, cannot be regarded as genuinely bilateral or stable. In the final analysis, only the white regimes of Rhodesia and Portuguese Africa were in a position to deal with South Africa on a broad base of mutual acceptability and reciprocity, although for many years prior to 1974 South Africa seems to have entertained some doubts regarding the viability and usefulness of its white supremacist partners.

South Africa's inability to conceptualize its national and foreign policy interests in terms that would make them broadly, rather than occasionally, compatible with those of other national actors accounts not only for its relative isolation (which in many ways reinforces an inbred feature of Afrikaner political culture), but also for its sensitiveness to relatively slight environmental changes and for the limited range (and thus comparative inflexibility) of its international response mechanisms.

FOREIGN POLICY ALTERNATIVES

The international environment in which South Africa tries to operate its foreign policy changed dramatically in the early 1960s, and now appears to be changing again in several, partly unrelated ways. Though less directly threatening to the security of the apartheid regime, the wave of wholesale decolonization initiated by Britain, France and Belgium during the 1950s and early 1960s was probably more traumatic and unsettling for Pretoria than the type of developments which are currently underway in southern Africa and in the Indian Ocean. For all its ambiguities, decolonization—or, perhaps more accurately, the planned transition from direct colonial rule to indirect neocolonial control—jettisoned the ideological baggage (stewardship, the white man's burden, etc.) on which the bulk of

South Africa's policies of racial stratification had been resting. Pretoria correctly perceived the fundamental incompatibility between its own domestic colonialism and African independence, even under the somewhat nominal form which actually prevailed in most of the ex-colonial territories. Adding to South Africa's dismay was the realization that it could not, and was not expected to have, any significant input in the decolonization process. The initial and still largely prevalent reaction by Pretoria to the developments of the late 1950s and early 1960s, therefore, was to view them as largely negative and basically hostile to its own security interests. Three days after the creation of OAU in May 1963, the South African minister of defense announced that "in the light of the threats which are now being made against our country at Addis Ababa, the question of ensuring that adequate training is given to our defense forces is even more important now.... I trust the time is not far off when we shall train every young man for military service" [3]. South Africa's buildup of its military capability during that period is reflected in the dramatic increase of its defense expenditures (see Table 9–1). Over the past 15 years, South Africa has developed a military arsenal which far outstrips the combined military capabilities of all black African states and which, by international standards, puts it in the lower level of the middle powers. We will later examine this military capability in a more detailed fashion, but it should be noted at this point that this capability was never fully directed against external attack or devised in the perspective of conventional warfare. Much of it was conceived in terms of curbing civilian unrest and of performing counterinsurgency tasks, whether domestically or regionally. At the same time, South Africa developed a set of nonmilitary responses which, it was hoped,

Table 9–1. South African Defense Expenditures
(Official estimates, in millions of Rands)

1960–1961	43.6
1961–1962	71.6
1962–1963	119.8
1963–1964	157.1
1964–1965	210.0
1965–1966	230.0
1966–1967	255.8
1967–1968	266.0
1968–1969	253.0
1969–1970	271.6
1970–1971	257.1
1971–1972	316.5
1972–1973	344.0
1973–1974	438.3
1974–1975	697.0
1975–1976	948.0

Source: Kenneth W. Grundy, *Confrontation and Accommodation in Southern Africa* (University of California Press, 1973), p. 217. *Africa Contemporary Record, 1973–1974; The Military Balance, 1974–75; South African Digest,* 4 April 1975.

could deal with some of the domestic and external factors of instability as perceived from Pretoria. Thus, while the conceptual roots of "separate development" as an ideological smokescreen for racial segregation and exploitation had been planted as soon as the Union of South Africa was founded (1910), and while its specific implementation in the form of "Bantu Homelands" had been carefully scrutinized during the 1950s, it was not until the early 1960s that it was developed into a relatively consistent whole, involving a narrowly controlled measure of African "self-government" and a deliberately ambiguous use of the term "independence" for external, as well as domestic consumption. In its present form, the policy of "separate development" permanently denies any political rights to the Republic's 16 million Africans outside the territorial limits of the former native reserves (dignified as "Homelands") where fewer than half of the Africans actually reside and whose land area (basically determined in 1936) represents less than 13 percent of South Africa's territory. In counterpart, Africans are scheduled to enjoy the benefits of full internal self-government within the "Homelands," up to and including, if they so desire, a somewhat nebulous form of "independence" which definitely includes the right to have one's own flag and national anthem (Transkei already has both) but becomes much hazier over such aspects of sovereignty as currency, diplomatic relations and, naturally, defense. For some time now, it has been clear that the South African government has been anxious to achieve maximum credibility, both domestic and international, for this latest form of "separate development," while those Africans who could be persuaded to take part in this elaborate charade have been equally anxious to avoid being cornered into the position where the exercise of "Bantu self-government" could be construed as a final settlement of all African claims (a preoccupation formulated most explicitly by Gatsha Buthelezi). At this point, eight "Homelands" have been advanced to various stages of self-government (Transkei, Ciskei, Kwazulu, Boputha Tswana, Lebowa, Gazankulu, Venda and Qwaqwa), and one of the major issues argued among Homeland leaders in several recent "summit conferences" has been whether or not it would be desirable at this stage for the Bantustans to organize a federation of black states in order to maximize their collective leverage while negotiating with the white governments the terms of their promised "independence."

The degree to which the policy of separate development is currently being promoted in Pretoria with an eye toward international opinion is further evidenced by its deliberate extension to Namibia as part of South Africa's alleged new flexibility on that sensitive issue.

Concomitantly with the refurbishing of separate development, South Africa gradually developed what has come to be referred to as its "outward" (or "outward-looking") policy [4]. The policy itself has been sufficiently analyzed and publicized to need no further description at this stage. Broadly speaking, it refers to the attempt by South Africa to break out of its isolated position in the world.

Somewhat more narrowly, in the African context, it applies to Pretoria's efforts to demonstrate (or at least to suggest) that mutually beneficial interaction could be achieved between South Africa and the black states of the continent or, at the very least, that the prospects for such interaction should be explored by way of a "dialogue" between the interested parties.

From the outset, however, the outward policy suffered from a serious credibility problem, part of which was due to the ambiguous context in which it was being formulated. While South Africa's increasingly strong military capability was not, properly speaking, an encouragement to "dialogue," it was at any rate an indirect incitement for African "realists" such as Houphouët–Boigny to regard it as a plausible alternative to hostile impotence. At the same time, however, the use of South Africa's military potential to bolster the white minority regimes in Rhodesia, Angola and Mozambique indicated that Pretoria's ideological commitment extended far beyond the innocuous "live and let live" attitude on which its outward policy was supposedly based. Furthermore, the concrete illustrations of goodwill which Pretoria offered, rather obtusely, as evidence of its benevolence all involved countries whose structural dependency on South Africa was such as to cast serious doubts on their capacity for autonomous action, thus suggesting that the only partners with which South Africa could actually interact were satellites or glorified Bantustans rather than genuinely independent African states [5].

In the final analysis, despite the conciliatory tone of the Lusaka manifesto issued at the April 1969 summit conference of East and Central African heads of state, the initial phase of South Africa's outward policy ended in partial failure. Not only was the principle of a dialogue decisively (if not unanimously) rejected by the OAU in 1971, but some of those African leaders who had initially been willing to lend their credit to that principle (or at least to the more circumspect perspective of a "dialogue about dialogue") began having second thoughts about the value of what South Africa could have to offer beyond the insulting prospect that African negotiators might be spared the indignities of petty apartheid, like Kamuzu Banda, should they decide to travel to Pretoria.

South Africa's reluctance to offer any significant concessions, like most rigid attitudes, was born in part from a confident assessment of its ability to withstand African pressures and possibly from an overestimation of the divisions within the ranks of the OAU, but also from its basic sense of insecurity and its attendent fear of initiating a process which it might not subsequently be able to contain. Much of South Africa's evaluation of the balance of power situation as it stood in the early 1970s was of course quite correct, and was reflected by American foreign policy analysts in the now famous premise to option 2 in the National Security Study Memorandum 39 (NSSM 39) of 1969, which read: "The whites are here to stay and the only way that constructive change can come about is through them. There is no hope for the blacks to gain the political rights they seek through violence" [6].

Today, the validity of this assessment clearly stands in doubt, though how much still remains unclear. South Africa's international environment is changing drastically, but in several directions, some of which may not be as unfavorable to Pretoria as seems to be commonly assumed. The April 1974 revolution in Portugal and its repercussions in Africa naturally remain the single most important set of changes affecting South Africa's position. African governments are now in charge in Angola and Mozambique. The advent of some form of African majority rule in Rhodesia and Namibia now seems increasingly probable, and is apparently being treated as such by South African policymakers. At the same time, however, the mounting strategic importance of the Indian Ocean and the intense jockeying for position which seems to be taking place all along its perimeter clearly enhances South Africa's importance—and thus its leverage—in the eyes of the Western powers. Pretoria has been remarkably swift to take stock of these two sets of seemingly contradictory developments, and our purpose will now be to trace their impact on the range of South Africa's foreign policy alternatives, with particular attention to that country's option of developing a nuclear capability.

From a strict geopolitical viewpoint, the collapse of colonial rule in Portuguese Africa clearly appears to alter the balance of forces in the subcontinent in favor of the black states. Not only does it remove two key bastions of white supremacy from South Africa's *cordon sanitaire*, but it also raises serious doubt about the viability and defensibility of the two remaining links in the glacis, Rhodesia and Namibia. This may have been one of the considerations which motivated Prime Minister Vorster to make his famous announcement of November 4, 1974, in which he dramatically asked the world to: "Give South Africa six months' chance by not making our road harder than it is: you will be surprised where we will then stand." In the following weeks, Pretoria initiated some tentative steps to disentangle itself from its vulnerable position in Namibia, and used its influence on the government of Ian Smith to secure the release of imprisoned black nationalist leaders. Although the negotiating process that was subsequently initiated in Lusaka soon ran into a confrontation of incompatible demands, there is no reason to doubt that the trend toward a phasing out of white supremacy in Rhodesia has not lost any of its momentum, and that South Africa appears reconciled with such a development.

At the same time, however, it may be argued that despite a significant change of style, South Africa's foreign policy will continue to rely on a combination of the same three components on which it has been based for a number of years: military preparedness, some form of "outward policy" toward the black states and a continuing attempt to achieve informal security arrangements with Western powers. It seems plausible to assume that the relative importance of each of these three components may shift significantly in a short term or intermediate term perspective, but in the absence of acceptable alternatives (acceptable, that is, to Pretoria) it does not appear likely that any one of them might be willingly abandoned. In fact, a reasonable case can be made that South Africa's apparent retrenchment may actually improve its capabilities on each of these three counts, whether temporarily or permanently.

In its original, pre-1974 version, South Africa's "outward policy" suffered, as we have noted, from a lack of credibility which ultimately stemmed from the fact that neither the African states nor the apartheid regime had any great incentives to encourage interaction, or any wide range of rewards or threats to be used as foreign policy instruments. While this may remain true with respect to most black African states, there is both the need and the opportunity for South Africa to develop a more constructive approach in its future dealings with the emerging governments of Angola and Mozambique (or, in due time, with those of Namibia and Rhodesia–Zimbabwe). Each one of these states has access to capabilities in which South Africa has some degree of interest. This is particularly true in the case of Mozambique, which controls the shortest rail outlet to the sea from the mining region of the Transvaal (through Lourenço Marques) as well as that from Rhodesia (through Beira). Mozambique also holds the key to one of Africa's most important sources of hydroelectric power, the Cabora Bassa Dam, for which South Africa is the major investor as well as the major potential market. Furthermore, an independent Mozambique will bring the reality of African rule much closer to South Africa's doorstep than any of the existing black enclaves ever could, as it will have a common boundary with Natal and the Transvaal, the two most vital demographic and economic centers of the Republic—and those where, parenthetically, most of its African population is concentrated. Mozambique, for its part, stands to benefit from South African contributions to its national income in four major areas: shipping and handling charges for all South African freight flowing through Lourençco Marques; income repatriation and bonuses accruing from some 140,000 Mozambican workers employed in South Africa; the tourist trade (which may be in partial jeopardy for some time); and, once the Cabora Bassa project is completed, revenue generated by the sale of hydroelectric power to South Africa. Militarily, Mozambique is of course highly vulnerable to South African pressure: its southern pedicle (south of the Limpopo River, around Delagoa Bay) would be indefensible in the event of armed conflict, and the capital city of Lourenço Marques lies within easy reach of South Africa, whether by land, by sea or by air. On the other hand, the FRELIMO nationalist elite of Mozambique, while perhaps not as cohesive as it may outwardly appear, seems to be less divided by factionalism than that of Angola or Zimbabwe and may thus be less susceptible to covert maneuverings from Pretoria. In sum, South Africa cannot ignore independent Mozambique any more than the latter can ignore South Africa, and the two governments have sufficiently balanced means of leverage upon one another to ensure that relations between them may be more genuinely bilateral than has been the case with Pretoria's earlier dealings with African states. The nature of South Africa's future relations with Angola is far more conjectural, partly because of the greater uncertainty which at this point surrounds the establishment of that country's political system, but also because of the far more limited incidence of interest intersection between these two territories. Significant bases for reciprocal involvement, although possibly of a more imbalanced type, also exist between South Africa and Namibia or Rhodesia.

Even in its dealings with the other states of black Africa, South Africa may now be in a position to develop more meaningful forms of interaction, at least temporarily. Though its economic and military potentials may be unimpaired in absolute terms, South Africa may now appear more vulnerable and less arrogant, and its appeals for dialogue less gratuitous or less cynical. Pretoria's mediation in Rhodesia, despite its limited effectiveness, is an illustration of its potential contribution to a détente in southern Africa. The black states are just as anxious as South Africa (though not necessarily for the same reasons) to achieve nonviolent settlement in the southern part of the continent and they can undoubtedly recognize the positive part which Pretoria can play in this process. Thus, while South Africa's earlier offer of a dialogue rapidly aborted on the recognition that there was not really much that the white regime cared to talk about, there is now clearly something worth talking about and an implicit admission on both sides that bargaining may be preferable to confrontation.

The limits of such incipient détente are all too obvious, however. For one thing, the African states are as suspicious of Pretoria's ulterior motives as South Africa is of theirs. For another, while Prime Minister Vorster may genuinely be interested in pursuing his unaccustomed role of mediator, he has repeatedly emphasized (partly for the benefit of his *verkrampte* critics, to be sure) that South Africa's pursuit of its policy of separate development should be regarded as non negotiable [7]. Such frankness may well lend more credibility to his attempts to strike a relatively flexible posture over the issues of Rhodesia and Namibia, but it also indicates that the partial convergence of interests between Pretoria and the African states remains purely tactical and basically ambiguous. In a fundamental sense, South Africa's goals appear to be unchanged, and it will take more than Vorster's cryptic promises of an impending metamorphosis to convince the African states that the leopard can indeed change its spots. In the meantime, and well beyond Vorster's six months deadline, there will be many concrete ways for Africans to evaluate the sincerity of South Africa's commitment to détente and bilateralism.

While South Africa may now be trying to walk more softly, it seems more determined than ever to carry a big stick. Defense expenditure multiplied over 16 times from 1960 through 1975 and more than doubled over the last two fiscal years alone. While defense now accounts for more than 10 percent of the South African budget, the level of military spending remains relatively modest when compared with the share of public expenditure allocated to defense by such countries as Israel (49 percent) or prerevolutionary Portugal (44 percent).

More significantly, South Africa's defense expenditures in 1974–1975 represented only 3.87 percent of that country's gross national product, which is more than the corresponding percentages for Uganda (2.72 percent), Tanzania (2.47 percent) or Zaire (3.35 percent), but less than those for Zambia (3.90 percent), Somalia (5.0 percent) or Nigeria (7.30 percent) [8]. It seems clear that, under escalating pressure, the apartheid regime could easily mobilize a much larger proportion of its resources for military purposes. This will have to be borne in mind when looking into South Africa's nuclear options. At this point, however, the Pretoria government undoubtedly has the capacity to deter any conventional attack that might be mounted against its territory by any combination of African states, and there is indeed little likelihood that such a confrontation might develop within the foreseeable future. Attempting to answer the question of "where South Africa stands in the balance of power in Africa," the defense correspondent of the Johannesburg *Star* recently described that country's arsenal as "a staggering amount of military hardware" [9]. By African standards, that characterization is hardly an overstatement. Tables 9-2 and 9-3 will offer a summary overview of South Africa's defense establishment. South Africa's 30,000 police also perform many duties falling within the broad purview of defense, particularly in terms of counterinsurgency measures. In addition, the government has recently initiated a cautious policy aimed at giving some military training to nonwhites (carefully organized into "Colored," Asian and "Bantu" units). South Africa's armaments superiority over its black neighbors is particularly awesome in terms of armored combat vehicles and of air power. It has five times as many tanks and armored cars as all the independent African nations south of the equator combined. The South African air force numbers more aircraft than are available to all other states south of the Sahara. When it comes to combat aircraft, Nigeria (42 combat planes), Ethiopia (40), Zaire (33, plus 17 on order), Somalia (31 on paper), Uganda (29), Tanzania (24) and Zambia (18) may appear to be more of a match for South Africa's 99 units, but many of the aircraft in African hands are of doubtful serviceability or all partly obsolescent models of the sort which South Africa has now relegated to the role of trainers. The recent purchase by South Africa of a number of sophisticated F-1 Super Mirage (basically the same plane which France recently tried to sell to the

Table 9-2. South African Military Personnel

	Regular	Conscripts	Reserves	Total
Army	7,000	27,500	60,000	94,500
Navy	3,200	1,250	9,000	13,450
Air Force	5,500	3,000	3,000	11,500
Paramilitary (Home Guard)				75,000
Totals:	15,700	31,750	72,000	194,450

Source: *The Military Balance, 1974-1975.*

Table 9-3. South African Armaments

Army:

Tanks: 100 Centurion Mk 5
        20 Comet medium tanks
Armed Cars: 1,000 AML–60 and AML–90 (Panhard, under French license)
        50 M–3
Scout Cars: 50 Ferret
Armed Personnel Carriers: 250 Saracen
        ca. 100 V–150 Commando
Artillery: 25–pounder gun howitzers; 155 mm. howitzers
Antiaircraft: 35 mm L–70/40 guns; 3.7 inch guns
        3 batteries of 18 Cactus (i.e. Crotale) SAMs (+7 on order)
Light aircraft: 1 squadron (Cessna 185A/D and A185E, to be replaced by AM3–C)

Navy:

Submarines: 3
Destroyers: 2 (with antisub. Wasp helicopters)
Frigates: 7 (3 with antisub. Wasp helicopters)
Minesweepers: 10 coastal minesweepers; 1 escort minesweeper
Seaward Defence: 5 boats
Support and Auxiliary: 5
Helicopters: 7 Wasp + 10 on order
On order: 6 corvettes (to be equipped with Exocet SSMs)

Air Force:

Bombers: 1 squadron (6 Canberra B (I) Mk 12; 3 Canberra T Mk 4)
Light Bombers: 1 squadron (10 Buccaneer S Mk 50)
Fighters: 2 squadrons (32 Mirage IIIEZ; 8 Mirage III DZ)
Fighter Recce: 1 squadron (16 Mirage IIICZ; 4 Mirage IIIBZ; 4 Mirage IIIRZ)
Marine Recce: 2 squadrons (7 Shackleton MR3; 9 Piaggio P–166S Albatross + 11 on order)
Transport: 4 squadrons (7 C–130B; 9 Transall C–190Z; 23 C–47; 5 C–54; 1 Viscount 781;
        4 HS–125 Mercurius
Helicopters: 4 squadrons (40 Alouette III; 20 SA–330 Puma; 15 SA–321L Super Frelon)
Trainers: 160 Impala MB–326M, some of which are armed for counterinsurgency missions +
        15 MB–326K on order; Vampire Fb Mk6, Mk9, T Mk55; TF–86; C–47; Harvard;
        Alouette II/III
Reserves: 20 Impala; 100 Harvard IIA, III, T–6G (Texan); 20 Cessna 185A/D, A 185E;
        12 Air Commando squadrons (private)

Source: *The Military Balance, 1974–1975.*

smaller European members of NATO) will further enhance the strike power of its air force and further widen the gap with the rest of Africa.

South Africa is also the only power in the subcontinent to possess an early warning system (deployed in the Transvaal), which in addition to detecting aerial intrusions can also guide interceptor aircraft response. This network, which was originally built by the British, was turned over to South Africa in 1965 and is now tied in with an underground air defense headquarters in the eastern Transvaal (completed in 1972) which is again the only operational center of its kind in the subcontinent.

In addition, the Pretoria government has been engaged in the development, testing and production of various missile systems. Apart from the Cactus surface-to-air missile (a South African version of the French Crotale), and the French surface-to-surface "Exocet" system, which equips the navy, South Africa reportedly holds air-to-ground missiles and has been testing a guided air-to-air missile (presumably similar to the French Matra R–550) with which it plans to equip its new F–1 Super Mirages.

South Africa's determination to achieve self-sufficiency in terms of the production of all of its major weapons systems has been apparent for more than a decade and is clearly related to the 1963 decision by the United Nations to initiate an arms embargo against the apartheid regime. The result has been the development of a local arms industry which now fills all of the country's needs in terms of conventional weaponry and also supplies an increasingly large share of its more sophisticated equipment. South Africa is now assembling some of its more advanced aircraft and manufacturing certain of their components and it will soon achieve full production capacity for certain models (Atlas Impala, Aermacchi AM3C, SA–330 Puma helicopters).

It is almost impossible to evaluate the adequacy of South Africa's military preparedness. For a country that tends to regard itself (sometimes with a certain perverse fondness) as being alone against the world, the answer to "how much is enough?" will obviously always be "just a little more." But part of the problem, in the case of South Africa, has to do with two sets of unresolved questions: (1) What type of conflict can or should it be prepared for? and (2) At what point does South Africa's military buildup become counterproductive in terms of its foreign policy goals—or in other words, At what point does the escalation of Pretoria's military capability generate more hostility than it is able to deter and invite great power intervention?

From a comparison of their respective military potentials, it seems clear beyond a reasonable doubt that South Africa could easily withstand a full scale armed confrontation with any combination of African states. Some African leaders probably share this view privately and a few, like Chief Jonathan of Lesotho or President Banda of Malawi, have actually expressed the opinion that South Africa could overwhelm the rest of the continent in a matter of days, if not hours [10]. Even discounting such excessive statements, the more sober evaluation of recent observers remains that "even if all the Black African states were to place all their military equipment under one united command, it is still debatable whether such a command would make a major dent on South Africa's military might" [11]. Portugal's pullback from Africa does not appear likely to make for drastic alterations of this assessment, at least in the short to intermediate range (the opinion quoted above was, in any case, published in early 1975). It might actually be argued that inasmuch as it has contributed to reducing South Africa's defense perimeter, and shortened its lines, the elimination of the republic's white bastions could make its whole defense posture more tenable. This may well have been one of the considerations accounting for Prime Minister

Vorster's conciliatory moves with respect to Namibia and Rhodesia. But if the sort of "peace" offered by Pretoria is merely "the continuation of war by other means," what may in fact develop in southern Africa could be a perpetuation of the present situation—namely, a low key form of hostility ranging from a neither-war-nor-peace type of tacit stalemate to a war of attrition involving only relatively insignificant levels of violence.

The third alternative (leaving out for the moment the hypothesis of an enlarged conflict) involves all forms of unconventional warfare, with or without coordinated African support and with any combination of a number of possible types of paramilitary activity. Though it is by no means the most improbable sort of contingency, it is also by definition the most elusive, and the one against which the concept of deterrence seems to be most irrelevant. Plans can of course be made against sabotage, hit-and-run attacks and outright guerrilla activity; in fact, a good deal of South Africa's military buildup over the past 15 years has been directed precisely at this type of contingency, but while the government in Pretoria may have the capability to contain or even to control such situations, it is doubtful that they can be deterred by the sheer piling up of weaponry. When it comes to the strategy of counterinsurgency, the resort to precedents can be enormously misleading. South Africa is not Indochina, nor is it like Malaya or Palestine. The nature of South Africa's terrain and the density of its communications network make it highly improbable that any sort of lasting armed insurgency could be sustained entirely from within the republic, even if equipment could be resupplied from outside with a fair amount of regularity. The most plausible type of scenario for unconventional warfare in southern Africa implies rapid incursions by relatively small forces, followed by retreat into neighboring territory. Such tactics actually involve enormous problems for the insurgents, as well as for the countries which could serve, more or less reluctantly, as sanctuaries, but they nevertheless offer the least unfavorable odds for the insurgents, if only because they would sooner or later trigger some kind of military action by South Africa against an independent African state, whether in the form of hot pursuit, pre-emptive strikes, retaliatory strikes, short term incursions or lasting occupation of border areas. Any one of these responses would automatically escalate South Africa's counterinsurgency efforts to the level of an international incident or crisis, depending on its magnitude and intensity.

The full range of plausible contingencies involving insurgency and counterinsurgency in southern Africa is far too wide to be analyzed in detail within the scope of this study [12]. Any serious contemplation of probable scenarios, however, must sooner or later come to grips with a certain number of nonmilitary or paramilitary variables. The most obvious of these considerations (and one which has been raised in nearly similar terms from Machiavelli to Spínola) is whether conventional weapons and methods can be the final answer to popular insurgency. In May 1973, the South African minister of police admitted that curbing "terrorism" might be more than a law and order problem and suggested that South Africa "may have to make use of improved human relations and international ties" in order to deal with such situations [13]. Apart from recognizing that insurgency has to be dealt with in political as well as military terms (the "hearts and minds" argument so frequently and so rhetorically put forth with respect to Indochina), what this statement also implies is that any type of insurgency situation in southern Africa is automatically invested with a conflictual potential of an international scope.

The prospects for international involvement in southern Africa have of course been recognized for years, but there is considerable range for differences of opinion as to the specific forms which such involvement might assume and, more importantly, as to their probable repercussions for South Africa and for the rest of the continent. Limited interaction involving only African actors is one such form of situation with which South Africa's advocacy of a "dialogue" (and, from a different perspective, its military preparedness) has been intended to deal. Concomitantly with this limited, regional perspective, South Africa has for many years maintained a much broader framework of analysis, in which the threats against the stability of the white supremacist regimes of southern Africa are presented (and quite probably perceived) in manichean, quasi-apocalyptic terms as only one facet of a global conflict between "freedom" and "communism." Such drastic dichotomies raise relatively few problems within the context of a system where the legal definition of "communism" can be extended to cover all forms of opposition to the regime, and where "freedom" is very much a residual category, but the least that can be said about their global validity is that its serviceability for purposes of international analysis is rather doubtful. Actually, Pretoria's devil theory of African liberation, while thriving on bipolarity, can easily be made to fit into a bi-multipolar or even a straight multipolar model: what matters, after all, is the alleged existence of a dark, looming menace hovering about the subcontinent, and Peking can fit that part just as well as Moscow. Of late, as a matter of fact, China has been increasingly identified as the chief villain. Writing in the September 1973 issue of *Paratus*, the defense force magazine, the South African Commandant-General, Admiral H. H. Beiermann, wrote that the "terrorist" threat against South Africa was formidable and real, adding: "Behind and within these fronts looms the dragon of Communist China" [14]. Such interpretations, with their crude ideological substratum, have obvious uses on the domestic scene, where they can not only help forge the widest possible consensus for the regime, but also cloak the cultural heterogeneity of the white society under a more respectable mantle than mere racial solidarity. In terms of international alignments, however, they also represent a permanent invitation for the Western nations to embrace the apartheid regime as their ideological kin and to include it within their collective security system. This search for some form of alliance with the West, which represents the third component of South Africa's international stance, has been going on for quite some time (longer, in fact than the other two) and probably stems in part from a sense of nostalgia for the relatively brief period (1919—1945) when South Africa was a respected member of the then predominantly white international system— a "golden age" highlighted by Jan Smuts' contribution to the genesis of the League of Nations and of the United Nations [15]. Over the years, it took such diverse forms as Foreign Minister Eric Louw's insistence that South Africa should be formally or informally associated with NATO, or the deliberate encouragement of American and European investments in the South African economy as a way of building up the Western nations' stakes in the stability of the white regime. During the 1960s and early 1970s, Portugal provided South Africa with a permanent illustration of the advantages to be derived from a Western alliance. Portugal was able to secure most of the weaponry it needed through its membership in NATO (a problem which South Africa has solved by other means) and it was effectively insulated against non-Western involvement in its colonial wars. At the same time, Portugal's allies clearly never intended to become themselves militarily involved in Portuguese Africa, so long as no other major power did, and the equanimity with which they witnessed the fall of the ultracolonialist faction in Lisbon should indicate to Pretoria than the degree of Western commitment to white rule in Africa can never be of the same order as its own. While pointing out the limitations of a Western alliance, these considerations have not deterred South Africa from seeking to share in some of its benefits. Yet, it might be argued that South Africa has actually enjoyed, on a de facto basis, most of the limited advantages which it could conceivably derive from a more explicit kind of Western support—a recent example being the triple veto whereby the US, Great Britain and France blocked the 1974 move to oust South Africa from the United Nations. Arms and military technology have continued to flow to South Africa, directly or indirectly, from a variety of Western sources. What capability not currently within Pretoria's reach could South Africa achieve from a more formal type of linkage?

The answer to this question is not an easy one, but it is reasonable to assume that it is related to South Africa's quest for absolute deterrence, not only against African states (which it already has) but also against any non–African power. This view is indirectly reflected in the opinion expressed in September 1973 by the South African Chief of the Army, Leitenant General Magnus Malan, who suggested that South Africa could not afford to lose a single battle and that its first defeat would also be its last. "African states can fight and lose," he said, "recover and fight again. But can we? I am afraid we can lose only once" [16].

In its attempts to become associated with a collective security system that would enable it to achieve such a level of deterrence, South Africa has been powerfully helped by the current debate in Western military circles over the strategic significance of the Indian and South Atlantic Oceans. It is not the purpose of this chapter to go into the details of a debate which has been simmering for nearly a decade and which has created deep rifts with the US defense community [17]. In the official view of the South African government, by contrast, there is no room for grey areas and the major themes of past and present propaganda are summoned to the rescue:

> In the event of a global clash with the West, the Communists would regard gaining control of the Cape sea route as a valuable prize, the Communist threat to Africa goes hand in glove with Red expansionism in the Indian Ocean area. Strategists in Europe and the United States have repeatedly called for joint action to secure the sea lanes of the Indian Ocean and the Cape, as they are their main trade lifelines. Comparatively recently, NATO realized that it could no longer ignore the situation and suggested that its area of operation be extended to South of the Tropic of Cancer.... The security of the Cape sea route and the southern Atlantic are vital not only for the safety of South Africa, but other countries as well [18].

Such views are widely shared in some African and British navy circles. Former US Navy Commander-in-Chief Pacific Command, Admiral John McCain recently expressed their sentiment by stating: "What has happened in Mozambique and Angola makes our possession of Diego Garcia more important than ever. But it also means that we absolutely need access to the South African naval facilities at Simonstown and Durban" [19].

The extensive and sophisticated naval facilities at Simonstown, which South Africa is currently expanding in obvious anticipation of future use by Western (particularly US) naval units, is clearly intended to serve as Pretoria's most substantial bid in its effort to become more formally associated with Western security arrangements. The base was originally built by the British and turned over to South Africa in 1957, but continued to be used by the Royal Navy until the Labour government decided to abrogate the agreement under which it was being used by the United Kingdom. Whether the US now decides to take advantage of this and other South African naval facilities (which include advanced maritime surveillance and communications systems) may well condition the extent to which the US will move from its current position of mild disapproval (closer, in practice, to benevolent neutrality) to one of de facto collaboration with South Africa [20].

ADVANTAGES AND DISADVANTAGES OF "GOING NUCLEAR"

Pretoria's apparent determination to develop a nuclear capability will now be examined against the background of its foreign policy alternatives and perceived deterrence needs. Let it be understood at this point, however, that the issue will be envisaged only from the viewpoint of South Africa itself. Whether American foreign policy, the stability of the African continent or the cause of world peace would be better served by a nuclear South Africa will not be dealt with in this study, not because these considerations are unimportant but simply for reasons of space. The probable reactions to South Africa's development of a nuclear capacity by the superpowers, by the African states and by other national or international actors will be briefly appraised, but only as variables to be evaluated by South Africa in measuring the advantages and disadvantages of "going nuclear."

The problem for South Africa actually involves two separate sets of questions. The first set of questions deals broadly with problems of capacity. Does South Africa have the necessary resources and technology to develop a strategic or tactical nuclear capability, including means of delivery adequate to ensure credible use? The second order of problems revolves around the issue of how the development of such a capability would affect the achievement of South Africa's foreign policy objectives, as well as the actual conceptualization of such objectives.

The answer to the first set of questions is partly academic and need not detain us very long. South Africa has long been regarded as one of the near-nuclear nations, along with Israel, Brazil and a number of others—one of which has recently joined the once exclusive club of nuclear powers. Dr. Louw Alberts, vice president of South Africa's Atomic Energy Board, stated bluntly that "our technology and science have advanced sufficiently for us to produce [an atom bomb] if we have to" [21]. Even though Pretoria has consistently maintained that its nuclear knowledge would be used for peaceful purposes only, few experts doubt that South Africa could indeed produce a nuclear device once it had determined to do so. South Africa is the third largest producer of uranium in the world and, as a country with limited sources of conventional energy, it has been pursuing the development of nuclear power for some years as a major priority. South Africa acquired a nuclear reactor from the United States in 1965 under the Atoms for Peace program and reportedly started work in 1970 on an original uranium enrichment process which became operational in 1975 with the opening of a pilot plant at Valindaba, near Pretoria [22]. Although Prime Minister Vorster claims that the uranium enrichment process was developed "without assistance from foreign countries," US Congressman Les Aspin (D—Wis.) revealed in April 1975 that US Nuclear Corporation of Oak Ridge, Tennessee, had provided enriched uranium for the project, while IBM and Foxboro Corporation of Foxboro, Massachusetts, had supplied some of the technological equipment for the plant [23]. South Africa now reportedly plans to develop another, larger enrichment plan "in cooperation with friendly outside interests" [24].

While the primary significance of these developments probably lies in their commercial application as industrialized countries begin to turn increasingly to nuclear energy, they nevertheless imply on the part of South Africa a manifest capability to produce weapons grade material. The manufacture of weapons from that point is a problem that South African technology can solve with ease. In fact, whether or not South Africa presently has nuclear weapons within its reach, it is perhaps equally important to observe that Pretoria has made no real effort to discourage speculation on this score. Commenting on Louw Alberts' claim that South Africa could produce an atom bomb if it had to, the President of the Atomic Energy Board, A. J. A. Roux, merely remarked that such a statement was "pointless," as South Africa had in fact "not yet" manufactured a nuclear weapon [25]. South Africa apparently feels that rumors about its impending capability may have some of the effects that it anticipates from the actual possession of weapons—namely, deterrence and increased credibility as a potential ally. And while it has declined to sign the Nonproliferation Treaty, Pretoria insists that it would be prepared to consider signing it if appropriate guarantees by the International Atomic Energy Agency were forthcoming [26].

But the mere possession of a nuclear device cannot be separated from the problem of its delivery. As observed by a South African political scientists, "[I]t serves no purpose to manufacture a bomb in a military vacuum in the hope that either the prestige or fear associated with it will make it worth the effort" [27]. The question of what delivery systems South Africa could use for its nuclear weapons actually tells us something about the kind of strategic context in which they are apparently being contemplated at this stage. In the absence of intercontinental ballistic missiles, and given South Africa's geographical location, the only conceivable targets for South Africa's nuclear weapons can be African countries and African coastal areas—a consideration which clearly predetermines some of the foreign policy implications of any future nuclear capability that South Africa might develop.

Writing in the early 1970s, George Quester could argue that "by manufacturing nuclear weapons itself, South Africa seemingly would stand to gain less than it would lose. Its conventional superiority over any political opponents in Africa is so clear that it would hardly seem advisable to change the rules of the game" [28]. The evidence offered in the earlier part of this chapter suggests that this evaluation remains basically valid today, at least in the hypothesis of conventional conflicts [29]. As a deterrent against conventional attack, nuclear weapons might be effective if we hypothesize a frontal drive by African forces against the South African territory. Such an offensive would obviously have to involve forces other than those of the neighboring black states, and would presumably take place under OAU or other collective auspices, with a minimum of two of Africa's largest military powers (e.g. Nigeria, Zaire or Ethiopia) as fully committed participants. As suggested earlier, it seems unlikely that such a concerted effort could be mounted in the foreseeable future, even assuming that the powers in question can deflect sufficient resources from more pressing priorities. Should such a combined offensive nevertheless materialize, South Africa's ability to repulse it through conventional means may be regarded as a virtual certainty. In this particular context, therefore, the deterrence value of nuclear weapons, while theoretically present, becomes largely irrelevant and falls, more properly, under the category of overkill.

In the perspective of counterinsurgency, the use of nuclear weapons runs into several problems too obvious to need belaboring, especially if insurgent activity takes place within South Africa itself. A limited number of situations may be conceived, however, where the threatened or actual use of nuclear weapons could be defended from a strict military viewpoint. Heavy concentration of guerrilla units in a neighboring territory (whether for training purposes or in a staging area) represents one such scenario. The threat of a nuclear strike to secure the passivity of an African state while South African forces move across the border to liquidate guerrilla bases is another such contingency. In either hypothesis, however, one fails to perceive what military objectives could be achieved through the use of nuclear weapons that could not just as effectively be secured by conventional means. As for the hypothesis of insurgents manufacturing a crude nuclear device and planting it in one of South Africa's urban centers, it is not a situation where the concepts of deterrence or retaliation appear to have much relevance. In sum, the types of insurgency which are likely to be impervious to South Africa's conventional capability for deterrence and retaliation are just as unlikely to be prevented through the availability of nuclear weapons.

It might be mentioned at this point that any use (or threatened use) of nuclear weapons by South Africa against an African state could well crystallize the sort of collective international reaction which Pretoria has been attempting to prevent for years through its "outward policy" and through its search for informal alliance. In such a situation, South Africa's "tacit allies" (especially the United States and Great Britain) would probably find it very difficult to effectively oppose the use of internationally concerted military sanctions against the Pretoria regime, particularly since South Africa would not be responding to nuclear blackmail or aggression in this hypothesis, but rather acting unilaterally in a first-strike or pre-emptive capacity.

The use (or threatened use) of nuclear weapons by South Africa against a non–African actor raises a somewhat different, though hardly more reassuring perspective. As indicated earlier, South Africa's lack of long range delivery capabilities postulates that such a move would be taking place along southern Africa's maritime approaches, including the Mozambique channel and the Namibian coastline. The scenarios usually developed in connection with this line of speculation are fairly predictable: they involve, for example, a UN–sponsored amphibious operation designed to secure control of Namibia in accordance with international rulings on the illegality of South Africa's presence in that area, the development of Soviet or Chinese naval bases along the coast of Mozambique, or again the use of Lourenço Marques or Beira as points of entry for the supply of military equipment to an African guerrilla force under the cover of the Mozambican army. Starting with the assumption of a South African nuclear strike in any one of these situations, the prospects of escalation rapidly become rather unattractive. Tactical retaliation against South Africa at that point certainly must be regarded as a very strong possibility, which in turn brings in the question of further escalation involving another nuclear superpower (practically speaking, the United States).

The credibility of a direct United States strike against the Soviet Union or China following retaliation by either of these powers against an initial South African strike seems rather dubious. It would appear rather implausible that the United States should allow any significant degree of nuclear escalation to take place under the hypothesis described above, given the risks to its own territory and the lack of a direct threat to its survival or vital interests. Thus the value of South African nuclear weapons in situations involving non-African powers would appear to be somewhat doubtful, whether these weapons are expected to function as a deterrent or in the more limited capacity of a "tripwire" intended to trigger major Western involvement.

It seems reasonable to conclude that, from a strict military viewpoint, South Africa's ability to deal with any of the types of challenges which it is likely to face within the foreseeable future would not be significantly enhanced through the development of nuclear weapons. What about the other components of South Africa's policy—dialogue and the quest for collective security? How would they be affected by the decision to achieve nuclear capability?

Positing that South Africa's "outward policy" toward the rest of the continent is predicated on the existence of a mutual willingness to "negotiate rather than fight," the problem is whether Pretoria's achievement of a nuclear capability is likely to improve its bargaining position vis-à-vis the African states. A reasonable cogent, though by no means compelling case can be made that it could. Because of its ominous and spectacular nature, the development of nuclear power by South Africa might serve to impress upon its black neighbors (and upon the world generally) Pretoria's determination not to negotiate under pressure, as well as its ability to mobilize its considerable potential. By providing the white South African community with the psychological reassurance that their government is not dealing from a position of weakness, it could also make it politically easier for Pretoria to offer some meaningful concessions to its African opponents.

There is no reason to believe, however, that African states would fail to perceive the drastic constraints under which South Africa's nuclear capability would be operating. The degree to which they would be impressed by South Africa's apparent resolve might accordingly be less than overwhelming. In fact, since we are ultimately dealing with a problem of credibility, it might be argued that whatever bargaining advantage Pretoria might be able to spin off from the exercise of its nuclear options could be derived more effectively (and is indeed already being derived) from publicizing its capacity to "go nuclear"—which is highly credible—rather than from the threat of subsequently using such weapons—which is demonstrably less credible.

It is perhaps within the context of South Africa's search for some sort of Western guarantee of its security that the pursuit of nuclear capability can be most validly justified. Paradoxically, however, Pretoria's nuclear potential may be more effective in this respect in terms of its nuisance value than because of any positive contribution which it could make to the global posture of the Western alliance.

The apartheid regime may be genuinely interested in demonstrating its attractiveness as a potential ally by building up a strong and well-equipped military establishment. In the final analysis, however, South Africa's value as an ally depends on its location, on its willingness to make naval, air and communications bases available to the West; on its ability to help control the southern oceanic routes; but not on its idiosyncratic brand of anticommunism or on the type of political order it stands for (both of which are either irrelevant or embarrassing from the perspective of Western diplomacy), nor certainly on its possession of nuclear weapons. Indeed, from the viewpoint of the United States, allies or would-be allies possessing an independent nuclear capability of limited credibility and proposing to use it in accordance with their own motivations represent an unwelcome factor of systemic disequilibrium, rather than of stability. Should South Africa, in view of these reservations, offer to place its nuclear forces-in-being under Western (i.e., United States) strategic command, it is doubtful that any great value would be attached to this modest addition to the West's ample nuclear arsenal, except precisely to the extent that this would remove an unpredictable variable from the "delicate balance of terror." It is not inconceivable, therefore that the United States might be the prime target of a mild form of nuclear blackmail whereby South Africa would exact some sort of guarantee of her security through a judicious mix of bomb-rattling and calculated orneriness. There are, in any case, enough voices in our defense establishment advocating (on rather questionable grounds) a closer cooperation with South Africa for Pretoria's "enfant terrible" gambit to bring off the desired result—i.e., a measure of American commitment to the preservation of its particular brand of order. In view of the consequences of some of our earlier commitments, however, there is more than enough reason for the United States to carefully look that kind of gift horse in the mouth.

NOTES TO CHAPTER NINE

1. See G. G. Lawrie, "South Africa's World Position," *Journal of Modern African Studies,* March 1964, p. 41–54.

2. A theme recently repeated by Prime Minister Vorster with reference to Namibia during his visit to Liberia (see *South Africa Digest,* February 21, 1975).

3. Quoted by Sammy Kum Buo, "Fortress South Africa," *Africa Report*, January–February 1975, p. 11.

4. Also described as "outward going" or, in its Afrikaans version, as an "outward movement" (*uitwaartse beweging*). See J. Barratt, "The Outward Movement in South Africa's Foreign Relations," *South African Institute of International Affairs Newsletter*, August 1969, pp. 15–16.

5. Lesotho, Swaziland and Botswana are tied to South Africa through a common currency and a customs union whose operation has clearly been detrimental to some of their interests. Malawi's economic dependence on the employment of a substantial fraction of its labor force inside South Africa is well documented. In all four countries, a significant number of white South African officials have occupied senior administrative positions (see Barbara B. Brown, "The Foreign Policies of Botswana, Lesotho and Swaziland toward South Africa: 'Constraint and Flexibility' " [Paper presented at the 16th Annual Meeting of the African Studies Association, 1973]).

6. On NSSM 39, see *Washington Post*, October 11 and 13, 1974; Bruce Oudes, "Southern Africa Policy Watershed," *Africa Report*, November–December 1974, pp. 20–23; *Southern Africa*, February 1975. The importance of this document stems from the fact that "option 2" was evidently selected as the basis of the policy implemented by the Nixon administration in southern Africa (see Edgar Lockwood, "NSSM 39 and the Future of U.S. Policy Toward Southern Africa," *Issue* IV (3), Fall 1974, 63–69.

7. See, for example, his statements to the South African press on returning from Monrovia in *South African Digest*, 28 February, 1975.

8. Calculated from data in *The Military Balance, 1974–1975*. London: International Institute for Strategic Studies.

9. *The Star*, December 21, 1974.

10. See their statements (in 1968) in *Africa Diary* VIII, 3736 and *Africa Research Bulletin* V, 1144.

11. Sammy Kum Buo, p. 16.

12. Such analyses have an unfortunate tendency to mix quantifiable data with imponderables and to view the field through a macroanalytical perspective. See, e.g., M. M. Yanchius et al., *Research Notes on Insurgency Potential in Africa South of the Sahara* (Washington, D.C.: SORO, 1965); Michael E. Sherman, *Racial War in Africa: A Peacekeeping Scenario* (Hudson Institute, December 1968). A more coherent, yet partly dated, picture is presented in Kenneth Grundy, *Guerrilla Struggle in Africa* (New York: Grossman, 1971); and in his *Confrontation and Accommodation*, Chapters 5 and 6.

13. Quoted in *Africa Contemporary Record 1973–1974*, B-437.

14. *Ibid.*

15. Smuts drafted the plan for the League of Nations and the preamble to the UN Charter.

16. Quoted in *Africa Contemporary Record 1973–1974*, B-437.

17. For some views relating to southern Africa, see Eugene S. Virpsha, *Southern Africa and the Indian Ocean: A Study in Power Strategy* (Brighton, U.K., Markham House, 1969); John E. Spence, *The Strategic Significance of*

Southern Africa (London, RUSI, 1970); *The Cape Route* (London, RUSI, 1970); *The Security of the Southern Oceans—Southern Africa the Key* (London, RUSI, 1972); David Johnson, "Troubles Waters for the U.S. Navy," *Africa Report*, January–February 1975, 8–10.

18. Quoted in *Africa Report*, January–February 1975, p. 20.
19. Quoted in Johnson, p. 10.
20. For some recent efforts by South Africa to influence the American public on this matter, see *Southern Africa* 8 (6), June 1975, 35–36.
21. *Rand Daily Mail*, July 11, 1974.
22. *South African Digest*, April 11, 1975.
23. *Washington Post*, April 14, 1975; *New York Times*, April 24, 1975. In the early 1970s, the plausibility of an uranium-for-expertise exchange between Israel and South Africa had also been raised by George Quester in *The Politics of Nuclear Proliferation* (Baltimore: Johns Hopkins Press, 1973), p. 98.
24. *The Star*, April 12, 1975.
25. *Oggendblad*, July 13, 1974.
26. J. B. Vorster in *Beeld*, October 11, 1970. For South Africa's objections to the NPT, see UN General Assembly, A/C.1/PV 1571, May 20, 1968.
27. M. Hough, "Selected excerpts regarding South Africa's position on nuclear energy,' *Politikon* (Pretoria) 1 (2), December 1974, 69.
28. Quester, pp. 201–202.
29. For some of the arguments against the need for South Africa to develop nuclear weapons, see Hough, pp. 70–73.

Chapter Ten

Japan's Response to Nuclear Developments: Beyond "Nuclear Allergy"

Yoshiyasu Sato

INTRODUCTION

On August 6, 1945 an atomic bomb (uranium 235) of 20 kilotons of TNT was exploded over Hiroshima. Three days later another atomic bomb (plutonium 239) of about the same power was exploded over Nagasaki. They were the only two atomic bombs that have ever been used. The destruction power—blast, heat and nuclear radiation—was enormous. Literally in a second the cities were burnt down and streets, paths and rivers became full of corpses. The development of a new weapon system had posed a challenge to mankind. The impact of the new weapon was not only upon the war leaders in Japan; it ushered in a new phase of world history. It triggered a chain reaction towards nuclear rivalry.

Thirty years have passed since then. Man has attempted to control the arms race and to promote peaceful uses of nuclear developments during the last three decades. And yet a new and serious look at nuclear developments seems to be needed in the light of deep concerns that the number of "threshold" countries may increase in the coming decade or so. The nuclear club, which recently counted only the United States, the Soviet Union, Great Britain, France and China among its members, is already losing its exclusivity. India detonated a nuclear device on May 18, 1974. The possibility of limiting the club to the five nations has now passed, despite the assertion that the Indian explosion was intended for nonmilitary uses. The Indian detonation and the growing interest on the part of developing countries in acquiring nuclear reactors have opened up a new period of nuclear developments and created great concern over the further spread of nuclear weapons. In the late 1960s the effort to ensure nonproliferation of nuclear technology for military uses was a major issue in international politics, and the issue will once more draw serious attention in the late 1970s. It is far more serious and complex now, however, simply because the number of potential nuclear powers has increased enormously. Weapons technology has also greatly developed since then and peaceful nuclear technology, which can be crucially linked with nuclear weapons, has widely spread.

Some expect that Japan will join the nuclear weapons club in the near future. In particular, Japan's hesitation in ratifying the Treaty on Nonproliferation of Nuclear Weapons (NPT) has created suspicions, as its technological developments and its economic progress have made Japan an economic giant ranking next to the United States in the free world. Herman Kahn predicted in his book *The Emerging Japanese Superstate,* published in 1970, that "within the next five or ten years the Japanese are likely to unequivocally start on the process of acquiring nuclear weapons" [1]. Will history bear him out? I personally doubt it. The arguments are that Japan will be able to develop nuclear weapons within a relatively short period of time; that she will justify them as "defense weapons" under the constitution; and that the government of Japan will decide to go for nuclear weapons out of consideration for her prestige, or because of a significant loss of credibility in the American nuclear deterrent for Japan, or a perception of nuclear threats from China or the Soviet Union, or a desire to strengthen Japan's bargaining power with other nations. I do not find these arguments persuasive.

My purpose in this chapter is to describe how the Japanese government and people have reacted to nuclear developments, and to present my own views and analysis that Japan will not and should not become a member of the nuclear weapons club. To do so would be neither in Japan's interest nor in the interest of world security.

EVOLUTION OF NUCLEAR POLICY

First of all, I would like to review some of the origins and basic concepts from which Japan's posture on nuclear developments has emerged since the end of the last war. The historical background of the last three decades does not necessarily present an absolute criterion for projecting the nation's orientation for many years to come. And yet Japan's rebirth out of the ashes has shown a remarkable growth under the unique setting in which she was placed during this period.

The Constitutional Framework
The starting point is the constitution, Article 9 of which reads as follows:

> Aspiring sincerely to an international peace based on justice and order, the Japanese people forever renounce war as a sovereign right of the nation and the threat or use of force as means of settling international disputes.
>
> In order to accomplish the aim of the preceding paragraph, land, sea, and air forces, as well as other war potential, will never be maintained. The right of belligerency of the state will not be recognized.

An extensive debate has been conducted with regard to the interpretation of this article and the legality of Japan's substantial military forces ever since the dilemma of the peace constitution and the reality of rearmament was born. At one time it was interpreted to mean that Japan was committed to peace without any military forces whatsoever.

The government at one time made an unsuccessful attempt to resolve the dilemma by advocating constitutional revision, in order to remove all doubts on the constitutionality of the Self Defense Forces (SDF). Despite the insistent efforts of opposition parties to bring the issue of SDF before the Supreme Court, the government's interpretation has now been established to the effect that self-defense is inherent in sovereign right, and thus the article does not necessarily prohibit the possession of military capability for defense purposes. This view was stated as early as November 1952, and this interpretation has been supported by the Supreme Court though it refused to rule on the constitutionality of SDF as not being pertinent.

The position of the government was restated in *The Defense of Japan,* its first defense white paper published in October 1970 [2]. The views in the white paper were welcomed as a reasonable framework of Japan's posture. It is fair to say that the status of SDF has steadily grown in acceptance over the years.

The problem of nuclear weapons has been argued within this concept of self-defense of the nation. An interpellation in the Diet on nuclear weapons dates back as early as the latter half of the 1950s, which was well before Peking began its nuclear testing program. On May 7, 1957 in the Budget Committee of the House of Councillors, the then Prime Minister Kishi said that nuclear weapons do not necessarily violate the Constitution.

This attitude of the Japanese government contrasts with the sharply divided opinions in the United States on nuclear testing during the same time period. The Japanese government was already conscious of the inevitable spread of nuclear knowledge and took a positive attitude toward peaceful uses of nuclear energy, yet carefully set aside the implications the technological development might have on Japan's defense capability.

The Chinese success with nuclear testing in October 1964 prompted renewed discussions of the implications for Japan's security. As in the 1950s, the government again evaluated nuclear weapons in the context of legitimate self-defense.

The 1970 defense white paper further developed the official view that small tactical nuclear weapons would not violate the Constitution [3].

With this interpretation, one might conclude that there is no constitutional reason why Japan cannot possess nuclear weapons to defend herself. The interpretation, however, must be matched by an attention to another system of laws as well as to a policy orientation of three non-nuclear principles.

The Atomic Energy Basic Law

With such evolution of the constitution as to peace and self-defense in the background, Japan started her "atoms for peace" program in 1954 by government initiative. "The Atomic Energy Basic Law" was enacted in 1955, providing the objective for Japan's basic policy on peaceful nuclear energy developments. Article 2 of the law stipulates that the research, development and application of atomic energy be conducted solely for peaceful purposes. The law explicitly prohibits Japan from manufacturing nuclear weapons on applying nuclear technologies for military purposes. The policy has since become the foundation of Japan's nuclear programs. Consequently, if a government should decide to have nuclear capabilities, even for self-defense, the law must be revised.

Article 2 also provides that the results of research, development and application be open to the public and conducive to the cause of international cooperation. If this is interpreted literally, it will be almost impossible for any government in Japan to divert her technologies for military purposes. Even in peaceful developments a certain degree of secrecy on technical know-how must be observed. In this connection the Atomic Energy Law of Japan has been regarded as a unique case. The "nuclear allergy" of the nation, still a prevailing national feeling at the time the law was enacted, did have an effect, inasmuch as it was felt that nothing should be hidden as far as nuclear developments are concerned.

The law was adopted by bipartisan action at the time to provide a framework for Japan's nuclear policy. Its principles have become so firmly established over the years that no political atmosphere will permit revision of them.

The Three Non-nuclear Principles

The Japanese government has repeatedly stated that it has no intention of arming with nuclear weapons. The policy includes three principles: neither acquire nuclear weapons, nor manufacture them, nor allow their introduction into the country by a foreign power. These have been repeatedly cited as the "three non-nuclear principles."

At a Diet interpellation on January 30, 1968 former Prime Minister Sato stated the four elements of the nuclear policy of his government—namely, to abide by the three principles already committed, to make efforts in the cause of nuclear disarmament, to depend on the American nuclear deterrent based on the Japan–US Security Treaty and to promote peaceful uses of nuclear energy.

The three non-nuclear principles have gained a strong consensus among the Japanese. However, the relationship between US deterrence and the non-introduction of nuclear weapons by a foreign power has left an area of contention. One such problem was port calls by the United States nuclear-powered submarines. This was not a problem of the introduction of weapons into Japan, but rather of safety from radioactivity leakages. However, nuclear-powered submarines are bound to be connected with the notion of war.

Another problem is the possible deployment of nuclear weapons in Japan, particularly on the island of Okinawa. It was assumed that the United States had deployed tactical nuclear forces on the islands while they were under its administration. There was a serious question as to whether the Japanese people could or could not be assured that Okinawa would revert to the nation *without* nuclear weapons. The hard-liners in Japan were of the view that they should not be withdrawn in the light of fluid international situations in the region. Quite a number argued that the return of Okinawa should be implemented on the same conditions the mainland of Japan enjoyed in the application of the Security Treaty. Okinawa was finally returned without nuclear weapons—an outcome which was assured by the exchange of diplomatic notes between the minister for foreign affairs of Japan and the secretary of state of the US to the effect that the special sentiment of the Japanese people concerning nuclear weapons will be respected, and that prior consultation under the Security Treaty will be observed, as it has been. Since the United States' policy concerning tactical weapons abroad has been to "neither confirm nor deny" their existence, these assurances were taken as the maximum the two governments could work out.

Problems were recently created by dissenting American soldiers or local Japanese employees, who provided Japanese opposition parties with information or documents circulated internally which indicated some kind of implication with nuclear forces. The government denied any deployment of US nuclear forces in Japan, in accordance with the assurances of the Security Treaty as well as the assurances exchanged at the time of the reversion of Okinawa. Another well-known challenge was the testimony of Admiral LaRoque before the Congress, stating that no removal operation of nuclear weapons would be conceivable without the ships loading them calling at foreign ports, including the case of Japan. The position the two governments took on this problem was no different from the one mentioned earlier.

Yet another challenge to the government was the demand of the opposition parties that the three non-nuclear principles should be adopted as a resolution of the Diet. The government was rather negative about this proposition. The reason was that the principles could only be realistic and operative on the assumption that Japan was dependent on the United States' nuclear deterrent, and that the Japanese government was not in a position to limit the nation's policy alternatives by separating the three principles from other elements of her national security. The pressures were so strong that on November 24, 1971 the House of Representatives passed a resolution to the effect that the government should respect "the three non-nuclear principles." It was supported by the Liberal Democrats, Komeito and the Democratic Socialists (the Socialists and the Communists were absent from the session). The government's commitment to the principles will continue to be challenged, as it has been, all the more critically because of this resolution, which will have a moral implication, if not legally binding force, on the behavior of the government and the psychology of the people.

Public Opinion

According to the various public polls in Japan, about 20 percent of the population supports a nuclear-armed Japan. They say that Japan should be nuclear-armed, or at any rate that she will be obliged to become so in the end.

The great majority, however, is against Japanese possession of nuclear weapons. Non-nuclear nationalism emerged after World War II and has by now become deeply engrained in the nation. It has been called a "nuclear allergy." Official documents issued between the governments of Japan and the United States repeatedly refer to it as "the special feeling of the Japanese people toward nuclear weapons." This is rooted in the memory of the atomic holocaust suffered by the Japanese people, so far the only victims of nuclear weapons in the world. The dislike of nuclear weapons and the strong advocacy for their abolition are felt by the whole Japanese people and are very deeply engrained emotionally in the minds of the nation. It is one expression of present-day Japanese nationalism. While this "nuclear allergy" may die down, it is inconceivable and improbable indeed to think of any situation in which Japan would dream of becoming a nuclear power. No government of Japan could survive if it made such a decision. It is a problem of the value system on which every aspect of modern Japan—political, economic, and social—is founded. It would not be possible to envisage a nuclear-armed Japan unless a Copernican revolution would take place in the Japanese value system. And this could only be caused by drastic changes in the international environment.

PEACEFUL USES OF NUCLEAR ENERGY

The "oil crisis" since October 1973 has drawn the world's attention to the diversification of energy resources. The development of peaceful uses of nuclear energy has intensified in major Western industrialized countries in order to overcome the high cost of oil. This is also the case in Japan. The Japanese government has placed its peaceful use program on a high priority in its energy policy. To a country like Japan which is so heavily dependent on the oil resources of the Middle Eastern countries as a source of energy, its efforts to lower the dependency on oil must be much more serious in the coming decade than those of any other country of the world. The government has emphasized the importance of developing new technologies such as nuclear fuel production skills and new types of nuclear reactors, including multipurpose gas cooled reactors, fast breeder reactors and so forth.

At the same time, it is now widely recognized that, as nuclear power plants are developed, the possibilities for nuclear weapon proliferation by governments are likely to increase dramatically. A critical link between peaceful uses of nuclear energy and manufacture of nuclear weapons lies in the development of nuclear reactors. Although a program of nuclear-fueled power generation has been developed as a means of exploiting new energy resources, a problem has now arisen because as uranium burns in a nuclear reactor, plutonium, an element not found in nature, is produced as an inevitable by-product. One isotope of plutonium, P-239, readily undergoes fission, as does U-235. Depending on the type of fuel cycle, a country which has developed a nuclear power reactor will have ever-increasing amounts of plutonium suitable for the construction of effective nuclear weapons and the capacity to produce plutonium of the type needed for the most efficient nuclear weapons. Some countries may wish to have nuclear power reactors in order to acquire the option of producing nuclear weapons. In the case of Japan, the objective has been purely peaceful, and yet, due to the magnitude of electrical power generated by nuclear reactors, attention has been drawn to the extraordinarily large production capacity and stockpiling of plutonium in Japan. The developments Japan has achieved in the field of peaceful uses of nuclear energy are now regarded in this context. How much potential does Japan have in reality?

Current Situations

Budgetary and Private Investments and Invested Manpower. The research program on nuclear energy developments was inaugurated in 1954 when the government appropriated 235 million yen (approximately $783,000 at current value) for it. The budget in fiscal year 1975 amounts to 85,590 million yen (approximately $285 million). During the last 21 years the government allocation has expanded substantially, as the figures show. The total budget between 1954 and 1975 amounts to 541,200 million yen (approximately $1,804 million).

The first investment by the private sector was made in 1956 when it amounted to 1,250 million yen (approximately $4 million). In 1973 it was 458,700 million yen (approximately $1,522 million), and thus the total amount of investment between 1956 and 1973 reached 1,584.3 billion yen (approximately $5.3 billion).

The investment made both by the government and the private sector has now exceeded 2,000 billion yen (approximately $6.7 billion). These figures may be contrasted with $1.7 billion invested for the "Manhattan Project."

The manpower which engages in nuclear energy developments now exceeds 54,000 persons, of which 28,000 belong to the private sector. These figures can be contrasted with the figures of 1,300 engineers and 500 scientists which were mentioned in a report of the United Nations in 1968 as the required figures for a nation to develop production facilities for nuclear weapons.

Nuclear Reactors. The first nuclear reactor in Japan was constructed in 1963, following West Germany (1960), Canada and Italy (1962), and Sweden (1963)—all of which embarked on purely peaceful nuclear power programs—and it was put into commercial use in July 1966. There are now eight nuclear power plants in commercial operation in Japan, the capacity of which amount to 3,893,000 kilowatt hours in total. As is the case of most countries, the type of reactor used in Japan is the light water reactor (LWR): one being a gas coolant reactor (GCR), four being boiling water reactors (BWR), and three being pressurized water reactors (PWR). Sixteen more plants are either under construction or planned.

As far as present capacity is concerned, Japan ranks third in the world, after the United States (28,300,000 kilowatt) and the United Kingdom (5,680,000 kilowatt). When all of the plants are completed the capacity will amount to 19,901,000 kilowatt. The rank, however, will by then be sixth, following the United States, the United Kingdom, France, West Germany and Spain. Japanese plans in this field seem to be rather modest in comparison with those of countries such as England and West Germany, which are smaller than Japan and whose densities of population are more or less the same as that of Japan.

According to a long term plan adopted by the Japanese Atomic Energy Commission in June 1972, which was well before the "oil crisis," the capacity is to be expanded to as much as 32 million kilowatt by 1980 and 60 million kilowatt by 1985, the idea being to supply one-third of Japan's total electric power needs by nuclear energy. In order to achieve the 1985 target, it is necessary to build 70 more nuclear power plants, each of which will be able to generate 600,000 kilowatt.

The projected plan to build more plants was very much behind schedule during 1974. Not only were there difficulties with the plan, but out of the eight nuclear reactors which went into commercial operation, six remained unused, one of them for more than seven months (as of the middle of March 1975), because of abnormal phenomena in the reactors themselves. Pessimism now prevails among the government and the power companies. Some business people have already indicated that the stated objective may not be achieved because of the program's slow implementation in 1974. The construction lead time being six to seven years, if the current pace of construction continues, capacity will only reach 22 million kilowatts by 1980 and 37 million kilowatts by 1985—just half the planned capacity. Some informed sources are predicting an even slower realization of the program.

Despite the assertion that nuclear power presents the safest method of generating electricity, the repeated shutdowns have increased popular doubt about the safety of the reactors. Escalating disputes between local residents and power companies over construction of nuclear facilities have now become a political issue. Reportedly some political leaders stated recently in their Diet interpellation that the long term program may need careful review in the light of environmental risks entailed by the construction of nuclear power plants.

The government's long term nuclear energy program is now endangered. The Ministry of International Trade and Industry has reportedly decided to cut drastically the 1985 target of generating 60 million kilowatt hours of electricity by power plants.

While the government's long term policy still remains to be seen, the Central Power Council (a business organization composed of nine regional power firms and the Electric Power Development Co., Ltd. announced its five year plan (fiscal 1975—1979) in April 1975. It, too, is far below the projection mentioned earlier. According to the council's announcement, the industry's investment in nuclear energy development during the five years will be to increase the capacity by 24,670,000 kilowatts. At the end of fiscal year 1979 the output facilities of nuclear energy will be 15,480,000 kilowatts. These figures are a significant contrast to the government's estimate, in terms of absolute quantity. It is recognized; however, that greater reliance on nuclear energy is projected in the council's proposition. Of the planned capital spending, the amount earmarked for nuclear energy during the five years, accounts for 43 percent, as against 40 percent for thermal power and 17 percent for hydropower. The ratio of nuclear energy facilities will stand at 14 percent at the end of fiscal year 1979, while the corresponding ratio at the end of fiscal year 1974 stood at only 5 percent. The government and business are in agreement as far as the long term trend is concerned. The problem at issue is how fast or how steadily nuclear energy resources should be developed. Both the government and business are aware of several advantages of nuclear power generation, despite the difficulties which are now exposed.

Uranium and Plutonium Resources. Japan's reserve of uranium ore is almost negligible. She has less than 5,000 short tons of U^3O^8 in her estimated reserve. She has to depend totally on foreign markets. She would have to import as much as 9,000 short tons in terms of U^3O^8 value when Japan's nuclear energy capacity reaches 30,000 to 40,000 kilowatt.

As the peaceful uses programs develop, Japan has so far acquired 391 tons of natural uranium, 83 tons of depleted uranium, 16.3 tons of U—235, 21.6 tons of trium and 1.9 tons of plutonium. One projection of annual plutonium production is as follows: 4.1 tons in 1980, 8.4 tons in 1985 and 14 tons in 1990. The stockpile of plutonium in this projection is thus: 12 tons in 1980, 40 tons in 1985 and 70 tons in 1990. The figures may be compared to the annual production in the United States, which is 10 tons at present and 25 tons in 1980. Even now Japan is capable of producing 280 of the Nagasaki-type atomic bombs annually. In five years (1980) the number will be 1,500, in ten years (1985), 5,000. These figures of production capabilities are derived from a projected stockpile of plutonium.

Enrichment Technologies and Reprocessing Facilities. While Japan is totally dependent on enrichment service of natural uranium by the United States, progress is being made in developing her own technology for enrichment. A special government agency for the development of reactors and nuclear materials built an enrichment facility by the centrifuge technology which has been in test operation since October 1974. It has been successful in producing a few grams of low enriched uranium. Another facility is under construction. After examining the data obtained through the test operation during 1976, the government plans to proceed with the building of a pilot enrichment facility plant in 1977. The centrifuge process is less costly and consumes less electricity than the gaseous fusion process, which is one reason why the Japanese government developed the former. Although Japan's development of enrichment technology is still in the infant stage, Japan aims at the production of enriched uranium by 1985.

A plutonium separation plant is also under construction. It is scheduled to go into test operation soon. The technical knowhow for this fuel recyling method has been imported from a French company. It has the capacity for processing 210 tons a year (0.7 tons a day) of used nuclear fuels. As a by-product of this reprocessing, approximately 230 kilograms of plutonium per year can be extracted, though the primary purpose is to separate residual uranium out of spent fuels. The nuclear power industry is studying plans to build another reprocessing facility to meet the growing need.

The Case of Mutsu

The Japanese government has had to face the enormous number of difficulties even in peaceful uses of nuclear energy. On August 26, 1974 the first Japanese nuclear-powered ship *Mutsu,* left port for her test navigation and returned on October 15 after drifting about for 50 days. During that period a leakage of radiation from the ship's power reactor developed. Strong resentment and fear then developed among the local inhabitants against the ship's return to its original port. A brief sketch of the background of the *Mutsu* plan shows some of the decisionmaking perspectives involved on the part of both Japanese government and business circles in building the ship. It seems to me that the plan was symbolic in two ways: namely, it was another significant expression of enthusiasm by the Japanese government for the peaceful use of nuclear energy, and the development of the ship's nuclear power generator would be Japan's first independent pursuit of nuclear technology. The failure of *Mutsu* was a shock to the authorities concerned, all the more because the plan was initiated with such hope and pride.

A program to build a nuclear-powered ship had been on the agenda together with nuclear power plants since the Atomic Energy Commission was established in 1956. In August 1963 the government established a special governmental agency to implement the program of constructing *Mutsu.* At that time the first nuclear-powered ship of Japan was the fourth ship of this type in the world. It was designed primarily for transporting special cargoes and training crews. According to the basic development plans, the ship should have sailed for its test voyage in 1973 and completed the two year test voyage by the end of 1974. The plan was behind schedule because of the fears of local residents regarding the safety of nuclear energy and the possible pollution of the sea resulting in damage to their fishing grounds.

The attitude of business circles to the plan was somewhat ambiguous. At an early stage, the shipbuilding industry and the maritime transportation industry earnestly supported the plan, expecting nuclear-fueled ships to be a major means of transportation in the future. The shipbuilding industry, which was leading the world market in its technology and production, could also see the plan as increasing its prestige.

This attitude changed, however, when Japanese industry obtained plenty of cheap oil. At the same time, they realized that the cost of building nuclear-powered ships would be extravagant. The cost problem has not yet been solved. There is no unified opinion throughout the world as to how nuclear-powered ships compare to conventional ships in terms of cost-effectiveness. In this respect one may recall the American nuclear-powered ship *Savannah,* which was deactivated after its maiden cruise in 1964 because of many mechanical troubles and financial burdens.

As a result of these uncertainties, by the time the plan was made public almost none of the shipping companies showed any interest. The plan was started rather prematurely in the sense that cooperation between government and business circles had not really been made firm. This was an important factor in the ship's failure. The reactor loaded on *Mutsu* is the only reactor that has ever been developed independently by Japanese engineers. They might have anticipated that the technological gap would further increase if research and development on nuclear-powered shipbuilding was delayed. It was natural that foreign countries were not all prepared to provide Japan with the technology for building nuclear-powered ships or with such an actual ship, fearing the possibility of military uses. Thus Japanese engineers took the path of independent development. It still remains to be seen what impact the failure of *Mutsu* will have on Japan's independent development of nuclear energy, which has been one of the government's policy objectives since the beginning of the peaceful use program.

The Promotion of Peaceful Uses of Nuclear Energy

Having observed the setbacks and problems to be overcome, Japan must still look toward a new era of energy development, so that the country will not have wasted the investment of more than 2,000 billion yen in the last 20 years. There is much justification for the development of nuclear energy resources in Japan. If this development is to be slowed down because of the 'nuclear allergy" it will be deplorable. To maintain the vitality of the Japanese economy and to ensure the progress of our society, the development of nuclear energy is essential. New types of technologies have always encountered popular opposition as the history of technologies shows. Leaders of a society are responsible for taking a long term view when it comes to opening up a new theater of progress.

There are three major considerations in advocating a more rapid thrust of nuclear energy development in Japan, as follows: (1) nuclear power is a source of energy independent of oil; (2) it is easily stored; and (3) the cost of development is relatively low.

Independent of Oil. Countries can avoid the threats of an "oil crisis" by diversifying their energy sources. If the world continues its recent rate of growth in the use of energy (6 percent) all of the world's oil will be gone by the year 2021, just 46 years from now. All conventional fuels will be gone by the middle of the next century. These facts will certainly force us to take a new and intelligent look at the problems of energy resources for the future, and nuclear energy will definitely enable the energy policy of the highly industrialized countries to be reoriented. In the United States Project Independence recommends that the capacity of nuclear energy supply should be 135 million kilowatts in 1980 and 505 million kilowatts in 1990. The percentage of contribution to the total supply of energy will be 42 percent in 1990. West Germany and France have also decided to put more emphasis on programs for the development of nuclear energy. It is quite natural that in order to maintain a better sense of security in this respect a new strategy for energy policy be pursued in Japan. Conservation efforts are not the only approach to the problem, but efforts to search for a new dimension must be made in this society where human activities are ever-growing.

Easy Storage. One gram of U–235 can produce energy comparable to that of three tons of coal. The thermal capacity is three million times that. To store 300 million tons of oil is impossible in Japan in terms of space, facilities, environmental effects and danger of fires. The storage of 100 tons of U–235, on the other hand, is easily possible.

Lower Cost of Development. The cost analysis of generating power by fission versus oil is complex. It is, however, a crucial factor in Japan. If it is proved that the cost of nuclear power generation is less costly than that of oil-fired generation, no time should be lost in making a shift so as to minimize losses.

The cost of ore, of the enrichment process and of all other parts of the fuel cycles must be taken into account. A number of uncertainties exist in evaluating each of those elements: uncertainty on uranium availability, uncertainty on the fast breeder reactor investment costs, and uncertainty on the discount rate for research and development investments. The capital cost of the nuclear-fired generation is higher than that of the oil-fired generation, and it will become more so in the light of required safety standards that are increasingly more stringent. The cost of uranium is also rising.

The nuclear electricity generating cost per kilowatt hour a few years ago was between 4 and 5 yen in Japan, while it is now between 6 and 8 yen. The axiom that power generation by fission energy is less costly than other sources of energy is now challenged. The point is whether the increased cost that is now anticipated would reach, in the future, a level where the comparative advantages of nuclear-fired generation may be discounted, and thus the incentives for it would become negative. With regard to the elasticity of the total nuclear fuel costs as against the increase in the cost of fuel enrichment, one analysis holds that a 20 percent increase in the latter cost would only add 7 percent to the former.

It seems to me that no definitive answers are possible at this moment. Given the factor that the oil resources of the world are limited, a country like Japan which is so heavily dependent on oil for her energy resources cannot help but look for alternative resources. Japan's efforts would be to make nuclear energy generation competitive with other sources of generation in long term planning. Technological developments and improvements may occur that have now been undertaken to build more efficient types of reactors. A standardization of nuclear reactors would also contribute to lowering the cost. However, Japan cannot waste the investment made so far in this field, and has not yet reached a stage where her nuclear energy policy should be given up because of rising costs. As long as reasonable prospects are still in sight, Japan has good reason to continue to develop nuclear energy resources in order to strengthen the structure of her economy.

The degree of dependence on imported resources is as high as 85 percent and the payment for petroleum alone now amounts to more than $20 billion a year. Despite the high price of oil, Japan must still continue to import it in order to sustain her economic and industrial activities which are projected to grow at a rate of between 5 and 10 percent per annum even in the age of the "energy crisis." Low cost energy developments will surely ease the nation's burdens and strengthen export competitiveness. Japanese development of related technologies for nuclear power plants and nuclear fuel recycling facilities which have more than a million parts as their components will also bring about multiple effects in restructuring Japanese industries toward more knowledge-consuming types of industry, independence in her energy policies and improving trade patterns. These are key points for Japan, which depends so much on trade for its existence.

ELEMENTS OF CONSTRAINTS

In previous sections I have tried to present some background on Japan's perception of nuclear developments since World War II. A commitment to becoming a "peace-loving nation" has taken deep root among the people, and the notion of dedication to "peace" has become everyone's jargon. Japan has been regarded as unique among the countries of the world with its strong flavor of peacefulness, and indeed the Japanese people have made efforts to maintain this uniqueness. And yet, following this evolution toward peacefulness, Japan is now regarded, with either admiration or skepticism, as a near-nuclear country with the requisite highly advanced technologies. It is argued that the policies so far evolved could be shifted in different directions, because of the temptations Japan might face on both her domestic and foreign fronts. Indeed, it would not be impossible for Japan to become the seventh nuclear power state if she so decides. Some Japanese nuclear physicists seem to be confident of producing an Indian type of explosive device within two months of such a decision. The problem is whether Japan can make this decision without experiencing any constraints and without endangering her security. It is obvious that whether or not to "go nuclear" is the major political decision Japan has to face. It must be a decision made through a complex of processes just as in any other democratic and open society. The decision will embrace not only political, economic and technical considerations, but also social, cultural and intellectual aspects of the mode of life of the Japanese people. Therefore, in order to analyze the likelihood that Japan will take up the option of nuclear armaments, one has to know whether she is really capable of doing so, technologically, economically and financially, whatever her intentions may be; through what processes the decision would be made; and whether a Japan armed with nuclear weapons would have the confidence to live peacefully and without friction with the rest of the world, on which it is so dependent for its life.

Technical Problems

No matter how determined any Japanese government might be to acquire nuclear weapons, to attempt to do so without solid technical capability would amount to suicide. Does Japan have such a capability?

Obtaining and Use of Resources. Japan's options are strictly conditioned by the problems of resources: (1) the scarcity of her own uranium ore reserve, and (2) her total dependence on the supply of materials from foreign countries, their use being strictly limited to peaceful purposes. The supply of enriched uranium is also dependent upon the services of the United States, which is the major contributor in this field. Also, Japan has not developed an uranium enrichment plant to produce the weapon grade enriched uranium, and the possibility of building such plants is not in sight. As has been mentioned, the stockpile of plutonium has grown as the number of nuclear power reactors has increased. A crucial point for Japan is a "cap" of peaceful uses of nuclear energy. Whatever quantity of natural uranium, enriched uranium and any other by-products such as plutonium she may acquire, and whatever technologies she may develop to process the materials, no other alternatives are permissible to her than the peaceful uses. Almost all fissionable materials in Japan are inspected under a severe safeguards system of the International Atomic Energy Agency (IAEA). Any diversion of the materials—natural uranium, enriched uranium or plutonium—to nonpeaceful uses will endanger the supply itself, which is detrimental to the Japanese industry and economy as both government and business in Japan are so firmly committed to the development of nuclear energy.

Warheads and Production of Weapons Systems. A simple explosive device could be produced in Japan in a short time. However, it is assumed that it would take at least ten years, judging from the cases of France or China, for the production of thermonuclear warheads deliverable by missiles, or miniature warheads for tactical use, to become an effective and credible weapons system. The difficulties for Japan are to find sites for building plant systems and for testing.

By this time it may be almost impossible to build weapons production plant systems in Japan. The capacity to accommodate industrial plants or factories is reaching its utmost limit. The flat lands of Japan, which represent only one-sixth of the total land area, are now full of houses and industrial complexes. An enrichment factory needs a vast supply of electricity, and must be located at a place of convenient access. It must be possible to protect it from natural disasters like earthquakes or typhoons. One can hardly find a location in Japan where the system of uranium enrichment, warhead and delivery system production could be established.

It is an even more difficult and probably insurmountable problem to find testing grounds in Japan. Extensive tests are essential. Underground tests can be a cause of earthquakes, and are even more difficult technically, given Japan's geological structure, than atmospheric tests. Japan has consistently opposed any form of testing. The development of nuclear weapons without test detonations may lack credibility, and may not produce the psychological impact which is of utmost importance in deterrence strategy.

Delivery and Radar Systems. Jet fighters currently owned by the SDF are not equipped to be used as a delivery system. The SDF would have to acquire aircraft which could carry tactical nuclear weapons. It is obvious, however, that these aircraft would not be sufficient to maintain a credible deterrent. If it is known that Japan's delivery system is limited to jet fighters or bombers, containment of a possible military operation by such vehicles would be easy. Dropping a nuclear bomb or bombs by aircraft would be a most dangerous gamble, if a massive missile attack by a prospective attacker is anticipated. For effective deterrence, missiles would thus be required.

While Japan has developed a propulsion system from which a Polaris type of rocket could be built in the future, the technologies for guidance systems (which are essential components for missiles) are extremely poor. Even the development of inertial guidance systems is still in an infant stage, and it would take some time before such technologies could be fully developed for military purposes.

The production of missile-launching nuclear submarines, which would constitute an essential element of deterrence to Japan, would also be a difficult matter. As we have seen in the case of *Mutsu,* the application of a nuclear propulsion system has not yet been fully developed. It would take ten years or more for the production of such submarines to become a possibility.

Another technological bottleneck for Japan would be the development of a warning system, which is a necessary component of deterrent forces. Radar systems, particularly OTHR (Over the Horizon Radar), to locate missiles as early as possible, the superspeed data handling necessary to identify them, and related communication systems cannot possibly be developed by the military technologies of present-day Japan.

The problems of technology may be comparatively easy for Japan compared to the problem of obtaining resources, and yet she still has a long way to go, as can be seen from the examples I have given. Moreover, it would be no easy matter to educate engineers and technicians to develop military technologies once it was known that the purpose of such education is military development.

Despite an early forecast by IAEA that Japan will be one of those countries which by 1970 would be able to produce plutonium as a by-product of nuclear power reactors in sufficient quantities and of sufficiently high grade to make warheads, it has already been proved that building a bomb, or even a minimal delivery system, is far more difficult for Japan than one can imagine. The sophistication and accuracy of nuclear weapons systems have been making such rapid progress that Japan's technology for producing plutonium cannot necessarily be regarded as a ticket to the nuclear club. The research and development of nuclear capability and the actual deployment or production of explosive devices are entirely different from each other. The application of technologies to a nuclear weapons system, which needs a high degree of sophistication and precision, is not as simple as the application of new antienvironmental devices to the automobile industry.

Financial Burdens

According to a 1968 United Nations report, it will cost $5.6 billion to have "small high quality nuclear forces," assuming that it will take ten years for their development. Therefore the annual cost is $560 million, or 168.3 billion yen ($1 = 300 yen). One might conclude by simple calculation that Japan can allocate such an amount for her nuclear development, and that with their large gross national product the Japanese are less constrained economically. A point at issue is how a Japanese government would allocate its given budget to a new policy program. Generally speaking, a Japanese budget has two components in the growth of expenditure, namely, "natural growth" and "policy growth." The former is the growth of expenditure which is allocated in an obligatory way without the discretion of the financial authorities, and the latter is determined by new policy requirements. In the past ten years or so, the rate of natural growth has been far higher than that of policy growth. It may safely be assumed that the rate of policy growth will be, on the average, only 5 to 8 percent a year, or even less. This may be characterized as limited financial resources, which make it almost impossible to pursue a new policy. In spite of the growing GNP, the percentage of the national budget against the GNP is relatively low in Japan, i.e., about 12 percent. In the United States, for instance, it is about 17.2 percent, 16.2 percent in West Germany, 20.6 percent in the United Kingdom and 19.9 percent in France. It is out of the question to allocate an amount of $560 million a year—perhaps more—for a period of ten years. A low rate of economic growth would make it still more difficult to allocate extensively for military establishments.

In this period when nuclear technology is developing so rapidly year after year, the halfway development of nuclear weapons will not really satisfy Japan's desire for security. After ten years, what was started today may well be outdated and obsolete. If the possession of nuclear weapons is not serving the cause of Japanese security, their acquisition is meaningless. The decision is indeed a grave one and one on which the opinions of the people will vary, but the financial scope of Japan just does not seem to allow the expenditure of such a huge amount of money. And once a country possesses nuclear weapons, the cost burden snowballs. In particular, the cost of delivery systems will be open-ended. Thus, from a financial point of view it seems inconceivable that Japan will be able to possess nuclear weapons in the foreseeable future.

Manpower and Organizational Problems

In the production of nuclear weapons, the number of engineers and amount of expertise are very important. Japan is lacking in those who are skilled in dealing with plutonium. Although a national consensus is needed to create strategic nuclear forces, the most difficult task faced by the government will be that of mobilizing engineers to work toward the stated objective, even if it should decide to create such forces. The lack of manpower will be a deterrent against such a policy alternative, and as far as I can see the organization for such a project could not be easily achieved against the hesitation of Japanese scientists, whose pacifism and reluctance to be involved in any programs related to war purposes or weapons systems is well known. Most of the peace appeals adopted by the Japan Academic Conference in the last 25 years have opposed atomic and hydrogen bomb testing, and favored the prohibition of chemical weapons. Since 1950 the conference has repeatedly taken a firm stand that "they will never participate in any scientific research programs whatsoever which aim at war." Under such circumstances one can easily imagine how difficult—almost impossible—it would be to gain the cooperation of scientists if Japan decided to arm herself with nuclear weapons.

Military and Strategic Vulnerability

Japan's demographic and geographic features make her strategic position inherently weak and vulnerable. The concentration of her population and industrial complex in a limited area and the lack of depth in defenses against an attack from the Northwest, together with the short distance from a prospective launching area, are the fatal constraints in deploying tactical nuclear weapons in Japan. Even though by the best of luck Japan could overcome the various technical difficulties mentioned earlier, one cannot think of any situation whereby Japan's deployment of such weapons could constitute an effective and meaningful deterrent to ensure her security. This is simply because Japan cannot change her given conditions.

More than 45 percent of Japan's population is concentrated in the major cities along the Pacific coast and about 25 percent live in metropolitan Tokyo and three adjacent prefectures. In contrast, only 25 percent of the Soviet Union's total population lives in the largest 100 cities, and only 11 percent of the Chinese population lives in the largest 1,000 cities. It is obvious from this fact that Japan needs a far greater number of bombs than are needed by the other two in order to achieve the same effect. The ratio of survivability in Japan could be extraordinarily small under such circumstances. It is said that 30 percent destruction of a nation's population would produce a fatal impact on the nation's survival. Under such extreme concentration of population Japan would not be able to rely on the ABM defense system, however effective the system might be, because even a very small number of hostile missiles could cause enormous casualties.

In recognition of such demographic peculiarities and of the growing vulnerability of land-based missiles, Japan would have to depend on sea-based missiles. In this respect another problem that Japan would have to face is the distance from the seas around Japan to Moscow—about 6,000 miles. It is such a long distance that Japanese military technologies could not possibly cope with it. Even a Trident submarine, which is to be completed for operation in 1978, can only cover a 4,000 mile span. It is unlikely that Japan could surpass the American submarine technology and that she could develop her own sea-based missiles beyond the order of Trident. Targets in the Soviet Union are too far away, and those in China are too many for Japanese missiles to attack. No strategists would feel comfortable in such a vulnerable position.

Because of the lack of defense depth, Japan has no areas in which to deploy the warning systems. In Japan's case, it is more probable that hostile missiles would be launched from under water, the sea or aircraft, rather than from land. Those launching bases are movable. Japan thus needs all-direction warning systems, the deployment and maintenance of which are prohibitively costly. Neither the United States nor the Soviet Union is deploying such warning systems. Japan cannot locate hostile missiles early enough to make her second strike credible. To defend herself against submarine-launched missiles would be virtually impossible.

Political Considerations

Over and above the previously mentioned unavoidable strategic weaknesses inherent in Japan's peculiar physical features, she would have to face more serious difficulties, externally and internally, once she decides to go nuclear. An adverse effect on Japan's diplomatic position as well as a domestic social instability would be inevitable, because of her great dependence on the rest of the world and because of her unique vulnerability—political, strategic and economic. The constraints here would also be far greater than for any other prospective nuclear power.

External Implications. The nuclear armament of Japan would, among other things, bring about Japan's isolation from the international community. The existing nuclear countries, as well as other Asian countries, would regard a nuclear-armed Japan as a destabilizing element in international relations despite the Japanese assertion that the development of a nuclear capability would be only to deter aggression against her own security. Such an anxiety and sense of threat might be amplified into the separation and isolation of Japan from international society. Japan should not only remember that her independent nuclear force will be regarded in its impact as something different from other cases, but she would also have to contemplate seriously how to overcome incipient vulnerability under such an isolated environment, which could last a decade or more. Certainly Japan, for obvious reasons, could not ask the Soviet Union for its nuclear protection. Cooperation of the United States with an independently nuclear Japan would only be possible if the United States demanded in return control over the Japanese nuclear force. Such United States intervention would almost certainly create anti-American resentment which might lead to political instability in Japan. Without the protection of the United States, Japan would surely feel very insecure.

A high degree of dependence on markets and resources abroad would be utilized against the interests of Japan by those countries which provide Japan with markets and natural resources. If Japan arms herself with nuclear weapons, the supply of resources for her nuclear energy industry as well as of technical know-how, would be terminated immediately. The nuclear energy industry, which is planned to supply one-third of the total electric power needed in 1985, would then have to face a most formidable consequence.

Domestic Implications. Vulnerability, international isolation and economic strangulation would lead to social uneasiness on Japan's domestic scene. The government would be criticized for its misjudged policy orientation. Nuclear forces development programs might be a target of sabotage or subversive activities, which probably would occur at a certain level of anxiety. Japanese policymakers should be aware that a state of disorder is one possible outcome of nuclear weapons development.

A Middle Class Non-Nuclear Japan

These enormous drawbacks that Japan would have to encounter in the process of developing her independent nuclear force would diminish the advantages to be gained from acquiring it. Even though "pacifism" and "nuclear allergy" among the population might dissipate over the years as the generations change, these fatal constraints would loom quite as large. The decision to become a nuclear power would bring down a number of governments.

THE NONPROLIFERATION TREATY AND JAPAN

Having stressed the constraints on Japan's going nuclear, despite her potentialities, we may now ask why the Japanese government has not ratified NPT as of April 1975, although she signed it on February 3, 1970. Will Japan not ratify because she wants a free hand for her nuclear development? How will Japan develop her peaceful use program without commitment to nonproliferation. In this section I would like to shed some light on how Japan has perceived the problem of non-proliferation.

Signing the Treaty

The first official comment on NPT was made by Prime Minister Sato on March 14, 1967 in the keynote speech in the Diet. He stated that the Japanese government would agree with the spirit of NPT in her appreciation that more efforts are needed toward a new international order and disarmament. "However, the opinions of non-nuclear nations," the government strongly insisted, "must be fully reflected and their legitimate interests must be fully respected in the process of drafting the agreement" [4]. Foreign Minister Miki (the present prime minister) elaborated further in his foreign policy statement at the same session. He mentioned the following four points:

1. In order to achieve the objectives of NPT, as many countries as possible should sign it and full attention should be paid to the security of non-nuclear countries so as to secure wider participation.

2. The nuclear countries should not only promote NPT, but should also make a sincere and clear intention towards conventional disarmament as well as nuclear disarmament. It is hoped that concrete measures for that purpose will be worked out.

3. The peaceful uses of nuclear energy should not be restricted at all.

4. In the field of peaceful usage there should be no inequality between the nuclear haves and the nuclear have nots [5].

This position of the government has been consistently restated. On signing the treaty on February 3, 1970, the Japanese government issued a statement of reservation, again making clear its concern with the inequalities in the NPT.

Japan's sensitivity concerning possible unequal treatment in this field despite the so-called "nuclear allergy" prevailing among the Japanese people was related to a growing confidence in her economic capability. Even if the majority did not wish to consider nuclear armament as a policy alternative of their choice, the government was sufficiently cautious not to be put into a situation which would limit its own maneuverability. It also attempted to draw international attention to its position.

Although the process of reaching agreement on NPT did not lead to a unanimous cohesiveness among nations, the points raised by the Japanese government represented a consensus among the nations which were doubtful or critical of NPT.

With respect to disarmament and national security problems, the Japanese government held that in order to attain the purposes of NPT the nuclear power states should achieve concrete measures for nuclear disarmament in pursuance of the treaty's Article VI; that no nuclear power states should use nuclear weapons or exercise the threat of their use against non-nuclear power states; that the Japanese government would attach importance to the UN Security Council Resolution 255 of June 19, 1968, hoping for further consideration to be continuously undertaken with respect to the national security of non-nuclear power states; that the Japanese government would take serious note, before ratifying the treaty, of how much progress would be made in disarmament negotiations as well as how credible the UN resolution would be; and that the Japanese government would take note of the provision of Article X of the treaty.

With regard to the equality in peaceful uses of nuclear energy, the Japanese government clarified that the ratification would be made by taking into full consideration the fact that the contents of the safeguard agreement Japan would conclude with the International Atomic Energy Agency in accordance with the treaty should not become substantially unequal compared with those of other treaty parties (particularly Euratom countries). Though the Japanese government tried very hard to incorporate the views of non-nuclear states in the process of drafting the treaty, the statement should not be construed as a negative attitude on the part of Japan toward nonproliferation of nuclear weapons. The Japanese government made very clear at the outset of the statement her conviction that the treaty will serve as a first step toward nuclear disarmament. Japan has not put an unnecessary emphasis on the inequality aspects of the treaty.

Ratifying the Treaty

It is a matter of regret that no significant and serious debate on the implications of the treaty for the future of Japan has taken place during the past five years since the signing of the treaty. No national consensus has been developed on the issue. Fortunately or unfortunately, international developments—including the "energy crisis," the Indian detonation and growing fears of nuclear proliferation—have now forced Japan to look more seriously at the issue. There is still a group of people who are dubious about ratification, holding that "freedom of nuclear armament" should not be abandoned so hastily and carelessly, as the world situation is still fluid despite détente between the United States and the Soviet Union. The argument has an element of truth, despite the special sentiment of the Japanese people in general toward nuclear weapons, in the sense that NPT requires the non-nuclear nations to accept serious restraints on their future conduct without imposing any significant limitations on the actions of the nuclear powers. After renewed efforts by the government, supported by the Liberal Democratic Party (the majority party in power), the nuclear industries and journalists, the NPT was finally, on April 25, 1975, scheduled for deliberation before the Diet. The deliberation will certainly cover a wide spectrum of Japan's future options.

Arms Control and Nuclear Disarmament. Important arms control agreements have been reached between the United States and the Soviet Union or in international organizations since NPT came into effect. SALT is undoubtedly the most significant response to the call for the superpowers' responsibility under NPT. The Treaty on the Limitation of Anti-Ballistic Missile Systems (1972 and 1974), the Interim Agreement on Certain Measures with respect to the Limitation of Strategic Offensive Arms and the Vladivostok agreement are the major achievements of the SALT negotiations. Outside the SALT forum, the two superpowers reached agreement on the Threshold Test Ban Treaty in 1974.

Can Japan or other non-nuclear power states consider these achievements as faithful implementation of Article VI by the nuclear power states? One argues that the nuclear power states should be urged, as further measures to meet their obligations, to move toward substantial reduction in numbers of strategic nuclear delivery vehicles, prohibition of further qualitative advances in the strategic arms race and a comprehensive test ban. These are the challenges that the United States and the Soviet Union are to face at the review conference in Geneva. Indeed, whether these are agreements to justify the arms race between the two superpowers or whether they are genuine agreements to minimize "human destructiveness" may be debated.

A long term objective of nuclear arms control is undoubtedly a process by which total disarmament may be reached. It is crystal clear, though, that instant liquidation of nuclear weapons would by no means be a realistic approach. Arms control has been directed to the containment of the possible pressures of nuclear proliferation. It can only be done over time through a process of continuous and attentive management. The process is sensitive, reflecting the power relationships between the two superpowers, and the international political process of the present-day world is so complex that enduring and unabated efforts are required to achieve the objectives. The apple of nuclear knowledge and technology has already been bitten. It would be impossible for those who enjoyed the taste to cast it away. The process of correction would be a long and patient one. Gradualism would be a practical reality in bringing about the results of disarmament. The Vladivostok agreement last fall would certainly not be the end of the game. Further efforts for reduction could be hoped for.

The Security of Japan. The problem of nuclear insecurity is the most difficult of all. The security interests of each party to the treaty are not identical, and thus the assurances by the nuclear power states must be tailored to the specific concerns each state or each group of states entertains. The UN Security Council Resolution 255 and the associated declarations on security assurances are now widely regarded as inadequate since they are no more than general statements, and international events since then have undermined the effectiveness of the resolution. The solution of this problem is not yet in sight. Moreover, while NPT is one of the major pillars to sustain the structure of international security, the problem of security specific to each nation would surely be dependent on the broader context of the international political process.

In the case of Japan, her security is assured by the security pact with the United States, which includes the latter's commitment to protect Japan against any nuclear hostilities or threats. The assurances have been repeated periodically, and the latest conversation (April 1975) of Mr. Miyazawa, the Japanese foreign minister, with President Ford and Secretary Kissinger, particularly focused on the US commitment to defend Japan against nuclear attacks or threats of nuclear attack in relation to Japan's readiness to ratify NPT. It has been commonly argued, despite such guarantees, that the credibility of the nuclear umbrella has now decreased because of the nuclear parity between the United States and the Soviet Union. Japan must pay careful attention to this problem, yet there is no reason for Japan to become independent of the pact because she can cope with the situation of "threats" only by relying on the US—Japanese security arrangements.

As Professor Wohlstetter points out:

If alliance guarantees are not completely certain, neither is any alternative to a guarantee. Alliance guarantees are matters of greater or lesser likelihood, not certainty. The question is only whether a guarantee against attack presents enough risks to the attacker, whether it has a large enough likelihood of being fulfilled to discourage attack. And it must do this sometimes when all alternatives seem bad to a prospective attacker [6].

Efforts have been made, and should continue to be made, toward developing a common US—Japanese perception of security problems in Asia and a common conception of the role of the security treaty so as to maintain the likelihood of the guarantee greater rather than lesser. I would argue that Japanese—US relations would quantitatively and qualitatively be viable enough to keep the likelihood greater, even though the relations might appear to run into some difficulties. Though it sometimes looks difficult, areas of coalition would and could be worked out between the two countries. Before it is hastily concluded that Japan should cease to have a security alliance with the United States because the credibility of the pact can no longer be relied on, a lot of diplomatic efforts should be made to promote the functions of the security pact. The pact certainly has deterrence value in ensuring Japan's national security.

Apart from the specific security interests to be protected, it is a very difficult task to solve the problem of "military disadvantages" and "vulnerability to nuclear intimidation" within the general framework of NPT. Any effort to frame provisions within the treaty that would meet this diversity of possible contingencies would be a matter of inordinate complexities. And yet without general efforts to promote the quality of NPT in this respect the frustration of non-nuclear sovereign states which have given up the nuclear weapons option would not be resolved. It could become even more serious.

In this connection two possible solutions have been cited: (1) a pledge by the nuclear powers never to be the first to use nuclear arms against a non-nuclear weapon state which is a party to the treaty unless that state is engaged in armed attack in concert with a nuclear weapon power; and (2) regional arrangements for nuclear-free zones. If these could be more than general statements and thus contribute to the specific security interests of the treaty states as well as to the general reduction of the danger of nuclear war, they should be the problems to be pursued by the initiative of the superpowers.

Peaceful Uses. Whether Japan can really be assured of substantial equality in peaceful uses of nuclear energy with other contracting countries of NPT has been a matter of the greatest domestic concern. Every possible measure must be taken to ensure Japan's continued and stable supply of nuclear fuel materials from foreign countries. The United States, on which Japan is dependent (and will be so for some time to come) for her supply of enriched uranium, has already notified Japan that it may cancel the supply of enriched uranium if the US government finds itself unable to observe Article III, paragraph 2, of NPT. It is obvious that Japan must eliminate such an unstable factor in the supply of materials.

The IAEA safeguard system agreed to by bilateral peaceful use agreements, and under which almost all nuclear materials in Japan have been inspected, is now too cumbersome to be compatible with the rapidly expanding nuclear power facilities in Japan. In 1971, 324 IAEA inspectors visited Japan. A simplified and rationalized safeguard system under NPT which is also designed to protect industrial secrets will certainly be more conducive to Japan's interests than the inspections as currently conducted. The Japanese government has finally reached a safeguard agreement with IAEA which was initialed on February 26, 1975. The protocol to the agreement assures most favored nation treatment to Japan, and Japan was assured that the Euratom nation formula of inspection would be applied.

It should be noted that, having been assured of substantial equality in the peaceful uses of nuclear energy (which are a matter of major concern to Japan), Japan's Atomic Industrial Forum, Inc., which has long hesitated to support the ratification of NPT, is now determined to press the government for early ratification of the treaty.

Commitment to Non-nuclear Option

On May 5, 1975 the NPT Review Conference convened in Geneva "to review the operation of this treaty with a view of assuring that the purposes of the Preamble and the provisions of the treaty are being realized" [7]. As a signatory, Japan participated but did not take part in formal decisionmaking since she has not completed the procedures of ratification and depository.

It remains to be seen how much the debate in Geneva will influence, in one way or the other, the deliberations in the Japanese Diet. Ambiguities in the prospects for arms control and uncertainties about national security are evoking caution from some. This group is basically different from those who oppose Japan's participation in the treaty on the grounds that it would make Japan a tool of the American imperialist nuclear strategy. What is important to the first group is a course of consensus-building in which they want to participate. The second group insists on a non-armed and non-nuclear neutral Japan but cannot support NPT. It is almost impossible to reconcile them on ideological grounds with the government's views though the bargaining tactics in the Diet are sometimes fruitful.

This writer is of the firm view that the time has now come for Japan to commit herself to a non-nuclear option based on clear state (not ideological) interests and on those considerations mentioned in earlier sections of this chapter. Japanese leaders—not necessarily only political, but those in business, journalism and academia—should discuss not whether the option for going nuclear would be available, but in what way Japan could contribute to solving the problems of the world without taking up that option, and by committing herself to the non-nuclear alternative.

Nothing is absolute and everything, therefore, is relative in this world. Thus, it seems to me inappropriate and unrealistic in the light of international political processes that Japan should hesitate to join NPT because of the lack of an absolute commitment by the two superpowers. In my view, the fundamental reality is that the nuclear superiority of the two superpowers versus other states does exist and will not change. The treaty cannot basically alter the superiority of the United States and the Soviet Union, which is inevitable. As any international legal system stems from the existing realities, NPT recognizes that nuclear states do exist. Once this is acknowledged, all nations have an interest in increasing their sense of stability and improving the international environment by limiting the number of nuclear countries. It is obvious how dangerous uncontrolled proliferation would be for everyone.

It is vitally important for Japan not to provoke the actions of another state, as her aim is exclusively to prevent aggression against herself. This is particularly so if an immediate and direct threat is not perceived by Japan.

Because of their military capabilities, their location and the character of their states, the Soviet Union and China would occupy a significant place in the foreign policy of Japan. Could they create a buildup of credible threats against her?

Certainly the overall military capability of the Soviet Union is a potential threat to Japan. It is another thing, however, whether the Soviet Union would be in a position to transform a potential threat into a real threat. The Soviet Union should remember the strong and traditional distrust that the Japanese have had of the Russians. The violation of the Neutrality Pact of 1941 in August 1945 was a clear example of the unfriendly acts which have caused Japan to regard the Soviet Union as unreliable. Any kind of nuclear blackmail would not only augment such distrust, but would also endanger the position of the USSR vis-à-vis China and other Asian countries. The Soviet Union has become an object of special concern in Japan's view of her relations with China, particularly after Japan normalized her relations with Peking in September 1972. Overt pressures of the Soviet Union against Japan will certainly affect the Chinese attitude toward Moscow and will further promote Chinese access to Japan. The recent development of relations between the Soviet Union and Japan indicates that Moscow has an interest in the improvement of relations with Japan, both politically and economically. As long as the basic structure assumed in this study exists, Moscow's interest will be to enjoy nonhostile relations with Japan rather than to adopt a policy of blackmail diplomacy, including the threat of nuclear weapons.

Japan's relations with China have improved to a great extent since normalization. Despite the growing Chinese nuclear capability, the Japanese people do not feel it as a threat to themselves. This was true even before the rapprochement between Tokyo and Peking. It may be worth noting that the Japanese people have entertained stronger attachments, both traditional and cultural, to China than to the Soviet Union.

No matter what the Chinese intentions may be on their nuclear development, China would ultimately have to face those problems that both the United States and the Soviet Union have faced in their process of nuclear weapons development, and thus her military as well as diplomatic alternatives would gradually be constrained. Furthermore, unless the Chinese are absolutely confident that, whatever action they may take, there would be no possibility of either the United States or the Soviet Union making a nuclear counterattack against China, the Chinese could not possibly resort to reckless acts. Therefore, the Chinese leaders will find a greater security interest in promoting better and friendlier relations, politically and economically, than in adopting a hostile attitude toward Japan.

Neither the Soviet Union nor China has an effective way of coercing Japan, and neither knows for sure in what cases a threat to Japan would be regarded as plausible, because of the lack of experience. They could only speculate about it; but a national security program cannot be built up on sheer speculation.

Thus, on the basis of the Japan–United States security treaty, Japan's national security is now relatively secure. If the worst and most unusual situations were to occur in the future, whereby Japan could only maintain her security by nuclear armament, it would be the end of NPT—which would certainly not be in the interests of Japan. Her interests clearly lie in being in the non-nuclear club, even if NPT is insufficient and incomplete.

All in all, it will only amplify mistrust and misunderstanding of Japan if she delays ratification of NPT any longer. It is almost inconceivable that Japan would be able to use "freedom of nuclear armament" as a diplomatic bargaining chip. Just as a British act of nuclear renunciation could have disclosed the fatuity of the argument that nuclear bombs are a key to world status, Japan today should be proud that she is uniquely positioned to reverse the trend toward proliferation. It seems to me that the NPT ratification problem is a symbolic condensation of what role Japan should play in the international community of nations. Japan is asked to make a more positive contribution toward progress and development of a new world order in the 1970s.

CONCLUSION

In the previous sections I have tried to explain as much as possible about the way in which the Japanese government and people have perceived the challenges they face in nuclear development—military or peaceful—and have advocated that Japan should ratify NPT as quickly as possible. Once Japan possesses nuclear weapons, she will become a potential target of the two superpowers and even of China. To a country like Japan, which is so dependent on harmonious relations with foreign countries and on world peace for her very existence, it is clear that she should not adopt such policies as may make many countries regard her with suspicion and mistrust. Japan's potential economic power is still growing, and the interdependence between Japan and the rest of the world will continue to develop. By contributing to the progress of the world economy through trade, investment and aid, Japan will also be contributing both to international order and to the strengthening of her own security. Now that nuclear weapons have become weapons which cannot be used in warfare and therefore have only very limited potential utility, Japan's efforts should be directed towards disassociating the possession of nuclear weapons from the notion of national prestige. The time has passed when only the nuclear-power states could be influential.

NOTES TO CHAPTER TEN

1. Herman Kahn, *The Emerging Japanese Superstate* (Englewood Cliffs, N.J.: Prentice–Hall, 1970).
2. Japanese Defense Agency, *The Defense of Japan* (Tokyo, October 1970).
3. Ibid.
4. *Diet Record* (in Japanese), March 14, 1967.
5. Ibid.
6. Albert Wohlstetter, "Japan's Security: Balancing After the Shocks," *Foreign Policy* 9 (Winter 1972–1973).
7. *Treaty on the Non-Proliferation of Nuclear Weapons,* 1970.

Chapter Eleven

Brazil's Nuclear Aspirations

H. Jon Rosenbaum

On June 27, 1975 Brazil and West Germany signed a 15 year, multibillion dollar nuclear technology agreement. It will bring West Germany at least somewhere between $4 billion and $8 billion in contracts for reactors, generators and a uranium enrichment system, as well as other items; it will also contribute not only to the fulfilment of Brazil's nuclear aspirations but perhaps to the spread of nuclear weapons as well. The accord provoked an international tempest [1] when news of the two countries' intentions was publicized in Washington [2]. Emotional, or at least undiplomatic, denunciations by several members of the United States Congress' Joint Atomic Energy Committee, including Senators Pastore, Symington, Ribicoff and Glenn, and critical editorials in the American press incensed many Brazilians and Germans [3]. The Brazilians were angered particularly by Senator Pastore's remark that by providing Brazil with access to the complete nuclear cycle Germany would be creating a nuclear peril "... in our own backyard, so to speak, while, at the same time the U.S. Government is heavily committed in West Germany's backyard to defend them against a likely peril" [4]. For their part, some of the Germans were resentful because they interpreted American opposition to the treaty as being motivated by a desire to prevent Germany from competing successfully with American firms in the nuclear export market [4A].

THE ACCORD

The Brazilian-German agreement that provoked this discord is an outgrowth of a broadly based scientific understanding signed in 1969 between the two nations. Negotiations continued, and the new agreement was reached on February 12, 1975.

According to the agreement [5], as many as eight 1200 megawatt pressurized water reactors of German design capable of generating 10,000 KW of electricity will be installed by 1990. The first two of these reactors will be built at Angra dos Reis, where a Westinghouse reactor, Brazil's first, is currently under construction [6]. This will increase the capacity of the facility at Itaorna beach by 2600 MW. The locations of the other reactors have not been chosen, but among the possibilities are the Vale da Ribeira in the state of São Paulo [7], near the banks of the Paraiba River, and the Angra dos Reis site. Construction of the first two reactors is planned to begin during 1977–1978; their completion is set for 1981–1988.

Mulheimer-Kraftwerk-Union (with the participation of Siemens and AEG) will assist in building the reactors, but fabrication will take place progressively in Brazil as the country's industries become familiar with German technology. It is estimated that Brazil has the technical capability to produce approximately 50 percent of the required equipment at present. It is also envisaged that private capital will be generated by the Brazilian stock market to finance the power plants [8]; the power plants alone will cost at least DM10,000 million.

The West German–Brazilian "package deal" specifies also that the two nations will mount a joint effort to prospect for uranium ore in Brazil. A mixed company will be created for this purpose with 51 percent Brazilian and 49 percent German financial participation [8A].

However, it is the inclusion of provisions for the construction of a fuel fabrication plant, a chemical reprocessing facility for extracting plutonium from spent fuel and a uranium enrichment factory for concentrating the fissionable isotope uranium 235 that has caused the greatest concern among those worried about nuclear proliferation [9]. These facilities will give Brazil the capability to produce nuclear weapons if she so desires. Additional anxiety has been produced by the subsequent announcement of a $2.5 million Brazilian–French contract for the development of a fast neutron or breeder reactor similar to the Rhapsodie and Phoenix reactors being built in France. This research reactor (called the Cobra, an ancronym for Cooperation Brazil), will be built by Techniatome and will produce more fuel than it consumes [10].

The German-designed enrichment facility is to be ready for operation by 1980 or 1981, cost $3 billion to construct and supply ten nuclear reactors with a minimum potential of 1,200 MW. Since the projected Jet Nozzle system requires large amounts of electricity, the plant will probably be located near one of Brazil's many hydroelectric complexes, and the enriched uranium will then be transported to the reactors, which will be situated closer to the country's industrial centers.

The German firm STEAG and NUCLEBRÁS (the Brazilian Nuclear Corporation) will establish a joint enterprise to develop the Jet Nozzle system. In the first stage Brazilian scientists will be sent to Germany to learn the German enrichment technology. In the second stage an experimental plant capable of producing nearly 200 tons of separation units a year will be built in Brazil. During the third stage a commercial plant capable of yielding one or two thousand tons of separation units per year will be constructed [11].

Aside from the German and French deals, new uranium discoveries have recently brought Brazil's nuclear aspirations yet another step closer to realization [12]. These finds, reported by a French geological team, also should help Brazil to meet its commitment to provide West Germany with sufficient uranium through 1990 as required by the accord.

Details of the French report are still secret. However, Project Radam (the Brazilian prospecting service) has also announced new uranium discoveries during the past several months. For example, rich deposits of uranium have been found in the northern part of Pará state near the Guianas. Some optimistic Brazilian officials are now suggesting that Brazil may have one of the largest uranium reserves in the world; current minimum estimates range from a minimum of 50 thousand tons to a maximum of 500 thousand tons (total confirmed world reserves are approximately one million tons).

Serious prospecting for uranium in Brazil did not begin until 1969, but since then a vigorous effort has been mounted, with $200,000 being spent for this purpose in 1975. Nineteen deposits of uranium have been found to date, but much of Brazil remains to be explored. Although the remote locations of many of these deposits will make exploitation difficult and costly, two mines, both in the state of Minas Gerais, are already in operation—Campo do Agostinho with reserves of one ton and Cercado with three tons [13]. Brazil also possesses significant amounts of thorium.

SAFEGUARDS

The primary criticism given by opponents of the Brazilian–German accord has been that the sale of uranium enrichment and plutonium extraction technology will allow Brazil to produce fissionable material that could be used in the manufacture of weapons. In response the Germans have argued that there is no need for apprehension since obligations of the Nuclear Nonproliferation Treaty will be respected (West Germany is a signatory but Brazil is not), and, moreover, Brazil will have to meet additional requirements.

The Germans have emphasized that the treaty between Germany and Brazil contains two requirements in particular that may establish precedents. First, technological information must be included in the agreement on safeguards that Brazil is required to conclude with the International Atomic Energy Agency (IAEA). Second, German approval must be obtained before any of the sensitive nuclear materials, installations and plants, or relevant technological information can be re-exported by Brazil [14].

In addition, the accord contains the following stipulations:

1. a general affirmation of the principle of nonproliferation of nuclear weapons;
2. an obligation by Brazil to submit all of the nuclear equipment, installations, materials and technological know-how it receives from Germany to IAEA safety controls;

3. a solemn commitment by Brazil not to use the nuclear equipment, installations, materials and technological information supplied by Germany for the production of either nuclear weapons or other nuclear explosives;
4. an agreement by Brazil to re-export such equipment, installations, materials and technological information only to countries that have concluded safeguard agreements with the IAEA;
5. a commitment to apply these safeguards for an indefinite period beyond that covered by the agreement (which has a minimum duration of 15 years with a provision for extension);
6. an assent by Brazil that the equipment and material will be physically protected from third parties; and
7. an agreement that the Federal Republic of Germany may be a party to the negotiations of the accord on safeguards to be concluded between Brazil and the IAEA [15].

In reply to critics of the accord the Germans also have stressed that the treaty's safeguards exceed those required by Canada for its exports of nuclear technology to India and that Brazil has the capability of eventually developing a nuclear industry without foreign assistance. If this occurred, the Germans have contended, there would be no international controls whatsoever on the Brazilian nuclear industry. Moreover, the German press has observed that France, which has not signed the NPT, is prepared to provide Brazil with a complete nuclear cycle as well. Had Germany not agreed to assist Brazil, a far greater danger of nuclear proliferation would have been posed [16].

It has been noted, in rebuttal, that the IAEA, given its small budget and staff, will be unable to monitor nuclear activities effectively in a country as big as Brazil [17]. In addition, doubt has been expressed about the willingness of future Brazilian leaders to abide by the treaty's stipulations [18]. Further, some commentators suspect that once Brazilians learn German technology, there will be no way to prevent the duplication, independently of the treaty's terms, of the facilities initially provided by the Germans.

Finally, it has been suggested that the German sale, the country's largest export deal to date, may lead to greater competition among exporters of nuclear technology and yet more nuclear proliferation [19]. It is unlikely that other developing nations, particularly Latin American countries such as Venezuela and Argentina, will be content to be surpassed by Brazil in the nuclear field, and in their desire to emulate the German example, exporters may compete by offering less and less stringent safeguards; thus, safeguards eventually could become so weak as to be entirely meaningless.

As was mentioned briefly at the outset, American reaction to the Brazilian–German accord has been especially disparaging. The accord has been branded as imprudent, dangerous and a peril to world peace by the American press, and an attempt was made by top Department of State officials to dissuade the Germans from signing it.

Naturally this was interpreted by the Brazilians as an example of unjustified interference in their internal affairs, a conclusion supported by Senator Pastore's unfortunate statement that "If this agreement goes through at this time in this fashion, it will make a mockery out of the Monroe Doctrine" [20]. As a consequence, nearly all segments of the Brazilian public have strongly supported the agreement.

Senator Paulo Brossard of the opposition MDB party termed Senator Pastore's comments an "abusive interference in matters which concern only Brazil" and asserted that "the time of the Big Stick is over, and Brazil will not allow itself to be treated like a banana republic" [21]. An opposition deputy, Antonio Carlos, asked the congresses of Argentina, Venezuela, Colombia, Mexico and Paraguay to support the accord on behalf of the technological interests of the continent [22]. Deputy Herbert Levy of the governing ARENA party declared: "The United States, which passively watched the installation of a communist dictatorship in Cuba and is now watching the strangulation of the Portuguese people at the hands of a minority of communist officers, should not be concerned over the signing of a nuclear agreement by an allied nation, aimed at peaceful objectives of development" [23]. Another ARENA deputy, Joaquim Coutinho, vice president of the Foreign Relations Committee of the Chamber of Deputies, stated: "The United States Senate also has its neurotic owners of the world's truth, who have forgotten about the genocidal practices in the world. They were the first to utilize atomic energy, not to build but to destroy" [24].

In a statement supported by Senators Rui Santos and Franco Montoro, the majority and minority leaders respectively of the Senate, opposition Senator Danton Jobim protested against US pressure to prevent the signing of the accord and said: "the answer we should give is that we are allies and friends of the United States and not puppets at its service" [25]. The entire Chamber of Deputies' Foreign Relations Committee charged Senator Pastore with "interference" in Brazil's affairs and "repudiated" his statements [26].

The Brazilian military, the dominant political force in the country, also registered its support for the treaty. For example, the commander of the First Army in Rio de Janeiro, General Reinaldo Melo de Almeida, argued that signing the accord "constitutes a decisive step that reinforces the country's sovereignty" [27].

Finally, the Brazilian press defended the treaty in numerous editorials. For instance, the *Jornal do Brasil* editorialized: "Only the hypocrisy of those who speak of nuclear control while increasing their stock of atomic weapons in quantities which can blow up the world explains the fact that a correct, natural, and sovereign agreement be dragged into the arena as a passionate polemic which only serves to feed extreme opinions and sentiments" [28]. Individual commentators also gave a great deal of attention to the accord. For example, Theophilo de Andrade, a columnist for the large Diários Associados newspaper chain, wrote: "Brazil is a country of peaceful tradition, being the great pioneer and champion of arbitration for the solution of questions that arise among the nations of the earth. It is not possible that it could use nuclear energy for military purposes" [29].

Despite the overwhelming support in Brazil for the accord, there has been some opposition to it as well, although few critics have dared to express their reservations publicly. Among the opponents are intellectuals who are worried about how the authoritarian military regime now governing the country might use nuclear technology since Brazil has not signed the NPT and the government is not subject to democratic restraints.

Some members of the scientific community also object to the treaty. They feel that Brazil should pursue an independent course in nuclear development and point out that by the time Brazil's German-designed enrichment plant begins production fast breeder reactors that can use thorium as fuel may be available. They note that Brazil, as far as is currently known, has much more accessible thorium than urnaium [30].

One dissenting scientist who has made his views known is Marcelo Damy de Sousa Santos of the Institute of Physics at the Catholic University of São Paulo. He continues to favor independent nuclear development, and he believes that Brazil should construct natural uranium and heavy water reactors and sign a cooperative pact with India, a nation with a comparable industrial base and scientific elite as well as immense thorium deposits. Germany, he insists, has a different cultural and scientific tradition from Brazil and, therefore, has less to offer the country [31].

During the past several years there has been a scientific debate within Brazil between those advocating that Brazil develop its own technology and those preferring that the country purchase nuclear plants on a turnkey basis from the industrialized countries. A 1973 study by the Bechtel Corporation, commissioned by Brazil, concluded that the country could produce 60 percent of the components for a light water and enriched fuel system by 1980, a finding that supported the views of certain military nationalists and scientists advocating the independent strategy. In a second, more recent, debate, it appears that scientists of the National Nuclear Energy Commission (Comissão Nacional de Energia Nuclear—CNEN) wanted the government to purchase nuclear plants from abroad on the basis of an international tender and became disgruntled when they learned that the finance minister, Mário Henrique Simonson, was negotiating only with Germany [32]. Of course, the West German deal has rendered these debates academic. Although there may be new controversies, it seems unlikely that the strategy for nuclear development can be reversed.

The devision to purchase the nuclear reactors from Germany was evidently made in November 1974 at a meeting attended by Hervásio de Carvalho, the head of CNEN; Mário Behring, the director of Electrobrás (the national electric company); and Shigeaki Ueki, the minister of mines and energy. This represented a compromise between those who wanted to pursue the independent course by purchasing a few reactors abroad while developing an autonomous capability and those, such as General Golbery do Couto e Silva, an advocate of close relations with the United States and an aide to President Ernesto Geisel, who favored dependence on external suppliers [33].

MOTIVES

Brazil's motives for signing the accord were numerous. First, there is a general belief in the country that the agreement will increase Brazil's international status. The foreign minister, Antonio Azeredo da Silveira, after signing the accord stated: "Brazil has gained new technological and political status on the world scene with the nuclear agreement" [34], and General Almeida declared that his country would "be transformed into a great power" [35].

Although the 1974 Indian test of a nuclear explosive probably reinforced the Brazilian belief that nuclear capabilities confer international status [36], the Brazilian government has been convinced since at least 1967 that nuclear power and international power are intimately related. This conviction has grown stronger in recent years [37].

The second motive was fear of falling behind Argentina. According to one source, confidential information reached Brasilia that Argentina was three years ahead of Brazil in the development of nuclear energy [38]. The Atucha plant had been dedicated, and Argentina was constructing several other units while Brazil's first nuclear power plant had not been completed [39].

The ancient rivalry with Argentina is a significant factor in the formation of Brazilian domestic and ·foreign policy, and relations between Argentina and Brazil were particularly strained when the planned accord with Germany was announced [40]. Nevertheless, the reported meeting of Brazilian national security advisors that decided to accelerate the pace of the two year old negotiations with Germany was probably not primarily due to accounts of Argentine success in the reactor field. It should, however, be noted that the new reactors being obtained from Germany are far more sophisticated than those built or being constructed in Argentina and may help Brazil to overcome Argentina's lead in this area of nuclear development.

News that France was planning to supply Argentina with a plant to produce plutonium, later denied by French sources [41], and that Argentina might construct a nuclear bomb clearly had a far greater impact on Brazilian policymakers [42]. Moreover, it can not have been overlooked by the Brazilians that Argentina signed a pact to exchange information with India in 1974, just after the Indian explosion.

During March 1975 Edgard Cossi Isasi, a member of a small provincial party within the Peronist movement, introduced a bill in the Argentine Congress calling upon the government to fabricate a nuclear bomb. Although the bill had little chance of passing, it did not go unnoticed in Brazil and no doubt also contributed to Brazilian fears about Argentine intentions [43]. Deputy Pedro

Faria (MDB), for example, alerted the government to what he termed "a nuclear threat in Latin America" and warned that the bill should be taken seriously and not underestimated [44]. A statement by the Argentine ambassador in Canada, Benito Llambis, added to Brazilian apprehension. He said that "all nations must enjoy the same rights, including the possibility of producing nuclear arms" [45].

The third motive for Brazilian interest in the German deal had to do with the country's increasing energy requirements, the rising price of imported petroleum and the belief that Brazil had only small fossil fuel reserves. Brazil's industry is expanding rapidly and so are the number of motor vehicles. However, when the decision was made to move forward with the German negotiations, Brazil was importing 80 percent of its petroleum and was concerned with the constantly increasing prices being charged by the OPEC nations. Moreover, Brazil relies heavily on hydroelectric power, and it was estimated that by 1985 the potential for expanding hydroelectric energy for the Center-South—where 45 percent of the country's population and 70 percent of its industry is located— would be exhausted, and wood (or alcohol), although widely burned in rural areas, certainly would not be able to alleviate the general energy deficit very much.

In recent months new petroleum reserves have been discovered in Brazil and off its coast, and there has been speculation that the finds off the state of Rio de Janeiro are substantial, although the government has cautioned that expectations should be restrained [46]. When the decision was made to negotiate the German agreement, Brazil was searching for new energy sources, and the German accord provided an opportunity to expand the country's nuclear generating capacity rapidly [47] and thereby help to satisfy the urgent need for electric power.

Ironically, instead of being concerned about OPEC's decisions, some Brazilians now are hypothesizing that their country will not only become a major world exporter of nuclear fuels but will be able to form a cartel with other nations having enrichment facilities in order to engage in boycott and pricing politics. However, this scenario appears highly unrealistic since cartel members probably would be unable to restrict access to enrichment technology.

Fourth, the agreement was attractive to the Brazilian government because, although negotiations were begun during the Médici years [48], President Geisel was able to take credit for the successful outcome. Certain circumstances had caused many Brazilians to lose confidence in Geisel and he was considered unlucky, a distinction that can be costly in Brazilian politics. Geisel needed a success badly, and the accord has helped him to reestablish his legitimacy. He has been able to portray himself as the liberator of Brazil's economy and defender of its sovereignty [49].

The government can assert that the economy has been liberated since Brazil is no longer dependent upon American nuclear technology. Moreover, the critical

American reaction to the accord and the United States' refusal to transfer nuclear technology have been overcome and this success has enabled Geisel to demonstrate that his government and Brazil are independent and no longer deserve to be characterized as America's "Junior Ally" in South America.

Finally, the Brazilians were prompted to seek the agreement with the Germans by their fascination with the peaceful uses of atomic energy. They were convinced that as long as they were dependent upon American nuclear technology they would not be allowed to apply this technology as they wished.

While the United States would not object to its technology being used for certain purposes, such as the generation of electricity or the treatment of disease, the Brazilians were not content to be restricted in this way, believing that the nations that use nuclear energy creatively will become the most developed in the future.

Clearly there was a danger that the United States would withdraw its assistance if the Brazilians began to use nuclear explosives for developmental purposes, and, although American experiments with peaceful nuclear explosives (PNEs) have not been particularly promising [50], PNEs intrigue many Brazilians. Proposals for extracting oil from the shale in the states of São Paulo and Paraná, the linking of the Plata, Amazon and Orinoco rivers to integrate South America, and the construction of ports and canals are just a few of the many benefits Brazil could realize from PNEs according to such Brazilians as Clóvis Ramalhete, one of the country's most distinguished lawyers [51]. The transfer of technology provided by the German accord will eventually allow the Brazilians to proceed with these and other schemes, should they wish, without fear of losing their access to American enriched uranium, for example, although perhaps the United States and even Germany might retaliate in other ways if Brazil were to do so.

CAPABILITIES

While the German agreement will contribute significantly to Brazil's future nuclear capabilities, the country, as this chapter has suggested previously, already has substantial nuclear assets. In fact, Theodore Taylor, an American physicist and authority on nuclear weapons, has indicated that Brazil's Angra dos Reis reactor will soon be able to produce enough plutonium for the country to construct every two weeks a bomb equal in destructive power to the one dropped on Hiroshima [52].

Brazil's nuclear energy program was initiated in 1951 by the National Research Council (Conselho Nacional de Pesquisas) [53]. Five years later a special commission, created to study nuclear energy policy, made 18 recommendations to the National Security Council, which approved the establishment of a National Atomic Energy Commission (Comissão Nacional de Energia Nuclear—CNEN), a National Fund for Nuclear Energy (Fundo Nacional de Energia Nuclear) and a program for the training of specialized personnel [54]. However, because of lack of resources and interest, the nuclear energy program received little financial support until 1961.

Three research reactors were constructed with the available funds. The first was inaugurated at the São Paulo Atomic Energy Institute (Instituto de Energia Atómica) in 1958; the others were built at the Belo Horizonte Institute of Radioactive Studies (Instituto de Pesquisas Radioativas) and the Rio de Janeiro Institute for Nuclear Engineering (Instituto de Engenharia Nuclear) in 1960 and 1965 respectively. A fourth research reactor, also located in Belo Horizonte, was completed in 1973 [55]. In addition, the CNEN installed three small, subcritical nuclear reactors for the study and production of radioisotopes in Recife, Belo Horizonte and Sao José dos Campos. Finally, since 1971 Professor José P. Sudano has been supervising construction of a thermonuclear reactor at the Institute of Aeronautic Technology in São Paulo, and a small fuel fabrication facility has been built in São Paulo as well.

In 1969 the CNEN included among its goals the discovery of uranium, training of personnel, basic research, international exchange, radioisotope development and mobilization of national industry for the manufacture of power reactors. Progress has been made in meeting each of these goals.

Mention already has been made of Brazil's uranium and thorium discoveries and reserves [56]. To date nearly 300 Brazilian scientists and engineers have received specialized education in nuclear energy, and 80 more are to complete their studies before the end of 1975. The construction and operation of the new nuclear power plants will require approximately 1,000 high level professionals; therefore, revised plans foresee the graduation of 150 new nuclear scientists and engineers annually beginning in 1976. This new goal probably is attainable since university enrollments and facilities have expanded dramatically in the last few years. Until these technicians are trained, however, Brazil will have to rely in part on foreign talent if the new reactors are to be installed on schedule.

Advances also have been made in basic research and radioisotope development. The applications of radioisotopes are being examined at the São Paulo Institute of Atomic Energy, the Center for the Development of Nuclear Technology of the Brazilian Nuclear Technology Company (CBTN) and other institutes [57], and the use of radioisotopes in medicine, agriculture, industry and engineering have received a great deal of attention. Basic research has focused on the fuel cycle, but work is being done in several other areas as well. These include nuclear astrophysics, nuclear structure, photonuclear reactions, nuclear fission, nuclear reactions produced by heavy ions and the displacement time of neutrons.

Although the agreement with West Germany is the most dramatic example of Brazilian nuclear cooperation with other nations, Brazil has also signed accords with several other nations. These nations include the United States, Bolivia, Israel, Peru, Chile, India, Canada, Italy, France, Paraguay, Portugal and Switzerland. A cooperative agreement also exists with Euratom [58]. Aside from the recent deals with France and Germany, an earlier agreement was signed with France in 1967 for joint research in thorium technology, the development of research and power reactors, and uranium exploration, while in 1969 Germany and Brazil agreed to joint consultation in nuclear power and gas centrifuge research. The latest accord with the United States was signed in 1972 and expires in 2002; in it the United States contracted to supply enriched uranium to the Angra dos Reis plants being constructed with the assistance of Westinghouse. It should be noted that this agreement does not specifically forbid Brazil from constructing PNEs. Moreover, during a 1967 visit to Brazil, former United States Atomic Energy Commission Chairman Glen Seaborg stated that enrichment services would be continued even if Brazil refused to abandon plans to develop PNEs [59]. Therefore, although the usual safeguards required by the United States are incorporated into the Brazil–United States agreement, Brazil might be able to divert material for the fabrication of PNEs from the Westinghouse reactor, or for that matter from its four research reactors which are also supplied with enriched uranium by the United States, without sacrificing the continued supply of American fuel; and, of course, once Brazil has its own enrichment facility, it no longer will need to fear the suspension of American fuel.

The mobilization of national industry for the development of power reactors has been stimulated by the creation of Nuclebrás. The national steel industry is capable of producing casings for turbines, and although Brazil can fabricate between 50 and 60 percent of the other equipment for nuclear reactors, it is unable to make cores [60]. However, due to the German accord, the main problem is no longer technological but financial. Brazil will need to spend an estimated $185 annually by 1985 for reactor fabrication, and this figure is expected to rise to $550 million by 1990 and $1.25 billion by 1995. Brazil can afford this, but whether it makes sense economically is another matter since petroleum resources now appear to be more abundant than they were previously believed to be and the transfer of technology supposedly will cost more than the importation of reactors. Politics seem to have weighed more heavily than economic considerations in the decision to launch the crash reactor program with German assistance.

Brazil has other capabilities that would be valuable, particularly if it intends to develop PNEs or a credible nuclear arsenal. For example, Brazil has vast, sparsely settled areas in remote parts of the country that could be used for testing nuclear explosives. These include several South Atlantic islands located far from shore such as Martin Vaz, Trinidade and the Fernando de Noronha group, and the Amazon and Center–West regions.

Brazil also has the most advanced missile capability and strongest air force in South America [61]. The Brazilian air force (FAB) is equipped mainly with Mirages and F–5s but has recently taken delivery of 42 new supersonic Northrop

F—5 "Tiger" fighters and has 48 Hercules C—130 transports on order. More importantly FAB is building its own fighter-bombers.

Since 1965 Brazil has launched over 400 missiles from a base near Natal, and it is increasing its investment in rocketry. The Brazilian National Commission for Space Activities (COBAE) and the Ministry of Aeronautics have established sites for launching sounding rockets and are developing the Sonda—1 two stage rocket. While the maximum altitude of this rocket is probably less than 50 miles (it is used for upper atmosphere research), some observers anticipate that Brazil will have developed intermediate range ballistic missiles by 1980 [62].

Whether Brazil is able to deliver nuclear weapons at present is difficult to determine, but its airplanes could conceivably deliver nuclear weapons to targets in neighboring countries. Brazil has air superiority in South America, and the neighboring nations have relatively primitive air defense systems. Although Brazil's present missiles are unsuitable for nuclear weapons delivery, by the early 1980s they also may be capable of carrying nuclear missiles to their targets with enough accuracy to pose a threat to the continent's other nations.

INTENTIONS

There seems little doubt that Brazil now has, or soon will have, the ability to fabricate nuclear explosives and that its capabilities in this field will improve as a result of the German accord. In fact, Brazil may already possess such devices or be engaged in their development. At least as long ago as 1967 the CNEN demonstrated enough interest in nuclear bombs to commission a study of Brazil's ability to build them. The study estimated that it would take between ten and 15 years and cost 1.5 billion cruzeiros to construct a nuclear bomb [63]. If a decision to develop nuclear explosives was made at that time and the estimate of the study group was correct, Brazil should now be close to obtaining at least one nuclear bomb. Of course, the 1967 estimate may have been incorrect or the government may have made no decision to launch a nuclear explosive development program in 1967 or at any other time. On the other hand, such a decision could have been made earlier than 1967.

Currently, it is impossible to know whether the Brazilians are building nuclear explosives and, if they are, how far they have progressed. Statements by Brazilian officials on this subject have been contradictory. The minister of foreign relations, Antonio Azevedo da Silveira, has declared that Brazil is not, and will not, fabricate nuclear explosives. However, his ministry's press secretary, Guy Brandão, has stated that "Brazil intends to detonate nuclear explosives for peaceful ends " [64].

Even without the confusion caused by contradictory government statements, it would be difficult to know Brazil's true intentions. As every reader is painfully aware, government officials do not always speak the truth. Officials may also make ambiguous statements. For example, the councillor of the Brazilian embassy in Tokyo, Franco Netto, is reported to have stated that Brazil "has no plans to conduct a peaceful nuclear explosion in the foreseeable future" [65]. This may be interpreted as meaning that Brazil does not have nuclear explosives yet, that it has them but is not going to test them or that it merely is not going to test them now. Moreover, when Franco Netto speaks of "foreseeable future," just how far can he foresee?

Speculation that Brazil does intend to gain access to nuclear explosives, if it has not already done so, has been stimulated by the remarks of Brazilian politicians, journalists and scientists, many of whom have urged the government to build these explosives. These proponents of a nuclear explosive program can be placed in three categories. First, there are those who feel Brazil needs nuclear weapons because they are convinced that Argentina has, or shortly will have, these weapons. Typical of these advocates is Deputy Pedro Faria (MDB). In his opinion Brazil has no other option than to "move quickly" and prepare its own nuclear arsenal "since never in our history has the adversary [Argentina] been so menacing while continually denying its military effort" [66].

These advocates fear that Argentina, seeing Brazil become more powerful, will build a bomb to freeze the power relationship between the two countries. They believe that the Argentinians see the bomb as the great equalizer and they are concerned about how an Argentine bomb might be used. Some feel that the Argentinians will threaten to use nuclear weapons when confronted by Brazilian policies that they find distasteful, such as the Brazilian decision (despite Argentine opposition) to proceed with the Itaipú hydroelectric project. Psychological terror, these Brazilians maintain, will be used to impede Brazil's development unless the nation is capable of nuclear retaliation, since the United States cannot be expected to defend Brazil against nuclear blackmail. Those anxious for Brazil to become nuclear further note that Argentina is politically unstable and has been governed by people they are afraid are mentally unbalanced, such as Lopez Rega, or politically naive, such as Isabel Peron. They wonder whether Argentina might launch a nuclear attack against Brazil by mistake in a period of political turmoil or as the result of irresponsible decisions by national leaders.

Just as there is no way to verify whether Brazil is building nuclear explosives or intends to build them, there is no way of knowing what Argentina really is doing in this field [67]. The statements of Brazilian bomb advocates, however, may help to convince Argentina to obtain nuclear weapons if such a decision has not been made already.

The Argentine–Brazilian rivalry [68], in some ways, seems reminiscent of the early Cold War days. Brazil feels that Argentina is trying to organize the other Spanish-speaking countries for the purpose of preventing Brazil from pursuing its legitimate developmental goals. On the other hand, Argentina believes that Brazil is expansionist and must be contained. The media in both countries have reinforced these perceptions. Although a simplification of the situation, in such an environment a nuclear arms race is quite possible and could be initiated by rumors that one country or the other had or was planning to obtain nuclear weapons.

Since Brazil is more powerful than Argentina and is likely to become increasingly more so, it would seem that Brazil would be foolish to announce or demonstrate that it had nuclear explosives. Such an action would only encourage the Argentinians (if they needed any encouragement) to develop their own. Similarly, public pleas by Brazilian nuclear weapons advocates would seem ill-advised. Brazil's superiority would be diminished if both countries had nuclear arms; therefore, probably recognizing this, the Brazilians may keep secret their intentions and capabilities, unless they are certain that Argentina has nuclear weapons or is constructing them.

The second group advocating Brazilian acquisition of nuclear explosives does so from pride. While they believe that Brazil is destined to become a great power eventually, they also feel that international recognition can be won sooner if the country has access to nuclear explosives. They observe that Britain, although a declining economic power, continues to play an important role in world affairs largely because it is a nuclear power and they presume that with nuclear weapons Brazil will be taken more seriously by the developed nations before its economic development alone would warrant such consideration.

This group also holds that the developed nations opposing nuclear proliferation are trying to prevent countries like Brazil from becoming world powers. An eminent Brazilian diplomat, J. A. de Araujo Castro, articulately expressed the sentiments of this group during a 1971 address given at the Brazilian National War College. Because his remarks represent the attitude of the current Brazilian government, they merit rather lengthy quotation. Araujo said, in part:

> On various occasions... Brazil has sought to characterize what is now clearly looming as a firm and undisguised trend towards the *freezing of world power*. And when we speak of Power, we speak not only of Military Power, but also of Political Power, of Economic Power, of Scientific and Technological Power. The Non-Proliferation Treaty... is the main instrument of this policy of freezing of World Power.... The Treaty ... established distinctive categories of nations: one comprising strong and therefore adult and responsible countries, and another comprising weak and therefore non-adult and non-responsible countries. Contrary to all historical evidence, the Treaty starts from the premise that prudence and moderation are built-in features of Power. It institutionalizes inequality between nations and apparently accepts the premises that the strong countries will become ever stronger and the weak ones will grow ever weaker [69].

Plainly the position of the second group of nuclear-explosive advocates is based on nationalism. If other nations can have the bomb, it is morally unjust for Brazil not to have it. If the superpowers are trying to prevent Brazil from obtaining nuclear explosives it is because the rich nations fear alteration of the current international system, which favors them. The contention of nuclear-explosive proponents is that Brazil must demonstrate its determination to control its destiny, and possession of nuclear explosives will symbolize Brazil's independence to the world.

Whether these views deserve to be criticized depends upon whether they are perceived as ultranationalistic. Will the acquisition of nuclear explosives really promote Brazil's national interests or endanger them? Will it serve Brazil's national interests if world peace and security are placed in greater jeopardy? Will a nuclear Brazil actually expose the world to a peril?

The final group of Brazilians championing the procurement of nuclear explosives does so for developmental reasons. They reason that Brazil should avoid nuclear explosives no more than it should refrain from using dynamite for peaceful purposes just because it can be used in war. They argue that nuclear explosives are similarly fundamental for Brazil's development and insist that they would be used only for peaceful purposes.

Representative of this group is the rector of the State University of Campinas (Universidade Estadual de Campinas), Professor Zeferino Vaz. He feels that the development of a Brazilian nuclear explosive is possible and necessary. It is needed because of the multiplicity of applications that it can have in a country with an immense territory to conquer [70].

Nuclear explosives may indeed be of assistance in the Brazilian development effort. However, even if they are of only marginal value they still might be used. Brazil's recent history is replete with examples of "show" projects—Brasilia, the Trans–Amazon road, the Rio–Niteroi bridge—that were undertaken, at least in part, to demonstrate Brazil's ability to achieve dramatic results in the face of adversity. Many of these projects were to be panaceas for pressing strategic, political and economic problems, and aroused exaggerated expectations. Unfortunately, some who want Brazil to obtain nuclear devices for peaceful purposes also are playing "panacea politics." While this can help to mobilize the Brazilian population for constructive purposes, failure to achieve all that is promised could eventually lead to cynicism and make people unwilling to work cooperatively in the national interest.

Finally, Brazil has a large underemployed labor force. It may be that many of the engineering feats projected for nuclear explosives could yield other benefits if human labor was used instead.

Before leaving this aspect of Brazil's nuclear aspirations, two qualifications should be made. First, the three categories used above to classify Brazilian nuclear explosives advocates are not only ideal types but tell little about each group's influence. No doubt some individuals who favor Brazil's development of nuclear explosives do so for more than one reason, and there may even be a few whose enthusiasm is based on other considerations entirely. For example, some members of the Brazilian scientific establishment are likely to have selfish motives for wanting their country to have a nuclear explosives program.

Measuring political influence is never an easy matter and is particularly difficult in authoritarian Brazil. Crucial decisions in Brazil are made by a small group of people who have no need to explain the reasons for their actions to the public. However, a fair guess would be that if the Brazilian government ever decides to fabricate nuclear explosives—which it may have done already—perceptions about Argentina's intentions will carry the most weight, with nationalism coming second. Nevertheless, the scientific argument should not be discounted; with the exception of the military, technocrats are the most important element in the Brazilian political system.

Second, the influence of each argument could rise or fall as a result of internal changes in Argentina and Brazil. It is possible that the Argentine threat may fade, at least temporarily. As this is being written there are strong indications that the Argentine military may seize power once again. If this occurs, cooperation rather than competition may characterize Argentine–Brazilian relations. Some of the military regimes that ruled Argentina in the 1960s had close relations with Brazil, and during the Ongania years rumors even circulated about a secret Argentine–Brazilian nuclear development plan. There is, however, no guarantee that a military regime in Argentina would wish to improve relations with Brazil, nor would improved relations necessarily cause Brazil to terminate an ongoing nuclear explosives program, although such a program could become less of a priority. The Brazilians may realize that military rule (or a new, moderate Peronist government) may be unable to deal effectively with Argentina's enormous problems. Therefore, the suspicion would remain that future Argentine governments might turn to a confrontation policy.

In contrast to Argentina, and to most other countries as well, Brazil appears to be exceptionally stable. While it seems unlikely that political "decompression" will take place very soon, more nationalistic military officers may take charge. In this event Brazil could pursue an even more independent course in foreign affairs, and national pride could become a more important motive for the development of nuclear explosives.

As for the scientific argument, it too could become more persuasive. If, for example, another country's experiments with peaceful nuclear explosives were highly successful, this would help to fortify the contention that nuclear explosives can be utilized constructively and that Brazil would benefit from their development.

CONCLUSIONS

To summarize, Brazil, with German assistance, is engaged in a major effort to develop its nuclear industry. While many Brazilian notables, and others, are encouraging the government to fabricate nuclear explosives, the popular enthusiasm probably was not a significant factor in Brazil's decision to negotiate the German accord. Brazil wants nuclear reactors for its own sake as well as for the transfer of technology that will lessen the country's dependence on the industrial countries. Although this technology could be used to fabricate nuclear weapons, Brazil probably had this capability prior to signing the agreement.

It is not known whether Brazil has built or intends to construct nuclear explosive devices. However, the guess is that Brazil will develop such explosives if it has not done so already [71]. These devices might be demonstrated in order to increase Brazil's international status and to accomplish certain developmental goals, or they might be concealed for two reasons: (1) to avoid encouraging the Argentinians and other Latin Americans to fabricate similar devices and (2) so as not to stimulate retaliation by the developed countries upon whom Brazil is still somewhat economically dependent.

Many fears have been expressed about Brazil's nuclear intentions, and the Brazilians have done little to allay them. Brazil has not signed the Nuclear Nonproliferation Treaty and has demonstrated no inclination of doing so. Similarly, the Brazilians have not shown very much interest in participating in multilateral arrangements with other Latin American countries to control or develop nuclear energy. Moreover, Brazil has indicated that it does not believe that the Tlatelolco Treaty prohibits the explosion of nuclear devices for peaceful purposes in Latin America [72].

The Brazilian–German deal represents the first sale of a complete nuclear fuel cycle system to a nonnuclear country [73]. This contract [74], together with the Indian nuclear test, has been responsible in large part for several initiatives designed to prevent further nuclear proliferation [75]. The major exporters of nuclear equipment—the United States, Great Britain, West Germany, Canada, France and the Soviet Union—have met secretly to try to develop methods of limiting the sale of enrichment and reprocessing plants [76]. The two meetings that these countries have had so far ultimately may prove to have been more significant than the recently held NPT Review Conference. However, these countries alone will not be able to prevent nuclear proliferation. If an export ban or limitation were imposed by the major nuclear equipment exporters, it could be broken by Brazil, India or China [77].

The Ford administration has suggested that private firms be allowed to build enrichment plants in the United States in order to increase the supply of enriched uranium. At present the export demand for enriched uranium cannot be met because of a production shortage, and the United States government fears that this will stimulate non-nuclear nations to develop their own enrichment facilities [78]. However, even without the supply problem of American enriched uranium, the Brazilians still would have sought to attain nuclear independence, and no doubt other countries also have aspirations that go beyond the assurance that they will receive a steady flow of enriched uranium from a foreign source.

Finally, there also has been renewed interest in the establishment of a system of worldwide international or regional organizations that could build and supervise nuclear generating, enrichment and reprocessing plants and provide PNE services [79]. In Latin America the Organization for the Prohibition of Nuclear Arms (OPANAL) might undertake these activities. Unfortunately, countries such as Brazil that are interested in acquiring status and perhaps even nuclear weapons as a by-product of their nuclear development programs have not been enthusiastic about this proposal.

To conclude, the Brazilian–German accord may be advantageous to Brazil and Germany, but it may also provide a further incentive for the development of new means to stem nuclear proliferation. Of course, the accord also could serve as a model and be but the first of many such agreements resulting from intense export competition and national aspirations. If this occurs, the accord will have contributed to nuclear proliferation rather than retarded it.

NOTES TO CHAPTER ELEVEN

1. The Brazilian nuclear energy program, initiated after World War II, has been a source of domestic controversy since its inception. The primary dispute has concerned the selection of an appropriate nuclear development strategy. While some Brazilian politicians and scientists have urged that a self-reliant course be followed, others have advocated at least some dependence upon nations willing and able to furnish the country with advanced nuclear technology and equipment. The history of the discord surrounding the Brazilian nuclear energy program is to be found in the following: Eduardo Pinto, "Energia Atômica: Uma Velha História do Brasil," *Jornal do Brasil,* June 8, 1975, p. 2, Special Section; General Juarez Távora, *Átomos para o Brasil* (Rio de Janeiro: José Olympio Editôra, 1958), p. 357; Olympio Guilherme, *O Brasil e a Era Atômica* (Rio de Janeiro: Editorial Vitória, 1957), p; 317; "Política Nuclear: Os projetos, as alternativas e o mistério," *Visão,* September 9, 1974, pp. 25–36; and "Brazil: nuclear controversies," *Latin America,* March 14, 1975, pp. 82 and 83.

2. See Robert Gillette, "Nuclear Proliferation: India, Germany May Accelerate the Process," *Science,* vol. 188, no. 4191 (May 30, 1975), pp. 911–914; and Lewis H. Diuguid, "Brazil Nuclear Deal Raises U.S. Concern," *Washington Post,* June 1, 1975.

3. *Pravda* also denounced the treaty and reported that the Soviet government had told the Germans of their opposition to the transfer of nuclear technology to Brazil. Curiously, Brazil's traditional South American rival, Argentina, remained publically silent, and even the sensationalistic Argentine press gave little coverage to the story. This was unusual since the Argentine press generally gives a great deal of attention to international developments relating to Brazil. Perhaps the Argentine media was preoccupied with the internal crisis of June 1975.

4. The statements of Senator Pastore and his colleagues are printed in

the *United States Congressional Record,* vol. 121, no. 85 (June 3, 1975), pp. S9312–S9327. The Brazilians were outraged by the use of the word "backyard," which was translated into Portuguese as *quintal,* meaning either a backyard for playing games or a vile one.

4A. In an interview with the writer, a senior member of the Joint Atomic Energy Committee staff insisted that the committee had not been influenced by American nuclear technology exporters such as Westinghouse or General Electric.

5. The text of the agreement was not available when this chapter was being written during June 1975. Therefore, the writer has had to rely on German and Brazilian press releases and news accounts, and interviews with government officials.

6. This 625 megawatt plant is scheduled to begin operation in 1977.

7. The São Paulo electric company (Centrais Eléctricas de Sao Paulo–CESP) has been anxious to obtain nuclear plants for several years.

8. Eduardo Pinto, "Acordo prevê transferência de tecnologia para o Brasil," *Jornal do Brasil,* June 6, 1975.

8A. A German banking consortim has agreed to provide Brazil with a one billion five hundred million dollar loan for nuclear research and development. *Boletim Especial,* Embassy of Brazil, Washington, D.C., No. 18, 1975.

9. Gillette, p. 188.

10. It is possible that France will become involved in several other areas of nuclear development in Brazil not covered by the Brazilian–German agreement. Before the German agreement was completed France suggested that it would participate if international bids were conducted for the Brazilian nuclear program. France has had broad experience prospecting for uranium in Nigeria and Gabon and has undertaken extensive industrial tests of the gaseous diffusion process for uranium enrichment. The French also have indicated that they would be willing to monitor the builder of the German reactors.

11. Arlette Chabrol, "Acordo nuclear com Alemanha é completo," *Jornal do Brasil,* June 12, 1975, p. 5.

12. Brazil's aspirations with regard to the peaceful uses of nuclear energy have steadily increased. An account of the country's aspirations during the mid–1960s is contained in Luis Cintra do Prado, *Perspectivas da Energia Atômica no Brasil* (São Paulo: EDWART Livraria Editôra, 1966), p. 245. Two official reports list the latest known targets; however, these targets have no doubt been revised as a result of recent developments. The reports are the "Annual Report of the Nuclear Technology Company," reprinted in the *Correio Brasiliense,* March 24, 1975, pp. 4 and 5; and the *II National Development Plan (1975–1979),* (Rio de Janeiro: Federative Republic of Brazil, 1974), p. 140.

13. "Brasil possui 4 mil t de urânio," *Folha de São Paulo,* June 5, 1975, p. 22. It is not known whether all these deposits are rich enough to merit the establishment of mining operations.

14. "Bonn Sources See Concerns About Nuclear Agreement With Brazil Unfounded," *Relay from Bonn,* vol. VI, no. 22 (June 11, 1975), press release from the embassy of the Federal Republic of Germany, Washington, D.C.

15. *Ibid.*
16. *German Press Review,* Press Office of the Embassy of the Federal Republic of Germany in Washington, D.C., June 11, 1975.
17. "... Halting Wider Danger," *New York Times,* June 29, 1975, p. 14E.
18. Letter to the Editor from Dr. Steven J. Baker, Center for International Studies, Cornell University, *New York Times,* June 15, 1975, p. 16E.
19. The accord also will allow Germany to diversify the sources of its uranium supplies, most of which now come from the United States. Coal is currently Germany's most important energy resource, but the country has an ambitious nuclear power program and should have a nuclear generating capacity of 30,000 to 50,000 MW by 1983. Paul Kemezis, "Bonn and Brazil in Uranium Deal," *New York Times,* May 3, 1975, p. 1. By having their uranium enriched in Brazil, the Germans will also be able to circumvent the prohibition against the production of enriched uranium that was imposed on the country by the victors after World War II. "Brazil spells out its plans for nuclear energy development," *Latin America Economic Report,* vol. III, no. 24 (June 20, 1975), p. 1.
20. *Congressional Record,* vol. 121, no. 85, p. S9323.
21. "Brossard investe contra Pastore," *Jornal do Brasil,* June 5, 1975, p. 4.
22. Broadcast by TELAM, Buenos Aires, June 21, 1975.
23. Broadcast by the Brasilia Domestic Service, June 11, 1975.
24. "Brossard investe contra Pastore."
25. Broadcast by the Brasilia Domestic Service, June 12, 1975.
26. Broadcast by ANSA, Buenos Aires, June 6, 1975.
27. Broadcast by TELAM, Buenos Aires, June 21, 1975.
28. "Átomo Inarredável," *Jornal do Brasil,* June 6, 1975, p. 6.
29. Theophilo de Andrade, "Questão de confiança," *Estado de Minas,* June 22, 1975, p. 6.
30. "Brazil spells out its plans for nuclear energy development."
31. "Damy continua favorável a programa independente," *Jornal do Brasil,* June 8, 1975, p. 26.
32. "Brazil: nuclear controversies," *Latin America,* vol. IX, no. 11 (March 14, 1975), pp. 82 and 84; and "Política Nuclear: Os projetos, as alternativas e o mistério," *Visão,* September 9, 1974, pp. 25–36.
33. Ibid. Part of the controversy concerned access to enriched fuel. One side was satisfied to rely upon American supplies, another proposed the development of a centrifugal enrichment facility and yet another preferred dependence upon the gas diffusion method.
34. Marvine Howe, "Brazil, Racing for Growth, Seeks to Rely Less on U.S.," *New York Times,* July 2, 1975, p. 2.
35. TELAM broadcast, Buenos Aires, June 21, 1975.
36. For an example of Brazilian editorial opinion on the Indian test see "Jogo Atomico," *Jornal do Brasil,* May 21, 1974, p. 6. The editorial says that India's actions were justified and that both India and Brazil need nuclear explosives for peaceful purposes.
37. For a more detailed analysis of Brazil's faith in nuclear technology see H. Jon Rosenbaum and Glenn Cooper, "Brazil and the Nuclear Non–Proliferation Treaty," *International Affairs,* vol. 46, no. 1 (January 1970), pp. 74–90.

38. Murilo Melo Filho, "O Poder Nacional Precisa do Poder Atômico," *Manchete*, June 21, 1975, p. 14. This information should have come as no surprise to Brazil's leaders since it was public knowledge.
39. A comparison of the two countries' nuclear programs is contained in Juan E. Guglialmelli, "Argentina, Brasil y la Bomba Atomica," *Estrategia*, no. 30, September–October 1974, pp. 1–15.
40. A detailed discussion of this rivalry is to be found in H. Jon Rosenbaum, "Argentine–Brazilian Relations: A Critical Juncture," *The World Today*, December 1973.
41. "Francia negó que ayude a la Argentina a producir plutonio," *La Opinion*, June 6, 1975.
42. Several unconfirmed reports give some credence to Brazilian fears in this regard. For example, after the accord was signed, a story was widely circulated that Argentina had removed 50 kilograms of plutonium waste from the Atucha plant without IAEA detection. Evidently this would be sufficient to construct five atomic bombs. Fay Willey, "Who's Going Nuclear?" *Newsweek*, July 7, 1975, p. 27.
43. Jonathan Kandell, "Argentines Assay Their Atom Potential," *New York Times*, April 2, 1975, p. 2.
44. "Deputado avisa de ameaça nuclear en América Latina," *Folha de São Paulo*, May 27, 1975, p. 4.
45. "Argentina defende armamento," *Folha de São Paulo*, June 12, 1975, p. 10. The effect of this statement on the Brazilian decision to seek the accord with Germany probably should be discounted since it was made after the accord was announced.
46. Petrobrás, Brazil's state petroleum company, traditionally is conservative when estimating the value of new discoveries.
47. Some reports state that when the German deal was concluded, Roberto Campos and Delfim Netto, two former cabinet ministers, were seeking nuclear energy agreements with Britain and France, where they currently are ambassadors. Both men are suspected of having presidential ambitions and supposedly were anxious to increase their standing with the armed forces by helping to resolve the country's energy problems so that they would be allowed to seek the office.
48. Negotiations were initiated by Paulo Nogueira Batista, a Brazilian diplomat who later became head of Nuclebrás.
49. "O temor á independéncia," *Veja*, June 11, 1975, pp. 18–21.
50. Similar Soviet experiments evidently have been more successful.
51. "Para Ramalhete, acordo beneficiará toda a AL," *Folha de São Paulo*, June 5, 1975, p. 23. Also see Flavio de Almeida Salles, "Por que o Brasil precisa do átomo," *Folha de São Paulo*, June 12, 1975, p. 29.
52. Quoted in Melo Filho, p. 8.
53. During the same year Brazil purchased equipment from Germany for the enrichment of uranium. Transfer of the equipment reportedly was delayed by British discovery, but the gas centrifuge is now stored, unused, in São, Paulo. Guglialmelli, p. 2.

54. The previous year Brazil signed a bilateral accord with the United States under the "Atoms for Peace" program.
55. These reactors are fueled with enriched uranium from the United States and are subject to IAEA safeguards.
56. Brazil has concluded an agreement with Gulf General Atomic Company for development of a thorium-fueled, gas cooled fast breeder reactor. *Nuclear Engineering International,* March 1974, pp. 8–9.
57. The CBTN was founded in 1970 and became part of Nuclebrás when the latter was established in 1975.
58. The agreement with Portugal is for the purchase of uranium from that country.
59. This paragraph relies heavily on a superb paper entitled "Nuclear Proliferation in Latin America." It was presented by John R. Redick at the Conference on Latin America in the World System sponsored by the Georgetown University Center for Strategic and International Studies and held in Williamsburg, Virginia on April 11–13, 1975.
60. The last two research reactors were totally constructed in Brazil by Brazilians.
61. This consensus was reached at a conference on Brazil's International Role in the Seventies. See "Brazil's International Role in the Seventies: A Conference Report," *Orbis,* vol. 16, no. 2, p. 553.
62. Norman A. Bailey and Ronald M. Schneider, "Brazil's Foreign Policy: A Case Study in Upward Mobility," *Inter–American Economic Affairs,* vol. XXVII, no. 4 (Spring 1974), p. 18.
63. "A bomba ao acesso de muitos," *Jornal do Brasil,* May 20, 1974.
64. "Negó Brasil que proyecte fabricar armas nucleares," *La Prensa* (Buenos Aires), June 4, 1975, p. 1.
65. Broadcast by KYODO, Tokyo, June 5, 1975.
66. "Deputado avisa . . ." *Folha de São Paulo,* May 27, 1975.
67. Argentina already has the atomic bomb according to some reports. See, for instance, "Afirman en Brasil que Tenemos la Bomba Atómica," *La Razon,* June 19, 1975. Argentina might not announce her possession of nuclear weapons, if it did obtain them, for fear of alienating the United States Department of State. The Argentine economy is in trouble, and the Department of State has been helpful in getting American banks to finance the Argentine debt.
68. This rivalry is discussed further in Chapter Thirteen by C. H. Waisman.
69. "The United Nations and the Freezing of Power," *United States Congressional Record,* August 4, 1971, pp. H8014–H8016. Although Ambassador Araujo Castro shares with some Brazilian nuclear advocates a concern about the freezing of world power, it is not known whether or not he personally favors the acquisition of nuclear explosives.
70. "Reitor crê em bomba brasileira," *Jornal do Brasil,* June 8, 1975, p. 26.
71. The writer casually polled several American specialists on Brazilian foreign policy; almost all felt Brazil would develop nuclear explosive devices.
72. Brazil has ratified the treaty, but does not recognize it as being in force yet.

73. Brazil could get the material necessary to fabricate a nuclear explosive from either the enrichment or the reprocessing facility it is obtaining from the Germans.

74. Actually two documents were signed by Brazil: an agreement on technical cooperation in the nuclear field pursuant to the five items of the program negotiated by the Brazilian Ministry of Mines and Energy and Nuclebrás, and a tripartite agreement by which Brazil and West Germany assume the commitment to maintain their plans of cooperation toward peaceful goals with the IAEA, thus formally incorporating the "fail-safe" provisions into the agreement. Supplementary documents outline the implementation of the agreements, providing the technical details regarding equipment and assistance which will be given to Brazil.

75. The Brazilian–German accord also seems to have contributed to the resignation from the Department of State of Dr. Dixie Ray, former head of the United States Atomic Energy Commission. Dr. Ray had wanted American firms to be able to compete with those of Germany for the lucrative Brazilian contract by being permitted to offer the entire nuclear fuel cycle for sale.

76. David Binder, "6 Nuclear Lands Meet on Controls," *New York Times*, June 18, 1975, p. 1.

77. Some Canadians, among others, have suggested a moratorium on the sale of enrichment and reprocessing facilities until new export rules can be established. They would also like to have the nuclear equipment exporters agree not to make sales to unstable countries. However, it will be difficult to establish standards for determining the stability of nations. On the other hand, safeguards may become more dependable as the result of technical innovations. Fiber optic seals and instruments are being developed for monitoring what becomes of the plutonium output from nuclear reactors.

78. Paul Lewis, "A Nuclear Peace–Profit Motive," *New York Times*, July 20, 1975, p. 6F.

79. The United States and the Soviet Union agreed, as part of the NPT, to make PNE benefits available on a nondiscriminatory basis to non-nuclear countries. An identical commitment by the United States is contained in the Tlatelolco Treaty. Nevertheless, neither the United States nor the Soviet Union have provided PNE services when requested.

Chapter Twelve

Incentives for Nuclear Proliferation: The Case of Argentina

C.H. Waisman

INTRODUCTION

In this chapter, Argentine nuclear prospects will be discussed. I will focus on the external and internal factors which might induce this country to follow "the Indian option"—i.e., the development of a policy aiming at the actualization of Argentina's nuclear potential at the highest possible level.

Argentina is a relevant case for the analysis of the incentives for nuclear proliferation, at least for four reasons:

In the first place, Argentina is a near-nuclear country that has achieved a relatively high level of nuclear capability, and where the development of either "peaceful nuclear devices" or of a modest weapons program seems to be, as I will show, entirely feasible.

Second, Argentine nuclear policy is an interesting case for analysis. This policy has been a consistent one for a relatively long period, and its aims have been two: the maximization of autonomy in the development and control of nuclear programs, and the careful assessment of commitments, in order to keep nuclear options open. The consequences of this policy are well known: Argentina has insisted on the building of power plants which operate on the basis of natural, rather than enriched, uranium; she has not signed the NPT, and has signed but not ratified the Tlatelolco treaty.

Third, the case of Argentina is pertinent for the study of external factors that might influence a decision to go nuclear. This is so because of the strategic context in which Argentine foreign policy operates: two central components of this strategic context are the traditional rivalry with Brazil—another near-nuclear nation—and the existence of moderate hegemonial aims in relation to smaller neighboring states.

Finally, Argentine society is an extremely interesting case for the examination of internal incentives for nuclear proliferation. Argentina is perhaps the near-nuclear country that has the highest level of political instability. I will focus on two determinants of political instability: the relatively low rate of growth, and a pattern of stalemate among the different social and political forces that compete for political power—what can be called the hegemonic crisis of Argentine society. Their potential effect on nuclear policy will be assessed.

I will start with a short discussion of Argentine capabilities and policy in the nuclear field. Following that, I will assess the nuclear options that Argentina has at her disposal, focusing on the interaction with Brazil. Finally, internal incentives will be considered.

ARGENTINE NUCLEAR CAPABILITIES

In this section, I will refer to the factors that determine Argentine capabilities in the nuclear field: the availability of uranium, reactors and power plants, plutonium, and the economic and technological infrastructure for the development of nuclear programs.

In the first place, let us consider uranium reserves. According to most estimates, these reserves seem to be abundant. Argentina is ranked by Barnaby as the ninth country in the world, in terms of "cheap uranium reserves" [1]—i.e., reserves that can be extracted at less than $10 per pound of uranium. In 1970, it has been estimated that reserves of 10,000 tons exist. And, more recently, new deposits have been discovered [2]. There are three extracting and refining plants in Argentina, and a production level of 100 tons per year has been reported [3].

According to Redick's useful summary presentation of the country's facilities, Argentina has five research reactors, and at least four of them are in operation. The first reactor, which was purchased from the US in 1956, went in operation in 1958. It should be noted that, as early as 1957, the decision was taken to develop research reactors indigenously. Therefore, the remaining reactors have been designed by the National Atomic Energy Commission. The most recent one, which has a power of 5 MW, is the largest one in Latin America. All these are enriched uranium reactors, and the fuel is obtained from the United States [4]. IAEA safeguards have been in effect since 1963.

Argentina entered the stage of nuclear power production when, in 1968, a contract for the construction of the first plant was awarded to Siemens. It has been reported that lower bids had been offered by Westinghouse and General Electric, but Siemens was selected because its project was based on a natural uranium reactor [5]. When this plant went into operation, it also was the first one in Latin America. Heavy water is obtained from the US, but there are reports that Argentina plans to build its own heavy water producing plant [6]. The plutonium-producing reactor has been placed under IAEA safeguards. This plant is only the first one of a series that Argentina is planning to build. As other energy resources are scarce in the country, it is understandable that Argentine planners have high expectations in relation to the contribution that atomic energy can make in order to further the country's industrialization. Therefore, two other power plants, of 600 and 1,000 MWE, are planned [7].

We now turn to the crucial item: plutonium. Estimates of the total production capacity for the second half of the seventies vary according to the source, but the minimum estimate is well beyond the crucial threshold for weapons production. In order to interpret these figures, it should be remembered that it is estimated that 5—10 Kg of plutonium are enough for the manufacturing of an atomic bomb, and that the minimum figure can be *much* lower.

On the other hand, it has been reported that, since 1967, Argentina has a reprocessing plant for the extraction of plutonium, the first one in Latin America [8]. Even though the Argentine chemical separation plant is very small—it reportedly produces 0.2—0.5 Kg per year [9], it should be noted that, besides the traditional nuclear nations and the internationally operated plant in Belgium, only Argentina, West Germany, India, Italy and Spain have plants of this type [10].

The feasibility of implementing a nuclear program is dependent not only on the availability of plutonium, but also, and more importantly, on the general economic and technological capabilities of a country. The question is, therefore, whether Argentina has enough trained manpower and economic resources to carry out the development of nuclear explosives.

Let us consider manpower first. According to a United Nations report, a modest weapons program requires for its implementation the cooperation of approximately 1,300 engineers and 500 scientists [11]. This is a rough estimate, and it does not take into consideration qualitative factors. However, it is adequate for a very general assessment. In principle, it seems to me that this quantity of manpower is well within the range of Argentina's capability. Argentina has a very high rate of university students per 1,000 of population: In the midsixties, this rate was higher than for any of the West European countries, with the exception of the Netherlands [12]. According to one source, in 1965

Table 12—1. **Argentine Production Capacity of Plutonium, According to Different Sources**

Year For Which Estimate is Made	Estimated Production Capacity (Kg)	Source
1975	100	Gilinsky[a]
1976	145	U.N. Association[b]
1977	400	SIPRI[c]
1980	300	Gilinsky[a]

[a]V. Gilinsky, "Military Potential of Civilian Nuclear Power," in B. Boskey and M. Willrich eds, *Nuclear Proliferation: Prospects for Control* (New York: Dunellen, 1970), p. 42.
[b]UNITED NATIONS Association. *Safeguarding the Atom* (1972), p. 15.
[c]SIPRI, *World Armaments and Disarmament* (New York: Humanities Press, 1972), p. 296.

engineering and natural sciences majors made up 22.3 percent of the total student body, nearly 200,000 students [13]. Probably most of them did not graduate—Argentina has a very high college dropout rate—and most who did do not have skills which are directly relevant for a nuclear development program. However, these figures support the statement that the required manpower pool is either available or can be attained without much effort. In relation to the specialized skills required in the atomic field, finally, it should be noted that the NAEC has had its own training program in operation for many years. Therefore, I would preliminarily conclude that the required quantity of personnel is not a constraint. Qualitative factors should also be considered, though, and I am not competent to make an assessment in this regard.

Finally, we may turn to the question of cost. In this area, I also think that the development of a nuclear explosives program by Argentina is feasible. Beaton estimated in 1966 that a minimum weapons program, consisting of the production of five bombs per year, without the development of a specialized delivery system, would require the expenditure of $450 million over a ten year period, or $45 million per year [14]. This is also a rough estimate. However, and allowing for inflation, it is my impression that a program of this type is also well within Argentine capability. Approximately at the date of Beaton's estimate, the expenditure of $45 million per year would have represented 15 percent of the country's military budget, and 0.15 percent of its GNP [15]. The cost of a comprehensive program, which includes an enrichment plant, a hexafluoride conversion plant, a plutonium reprocessing plant, the manufacturing and possibly testing of weapons, and a delivery system, was estimated by the International Institute of Strategic Studies to be $1,750 million—1972 prices—for a force of 100 weapons [16]. If this expenditure is spread over the rest of the decade, it was estimated that Argentina is one of the countries that could afford a program of this type by diverting less than 1 percent of its GNP [17]. In summary: On the basis of the previous discussion, it seems safe to conclude that Argentina's capability in the nuclear field is such that the development of a modest explosives program is a feasible option. We may turn now to the country's nuclear policy.

ARGENTINE NUCLEAR POLICY

Argentina's interest in the nuclear field goes back 25 years. In 1950, the National Atomic Energy Commission was created. In 1951 a rather curious episode brought the intensity of Argentine nuclear efforts to world wide attention. R. Richter, an Austrian physicist who had worked on nuclear fusion in Nazi Germany, migrated to Argentina after World War II, and was appointed as head of a research facility built in the south of the country. In March of 1951, President Peron reported the discovery of a simple, cheap method to produce energy without uranium, by the controlled release of "thermonuclear reactions, identical to sun's processes" [18], and Richter announced that Argentina had discovered the H Bomb. The international scientific community was skeptical, and at the end it was discovered that the "discovery" was a bluff, and it is still unclear who was responsible for it. The following year, it was reported that Richter had been arrested, and that 300 staff members at the research facility had been dismissed [19].

After Peron's overthrow in 1955, there was a change of style in Argentine nuclear policy. Restraint and the stability of goals and leadership in the NAEC are the two remarkable new aspects of this policy. Stability of leadership is particularly impressive, in the light of the political instability that plagued the country in the post-1955 era. Perhaps this stability is due to the fact that the NAEC was placed under strong military influence. Another reason is possibly related to the technical constraints of a nuclear program: the successful operation of an agency of this type requires more long range planning and responsible leadership than most other agencies and programs. Hirschman has remarked that less developed countries are more efficient when operating complex industries, in which the tolerance for poor coordination or maintenance is relatively low [20]. In nuclear affairs, the penalty for poor performance is more serious, especially to a military mind, than in most other policy areas, so that the imperious need for the protection of the agency leadership from the vagaries of political instability is clearly perceived.

Perhaps the simplest characterization of Argentine policy can be made in terms of three traits: emphasis on the goal of peaceful application of atomic energy, development of an independent nuclear capability and careful assessment of international commitments in order to keep the country's options open.

The decision to select a natural uranium power plant provides a significant instance for the evaluation of the aims of Argentine nuclear policy. At that time, the decision to purchase the Siemens reactor—which, according to one source, was still untested [21]—was explicitly justified in terms of protecting national independence.

Another instance is, of course, Argentina's refusal to sign the NPT and to ratify the Tlatelolco Treaty. Argentine arguments—which, incidentally, are not very different than those advanced by Brazil—have stressed the need to have access to all the means that can enhance national security, and to preserve the country's ability to develop all the types of technology which might contribute to furthering its economic progress. The Argentine ambassador to the UN, who participated in the discussion of the NPT, has summarized the Argentine point of view in the following terms:

> The first value which cannot be overlooked is the protection of the security of each member of the international community. In the present state of the international panorama, with nuclear and non-nuclear countries, *nonproliferation freezes the existing situation.* Consequently, it is

necessary that the states which do not have nuclear weapons are given effective guarantees by those who, because of their greater military power, have a primordial responsibility in the nuclear sphere.

The second fundamental value that must be protected when nonproliferation is reached, is technological progress, particularly in the developing countries, because this progress is the key to all economic and social development [22]. (Emphasis added)

And, in the UN debate, the ambassador said, "[Argentina] ... cannot accept remaining in a subordinate position [in nuclear technology] especially when the bases of a nuclear technology, which our economic development requires, already exist in the country" [23]. Argentina stated several specific objections to the text of the NPT. The most important one has to do with the banning of "peaceful explosions." However, as stated above, Argentina has not ratified the Tlatelolco Treaty, which does permit these explosions. According to the Article 18 of this treaty, the contracting parties are allowed to "carry out explosions of nuclear devices for peaceful purposes, including explosions which involve devices similar to those involved in nuclear weapons" [24].

Finally, from time to time there are rumors that Argentina is carrying out a weapons program. For instance, at the 1969 Pugwash Conference it was reported by a participant that Argentina was mobilizing her nuclear physicists in a crash program for the production of weapons [25]. No confirming evidence has been presented so far.

In completing these references to Argentine nuclear policy, it is important to point out that Brazilian policy is not very different, particularly in relation to the establishment of international commitments. Brazil has not signed the NPT either, and is not bound by the provisions of the Talatelolco Treaty, in spite of having signed and ratified it. Unlike the other signatories, Brazil has not waived the requirements of Article 28, so that the treaty will not come into force for Brazil before all the Latin American nations have ratified it and the nuclear nations have ratified the additional protocols [26].

Having established that Argentina has the capability for the production of nuclear devices, and that her nuclear policy has consistently sought to enhance this capability, we may now turn to the question of the incentives that might prompt Argentina to follow "the Indian road." I will examine both the external factors—focusing on the interaction with Brazil—and the internal ones. In considering external incentives, it will be useful to discuss the theoretical options that are available to a near-nuclear country. After that, the potential impact of the competition with Brazil on Argentine nuclear choices will be assessed.

POLICY OPTIONS FOR NEAR-NUCLEAR COUNTRIES

For a near-nuclear country, the decision to develop, or abstain from developing, nuclear weapons becomes one of the central choices, to which most other decisions are subordinated. The possible options can be conveniently discussed as different degrees of actualization of the nuclear potential. In principle, it seems that three levels of actualization of this potential can be distinguished, and each of them becomes an "option." These three alternatives can be labeled the Canadian option, the Israeli option and the Indian option.

The Canadian option consists simply in the public and unconditional renounciation of the development of nuclear weapons. The nuclear potential is not actualized—except, of course, for nonmilitary applications. The maximum level of actualization, on the other hand, corresponds to what has been called the Indian option: the country that selects this policy aims at assembling and testing explosive devices—"peaceful" or otherwise—demonstrating in this way its nuclear capability. Finally, the Israeli option represents an intermediate strategy: it consists of the creation of all the material prerequisites for the production of weapons, short of the final stages of assembly—or, perhaps, including assembly and testing but (and this is the basic difference from the Indian option) without acknowledging that this decisive threshold has been passed.

Thus, the Canadian option consists of the renounciation both of producing weapons and of demonstrating them; the Israeli option renounces demonstrating the nuclear capability; and the Indian option utilizes both production and demonstration as instruments of policy. However, the use of the nuclear potential in order to influence the behavior of other countries is not strictly correlated with the degree of actualization of this potential. It is true that a country that selects the Canadian option is foregoing the possibility of utilizing its nuclear capability as an instrument of policy—except in a "moral" sense, by setting an "example" or, paradoxically, as in the case of Canada, by indirectly facilitating nuclear proliferation by disseminating atomic technology. But it is the possibility of utilizing nuclear power as a negative sanction that is forsworn. On the other hand, a country that follows the Indian option is, of course, able to affect the behavior of other states through the threat of utilization of nuclear sanctions. However, the reaction of other states potentially affected by the entry of the state in question into the nuclear club is more or less predictable: They will try to increase the level of their own defenses and/or look for a nuclear umbrella provided by a superpower. What characterizes these two polar options is the element of certainty they introduce into the international system: the rest of the world knows that Canada does not and India does have explosive devices, and the affected parties will react consequently. On the contrary, what is characteristic of the Israeli option is the uncertainty it introduces—and, consequently, the greater ability it gives to the country in question to control the behavior of other states. Potential antagonists know that the Indian option is still available to the country in question, and that the transition to it is contingent upon their behavior. Escalation is also a consequence, but the competition looks more like a chess match than an all-out race. The "Indianization" of the players is a possible outcome, but by no means the only one, for a "Canadian" alternative is possible as well.

These are, of course, three abstract options. The feasibility of each varies according to the constraints and opportunities provided by the strategic context faced by each country. Three crucial dimensions along which these context can be classified are the intensity of conflict, the presence or absence of hegemonial aims in relation to other countries and the existence or inexistence of credible third party protection. Thus, the Canadian option is facilitated by the context of Canada's foreign policy and is, within that context, the most rational one: it is a context characterized by the absence of direct adversary relationships with any other nation, the absence—and unfeasibility—of direct hegemonial aims in relation to other states and the reasonable certainty that Canada is covered by the US nuclear umbrella.

On the other hand, the Israeli and Indian options make sense in contexts characterized by higher degrees of conflict with other nations, environments in which hegemonial aims—either by the countries in question or by their antagonists—are possible or actual and by the absence of the reasonable expectation that absolute nuclear protection by a superpower is available. In addition, India is engaged in a direct adversary relationship with a nuclear power, so that the course of action she has followed could have been predicted for any country in that situation.

In the light of the previous discussion, it would seem that Argentine nuclear policy corresponds more to the model that has been labeled "the Israeli option" than to the other two. We should now examine the environment of Argentine foreign policy, and try to assess whether this option is the most adequate one. As stated above, we will focus on the interaction with Brazil.

THE ARGENTINE–BRAZILIAN INTERACTION

There has been a traditional rivalry between the two countries. Even though some direct friction exists—for instance, in relation to the control of the waters of the Parana river, which flows from Brazil into Argentina—the confrontation is mainly an indirect one. It is primarily a competition for influence on smaller nations: the buffer states that separate Argentina and Brazil—Bolivia, Paraguay and Uruguay—and on the Andean countries, especially Chile and Peru.

Sometimes, this rivalry is perceived as an autonomous one between two regional powers. On the other hand, some writers view it as a part of a broader confrontation. Thus, leading Argentine writers on international affairs have described Brazil as the *pais llave,* or the "key," of American influence in Latin America [27], while Argentina has been more traditionally linked to Europe. Among leftist writers, the application of the term "subimperialism" to describe the relationship between the United States and Brazil is standard. Brazil is thus seen as the main representative of American influence in the area.

Traditionally, the rivalry was geared toward the achievement of political and ideological influence over the smaller states. At different points in their recent history, both countries attempted to "export" some characteristics of their political system. During Peron's regime, from 1946 to 1955, the Argentine government tried to penetrate the labor and populist movements in neighboring countries. Similarly, today Brazil has established strong ties with military and authoritarian regimes in the area.

With the industrialization of both countries, this rivalry has acquired an increasingly economic content: smaller nations, besides being arenas for political influence, started to be seen as sources of raw materials and energy resources, and as markets for manufactured goods. This shift has been apparent especially in relation to Bolivia and Paraguay.

At this point, it might be useful to make some references to the balance of power between Argentina and Brazil, and to the shifts that occurred in the recent period.

In the past, and until Brazil's industrialization spurt in the late sixties, Argentina's high level of modernization roughly compensated for Brazil's greater size. Table 12-2 shows some basic indicators in the midsixties.

However, the balance has been shifting in Brazil's favor since the late sixties, due to the latter's spectacular rate of economic growth. While Argentina has not precisely been stagnated in the recent period, her rate of growth by no means matches that of Brazil. What is most important, Brazil's growth has been based on the expansion of the manufacturing sector. The relevant indicators are shown shown in Table 13-3.

Thus, Brazil's GNP in 1965 was 1.29 times larger than that of Argentina, and was 1.55 times larger in 1972 [28]. Brazil's population growth is an additional factor that should be taken into consideration, and it is particularly worrisome to Argentine policymakers. Due to the different levels of modernization in both societies, their demographic behavior is also very different: in 1960-1965, Argentine population grew at a rate of 1.6 percent per year, versus 3.3 percent

Table 12-2. Indicators of Modernization in Argentina and Brazil, Circa 1965

	Argentina	Brazil
Population	22 million	82 million
Total GNP	$17 billion	$22 billion
PK GNP	$770	$267
Percent Labor Force in Manufacturing	26	11
Percent Industrial Share of GDP	37	28
Energy Consumption	30 million tons	29 million tons
Literacy (percent)	91	61
Enrollment in Higher Education	11 per thousand	2 per thousand

Source: compiled from C. L. Taylor and M. C. Hudson, *World Handbook of Political and Social Indicators,* 2nd ed., (New Haven: Yale University Press, 1972).

Table 12-3. Annual Growth Rates of GDP and Manufacturing Sector, 1965-1973

	Growth Rate GDP			
	1965–1970	1971	1972	1973
Argentina	4.1	3.7	3.8	4.8
Brazil	7.5	11.3	10.4	11.4
	Growth Rate Manufacturing Sector			
Argentina	5.0	7.1	7.2	6.8
Brazil	10.3	11.3	14.1	15.8

Source: UN–ECLA. *Economic Survey of Latin America, 1973*.

for Brazil's. As a consequence, while in 1965 the total population of Brazil was 3.72 times larger than that of Argentina, the ratio between the two was 4.20 in 1972.

These shifts are viewed by Argentine analysts with the utmost concern. In 1970, the publishers of the leading journal on strategic affairs wrote: "A realistic analysis of the Argentine–Brazilian situation cannot hide that, in the long run, a disequilibrium in Brazil's favor can be produced, which may become the main threat to our national security" [29].

In addition, the establishment of military regimes in most countries of the area—at this moment *all* the smaller states mentioned previously have military or authoritarian regimes of different types—has been conducive to the development of economic and ideological links between Brazil and most of these countries. Therefore, the development of a siege mentality in the Argentine ruling strata is a strong possibility.

Given the characteristics of the situation that has been sketched above, it is not surprising that the "Israeli option" is the alternative that seems most attractive to Argentine—and perhaps Brazilian—policymakers. And this is really the option that best fits the interests of both countries. This is especially apparent if the strategic context is described in terms of continuous variables, rather than in terms of attributes. There is a latent conflict between Brazil and Argentina, but the level of conflict is low: a major direct confrontation between both countries is unthinkable in the foreseeable future. Besides, both countries' hegemonial aims are moderate—among other things, because their means are moderate at the world level and will remain so in the foreseeable future. Therefore, even if Argentina and Brazil adopt the Israeli option, neither country needs to reach the level of nuclear preparedness that Israel has reportedly achieved. Probably they will not go as far along the nuclear road: they can afford to be months away from the final stage of the production of weapons, rather than days away, or hours away, as in Israel's case.

Of course, this is still a delicate balance, and it is intrinsically unstable. An escalation can be triggered by minor changes in one country's perception of the level of preparedness of the other. However, it is probably clear to policymakers in both countries that a nuclear race, which would consume a progressively larger amount of economic resources, is not in the interest of either. In this connection, Argentine circumspection in nuclear matters seems particularly adequate.

INTERNAL INCENTIVES

In this section, the potential effect of two internal factors which produce political instability—the low rate of growth and the pattern of stalemate among different social forces—will be explored. It should be clear that the distinction between external and internal factors is artificial, for both are interrelated.

In discussing the internal inducements, the rivalry with Brazil should be seen not only as a background, but also as a potential scenario for the externalization of internal conflicts.

We may start with the rate of growth. In the first place, it should be clear, on the basis of the figures presented in Table 12—3, that Argentina is by no means a stagnated country. Moreover, if the low rate of population growth is taken into consideration, it turns out that the rate of per capita growth is relatively high. However, the political impact of these rates is contingent upon the relationship between aspirations and satisfaction in the population. A very low rate of growth can have no destabilizing consequences, if the population is "satisfied," either at a low or at a high level of expectations and achievements. On the other hand, a relatively high rate of growth may be associated with a high level of discontent, either because the increase in the amount of wealth is not large enough to match the expectations of the population, or because the distribution of wealth is very unequal. This is what happens in Argentina. It is a highly mobilized society in which many groups are dissatisfied with their level of reward. This discontent is aggravated by the relatively skewed pattern of distribution of wealth, which is caused by the antiquated land tenure system and and by other factors. The state of collective discontent which ensues is one of the direct determinants of political instability. However, before we consider the other factor—political stalemate—it might be useful to discuss the relevance of the low rate of growth in the light of the conflict with Brazil.

Argentina, from being considered one of the richest countries in the world in the pre–1930 period, slowly acquired consciousness of her status as a "less developed country" in today's world. According to Diaz–Alejandro, "As early as 1895, the Argentine per capita income was about the same as those of Germany, Holland, and Belgium, and higher than those of Austria, Spain, Italy, Switzerland, Sweden, and Norway" [30]. On the other hand, in the midsixties Argentina's per capita income ranked thirty-second in the world, and it was lower than that of any European country mentioned above, with the exception of Spain [31].

This relative retrogression should be seen in the light of the high level of aspiration of most groups in the society. This level of aspiration stems from two facts: in the first place, Argentina is to a large extent a nation of European immigrants who impregnated the culture of the country not only with a high level of expectations, but also with a high level of sensitivity to international comparisons of standards of living. Second, Argentine society had, until the fifties, a very high rate of mobility. It was not only a country where immigrants with high expectations migrated in search of opportunity, but also, to a large extent, a land of opportunity.

How can this state of collective dissatisfaction be related to nuclear policy? It could be argued that acute collective discontent might trigger a process of compensation: the country would try to compensate for her decreased economic and military capabilities—in relation to those of Brazil—by increasing its coercive potential, and by basing its foreign policy on negative, rather than positive, sanctions. If a mechanism of this sort operates, "going nuclear" would become a very attractive option. After all, there are other cases on record of countries that have tried to retain their prestige through the development of nuclear power.

An outcome of this sort would be consistent with a simple interpretation in terms of relative deprivation. In the long run, individuals and countries adjust their aspirations to their capabilities. In the realm of foreign policy, this means that the actual conduct of foreign policy tends to pursue goals that, in the long run, are attainable with the means at the disposal of the country. However, it is in the short run that gaps between capabilities and aspirations develop, and one of the consequences of these gaps is that they trigger international processes which make a re-equilibration between means and goals very difficult. Usually, an absolute or relative decrease in the level of capability is not immediately followed by an adjustment of the level of expectations. These are the situations in which uninstitutionalized behavior—i.e., behavior that does not correspond to normative expectations—tends to occur, both at the individual and systemic levels [32]. In foreign policy, these periods of discrepancy between means and goals are likely to trigger nonrational courses of action. And it might be argued that Argentina is, or in the near future will be, caught in this "relative deprivation trap."

How valid is a hypothesis of this type? In principle, an outcome of this sort cannot be dismissed, especially if the intensity of relative deprivation increases in the future. However, up to now the evidence is not consistent with the hypothesis. Slow growth and collective dissatisfaction have gone on for many years now, and Argentine foreign policy, at least as far as military and nuclear matters are concerned, has been a moderate one. Perhaps this is due to the slow and gradual character of the relative decrease in the level of capability. The evidence on the impact of relative deprivation consistently shows that it is when a sudden, drastic decrease occurs that uninstitutionalized behavior tends to take place

[33]. A gradual loss of capability, on the other hand, allows for a gradual readjustment in the level of expectations.

We may now turn to the consideration of the impact of the second determinant of political instability—the hegemonic crisis.

Argentine society has been described as a case of stalemate [34]. In a context of this type, different social forces of roughly similar power—armed forces, industrialists, big farmers, industrial workers—contend for political power in such a way that there are no stable winning coalitions: no social force or coalition of social forces is able to impose its hegemony for a long period, and each social force has a "veto power," for it has enough strength to overthrow the coalition that in a given moment controls the state apparatus.

This pattern results from the combination of high mobilization and the relatively low rate of growth. In a context of this sort, conflicts over the distribution of wealth and the control of political power become zero-sum situations. Most significant social groups feel deprived, and they confront each other in a naked conflict in which all resources are legitimate. The combination of high participation and low institutionalization that Huntington [35] has called "praetorian society" is the consequence: Argentina has had 13 presidents in the past 20 years. In this period, there was only one instance of constitutional transfer of power, and this was due to the death of the incumbent, Peron. Five presidents have been overthrown by military coups and there have been three transfers of power from military regimes to civilians.

In a context of this type, it might be argued, it is not inconceivable that a weak government—as all the Argentine governments in the recent period, both civilian and military, have been—might be tempted to utilize foreign policy for domestic purposes and increase the level of external conflict in order to increase internal cohesion. Particularly, the development of a weapons program would accomplish two objectives: it would dramatically enhance the legitimacy of the powerholders, and it would increase the costs of opposition to the regime, given the increased level of tension in the international environment in which the country operates.

Again, even though the possibility of a course of action of this type cannot be discounted, the likelihood of its occurrence is not validated by the record. Forty-five years of political instability in Argentina have not produced an adventurous foreign policy. Neither weak populist governments nor authoritarian military regimes have carried out policies of this type, even though opportunities in which these policies would have been effective certainly existed. The moderation of Argentine foreign policy can be seen in relation to what amounts to an extreme example—the case of the Malvinas Islands, called Falkland by the British. These are a group of islands off the Argentine Patagonia which are claimed by Argentina and occupied by Britain. The islands probably have a low economic and strategic value for Britain—especially in view of the reduction of her international commitments—and high emotional value for the Argentines. In addition, Argentina enjoys considerable international support for her claims, and the economic costs of a confrontation with Britain would not be very high for Argentina, due to the diminishing British role in Argentine industry and international trade. In this situation, and taking into consideration the redistribution of power that is taking place in the international system, a more belligerent Argentine policy would not be entirely ineffective. Further, the internal consequences should have been extremely attractive to many weak and illegitimate Argentine governments. However, a strategy of this sort has been avoided, and all the Argentine governments have instead followed a policy of quiet diplomacy.

The conclusion from this analysis, therefore, is that internal factors are not likely to induce a selection of the Indian option in Argentine foreign policy. The evidence suggests that there is a discontinuity between domestic and foreign policy, perhaps because the stakes are greater in the latter, and there is a clear perception that many of the crucial factors involved are beyond the country's control. Therefore, factors that are consequential at the internal level do not necessarily produce changes at the external one.

Finally, the discussion of the interactions between Argentina and Brazil suggests that external incentives for proliferation are potentially more significant than the internal ones. In the particular case of two rivals who select the "Israeli option," reciprocal uncertainty can have two opposite effects. On the one hand, it can be conducive to proliferation, for the perception of disequilibria can trigger a process of escalation whose control can be difficult. However, on the other hand, reciprocal uncertainty creates an environment which is conducive to the maximization of rationality, and the outcome could be the one that fits best the interests of both parties.

NOTES TO CHAPTER TWELVE

1. C. F. Barnaby ed, *Preventing the Spread of Nuclear Weapons* (London: Souvenir Press, 1969), Introduction.

3. T. R. Redick, *Military Potential of Latin American Nuclear Energy Programs* (Sage Professional Papers. International Studies Series, 1972), p. 13.

3. *1974 Commodity Yearbook* (New York: Commodity Research Bureau, 1974).

4. Redick, p. 11.

5. Ibid. p. 13; and *New York Times,* February 23, 1968.

6. *New York Times,* March 15–16, 1972, cited by Redick.

7. Redick, p. 13–14.

8. Redick, p. 18.

9. International Institute for Strategic Studies, *Strategic Survey 1972* (London, 1973), p. 73.

10. Ibid.

11. Barnaby, p. 28.
12. C. L. Taylor and M. C. Hudson, *World Handbook of Political and Social Indicators*, 2nd. ed. (New Haven: Yale University Press, 1972).
13. Oficina de Estudios Para la Colaboracion Económica Internacional, *Argentina económica y Financiera* (Buenos Aires, 1969), p. 49.
14. L. Beaton, "Capabilities of Non-Nuclear Powers," in Buchan A. ed., *A World of Nuclear Powers?* (Englewood Cliffs, N.J.: Prentice–Hall, 1966), p. 32.
15. Taylor and Hudson.
16. International Institute of Strategic Studies, p. 74.
17. Ibid.
18. *New York Times*, March 25, 1951.
19. *New York Times*, December 5, 1952.
20. A. O. Hirschman, *The Strategy of Economic Development* (New Haven: Yale University Press, 1958), p. 144.
21. Redick, p. 15.
22. J. M. Ruda, "La posición argentina en cuanto al Tratado sobre la No Proliferación de las Armas Nucleares," *Estrategia* (2, 9, 75–80), p. 77.
23. Ibid.
24. The test of the Tlatelolco Treaty can be found in G. Fischer, *The Non-Proliferation of Nuclear Weapons* (London: Europa Publications, 1971), pp. 249–265. Cfr. also A. Garcia Robles, *The Denuclearization of Latin America* (Carnegie Endowment for International Peace, 1967).
25. *New York Times*, October 27, 1969.
26. Redick, p. 28.
27. Cfr. for instance: Dirección de Estrategia, "Relaciones Argentino–brasileñas," *Estrategia* (1, 5, 48–57); and O. Camilion, "Relaciones Argentino–brasileñas," *Estrategia* (4, 21, 43–48).
28. Taylor and Hudson; and Publishing Sciences Group, *The World Factbook, 1974* (Acton, Mass., 1974).
29. Dirección de Estrategia, p. 51.
30. C. Diaz–Alejandro, *Essays on the Economic History of the Argentine Republic* (New Haven: Yale University Press, 1970), p. 1 *n*.
31. Taylor and Hudson.
32. On relative deprivation models, cfr. T. R. Gurr, *Why Men Rebel* (Princeton, N.J.: Princeton University Press, 1970), chapter 2.
33. Ibid.
34. For general discussions of the Argentine political system and the pattern of stalemate, cfr. T. di Tella, "Stalemate or Coexistence in Argentina," in J. Petras, and M. Zeitlin eds., *Latin America Reform or Revolution?* (New York: Fawcett, 1968); and G. A. O'Donnell, *Modernization and Bureaucratic–Authoritarianism: Studies in South American Politics* (Institute of International Studies, University of California, 1973).
35. S. P. Huntington, *Political Order in Changing Societies* (New Haven: Yale University Press, 1968).

Discussion Essay: The Typical Nuclear Proliferator

The realities of nuclear proliferation have encouraged the early, somewhat theoretical, discussions about the nature of a multinuclear world to become more specific—to confront actual cases of new or potentially new nuclear countries with their unique foreign policy goals, economic and technological capacities, and positions in the emergent strategic system.* The preceding case studies focus attention upon these attributes of the nuclear powers-to-be.

Discussions of the case studies at the conference were largely country-specific. Time constraints limited generalizations about "the proliferating country." In many ways this specificity mirrors the real world. Early references to the "Nth" nuclear power were appropriate so long as proliferation was just a possibility. Once the near-nuclear countries became identifiable, their unique attributes became more evident.

The question asked most frequently in the case studies, as by others who have written about proliferation, is "What is the probability that country 'X' will 'go nuclear'?" Following from that are the further questions of "What level of nuclear capability will be developed?" and "What military options will become available with nuclear weapons?" The three questions are interrelated. A nuclear power with a sophisticated delivery and storage system will use nuclear weapons in a different way than one with a vulnerable, first strike capacity (Rosen). An elementary classification scheme of nuclear states which relates weapons levels and types and strategy was suggested. Beyond these few classes, however, conference participants advocated still more specificity.

The cases illustrate the varying experiences and environments of the near-nuclear countries. No typical "Nth" country exists in terms of (1) a technical

---

*Three recently published volumes are exceptions to the general discussions of proliferation. They are: Robert Lawrence and Joel Larus, *Nuclear Proliferation: Phase Two* (Lawrence: University of Kansas Press, 1974); Geoffrey L. Kemp, Robert Pfaltzgraff and Uri Ra'anan, *The Superpowers in a Multinuclear World* (Lexington, Mass.: Lexington Books, 1974); and George H. Quester, *The Politics of Nuclear Proliferation* (Baltimore: The Johns Hopkins Press, 1973).

capacity to develop an indigenous nuclear weapons capability within a specific time period; (2) military-strategic needs leading to a nuclear conclusion; (3) the impact of internal politics upon its nuclear weapons programs; and (4) the command and control and safety and safeguards issues that a nuclear capability raises in a country-specific context. Discussion centered around the first three issues.

TECHNOLOGY AND MANPOWER

Whether or not a country develops the technology for a nuclear weapons program depends largely upon its ability to bring together engineers and technicians with the requisite skills ideally to reproduce the entire nuclear fuel cycle. In the near-nuclear countries where this capacity exists, neither resources nor information are overwhelming barriers. Natural uranium is widely distributed throughout the world and information about nuclear fuel cycles and reactor technology is widely published. In this environment, available skilled manpower can virtually ensure a nation's entry into the "nuclear club."

The necessary skills can be either borrowed or trained. Large numbers of scientists do cross national boundaries and make their skills available to states of which they are not nationals. Alternatively, potential nuclear states can and are taking advantage of training programs within the existing nuclear states. The second path has much to recommend it to a state which intends to develop nuclear weapons because the allegiance of the engineers becomes critical in a weapons program (Hohenemser).

The development of an indigenous group of nuclear engineers is a lengthy process, however, both in terms of numbers to be trained and of the organization of the engineers into a nuclear science program (Subrahmanyam). India's 1974 nuclear explosion came nearly 20 years after the beginning of the Indian Atomic Energy Commission.

Severe internal political conflict also inhibits indigenous training programs. South Africa, which has benefited both from borrowed talent and from training in foreign programs, is a case in point. The racial conflict within the country has implications for skilled manpower development. Projections of South Africa's skilled manpower needs indicate that increasing proportions of South Africa's pool of skilled manpower will be black (Feit). In 1970, whites composed 70 percent of the country's professional class. Assuming a professional class of about 20 percent of the work force in the year 2000, the total number of whites in the working population would fall short of meeting professional manpower needs by more than one-third (or 1.2 million persons). The strong link between foreign and domestic sources of conflict in South Africa is likely to cause white leaders to be hesitant of beginning a nuclear weapons program which might be staffed partly by blacks if the rest of the economy were not to be denuded of technicians (Feit).

Iran was the only country discussed in which manpower limitations were likely to be a decisive factor (Subrahmanyam, Hohenemser). Japan's technological capabilities could be developed rapidly (Gordon); Argentina's are presently being developed. Moreover, both Israel and South Africa are now capable of producing their own missile delivery vehicles (Rosen, Mittelman). The major determinants of future nuclear options thus become internal and external political constraints and opportunities.

STRATEGIC NEEDS AND PROLIFERATION

The major strategic issues revolve around (1) the position of the near-nuclear country vis-à-vis present nuclear states (for example, whether or not the near-nuclear country is under a nuclear umbrella in which it has a reasonable degree of confidence), and (2) the nature of any perceived threat(s) to its national security, or, conversely, the extent of the country's international ambitions."

The Proliferator and the Superpowers

Three of the countries occupy crucial positions in the strategic framework of the United States—Japan, South Africa and Iran. The strength of the United States' commitment to Japan lends credibility to US nuclear guarantees and, thus, operates as a disincentive to a Japanese weapons program (Gordon). The same argument was made for South Africa to the extent that: its ports serve as alternatives to those in Angola and Mozambique for Indian Ocean naval operations; it faces on the Cape of Good Hope; and it supplies mineral raw materials to the EEC (Mittelman).

Iran is a major focus of US interests among both the Indian Ocean states and the oil-producing nations. However, the Iranian government is not especially sanguine about the American umbrella (Chubin). American guarantees depend partly on the internal stability of each of the "protected" countries, and political leaders in the protected countries are aware of this (Feit). Iran is concerned about its indigenous defense capabilities in the regional context of the Arab-Israeli arms race and Soviet overflights.

South African observers point to the United States' "selective relaxation" policy toward the white regime and its willingness to provide South Africa with uranium enrichment facilities as evidence that South Africa's friendship is regarded as vital to the United States (Mittelman). But whether the posture is a sign that US nuclear forces would come to the support of South Africa in the event of a threat to the regime's security is another question.

To neighboring states, the presence of external nuclear forces in a non-nuclear state may be regarded as equivalent to that of an indigenous nuclear capacity (Subrahmanyam, referring to US forces in Japan). However, to the "protected" country, the cases are not so similar. For South Africa, nuclear weapons are not uniquely useful in terms of regional conflicts and the credibility of US nuclear protection relates to any future Soviet threat. To Japan, the distinction between indigenous capacity and external protection may become crucial because neighboring states have the possibility of developing nuclear weapons. Thus, on the one hand, the Japanese government's three non-nuclear principles and accepting external nuclear weapons on its territory are mutually compatible positions (Gordon). On the other hand, the Japanese government faces the prospect of eroding US support if it does develop a nuclear weapons capacity.

Which option the Japanese government chooses in the future depends largely upon what happens in the region—the Japanese government will "want to watch what happens in Korea and Taiwan" (Sato). If Japan does opt for indigenous nuclear weapons, it will probably be for a major second-strike capability as contrasted with India's low level ambiguity (Gordon).

Sources of Conflict

Questions of external protection inevitably lead into the identification of potential conflicts. The country case studies indicate wide variation in the types of major conflicts envisaged by the near-nuclear states. South Africa and Israel are both prepared for nonconventional warfare and regional conventional conflicts as well as the remote possibility of confrontation with a nuclear superpower.

Iran, Japan, Argentina and Brazil all tend to identify the potential sources of conflict as regional, and for the latter three those regional conflicts could potentially be nuclear exchanges. Iran is the only state among these to have signed the NPT. India is the only nuclear state in the region and neither Iran nor India publicly admits to potential conflict between them. Strategists argue that Iran's regional needs, primarily dominance in the Persian Gulf, can be met without nuclear weapons.

As long as South Africa and Israel are operating in nuclear-free regions, their conventional forces would be sufficient deterrent to likely opponents. Strategic debates within Israel center upon that country's conventional superiority (Kemp). In the foreseeable future, South Africa's likely protagonists among the black African states probably will not develop either a unified military policy toward South Africa nor a nuclear capacity singly or jointly. Moreover, the threat of regional conventional conflict may diminish in both southern Africa and the Arab–Israeli arenas (Feit, Tahtinen). If, instead of being surrounded by the "cordon sanitaire" of white states, South Africa is bordered by black states which are economically dependent upon it, informal settlements may be made (Feit). Similarly, if a limited accommodation is achieved in the Middle East, Israel's nuclear capability will no longer be useful in the regional context. In fact, even without a limited accommodation, the use of nuclear weapons on either side is most likely as a result of an irrational decision (Tahtinen), not from a deliberate calculation of tactical advantage.

INTERNAL POLITICS

Internal politics is perhaps the least clear of the factors leading to nuclear weapons decisions. Japanese public opinion has been polled regularly on the issue of exercising a nuclear weapons option and has shifted over time toward support for an independent nuclear capability (Gordon). The trend of Indian opinion also has been toward support for nuclear weapons.

The extent of public support for nuclear weapons development is not so clear in other cases. The Argentinian and Iranian regimes might decide to develop a nuclear capacity in the belief that it would create internal support (Kazemi, Waisman). Both regimes face serious internal opposition which might be divided by appealing to national pride through military strength. But these predictions are more hunch than science.

"Bomb lobbies" are another aspect of public support for nuclear weapons. Such a group exists in Japan (Gordon) and could develop elsewhere. For example, although an Iranian decision to go nuclear in the near future would probably be made by the shah, with minimal influence from specific groups, over time a nuclear lobby might well grow out of a nuclear energy program (Leitenberg).

The case studies each have weighed the acquisition of nuclear weapons by second order states from the standpoint of military benefits to be gained. To several, the costs of developing the potential to produce nuclear explosives appear to outweigh those benefits; to others, developing a nuclear capacity may seem the wisest path. To each, national interests rather than international protestations will be the determining elements in a nuclear decision. The character of the multinuclear world which is emerging, its military structure and overlapping patterns of strategic perspectives form a critical agenda.

Epilogue: The NPT Review Conference, Geneva, 1975 [1]

Onkar Marwah

According to Article VIII, paragraph 3, of the NPT which came into force on March 5, 1970:

> "Five years after the entry into force of this Treaty, a conference of the Parties to the Treaty shall be held in Geneva, Switzerland, in order to review the operation of this Treaty with a view to assuring that the purposes of the Preamble and the provisions of the Treaty are being realized...".

With preparatory work for the Conference undertaken during the early part of 1975, fifty-eight (out of ninety-four) states parties to the NPT convened at Geneva from May 5 to May 30, 1975. Among them, the following special categories of participants were in attendance at the NPT review [2]:

1. Seven signatory but non-ratifying states parties to the NPT (Egypt, Japan, Panama, Switzerland, Trinidad and Tobago, Turkey and Venezuela).
2. Seven nonsignatory, non-ratifying states in the capacity of Observers (Algeria, Argentina, Brazil, Cuba, Israel, South Africa and Spain).
3. Two regional organizations in the capacity of Observer Agencies (the Agency for the Prohibition of Nuclear Weapons in Latin America (OPANAL), and the League of Arab States).
4. The United Nations Organization and the International Atomic Energy Agency.
5. Several national and international non-governmental organizations (NGOs).

Thirty-six NPT adherents did not attend the Review Conference. However, the major potential nuclear weapons states such as Argentina, Brazil, Iran, Israel, Japan and South Africa did go to Geneva in one capacity or another. Formally and informally, several of these states raised their concerns about the NPT regime's functioning with the Treaty's co-sponsors (the USA, the USSR, and Britain). The three co-sponsors, in turn, restated their unwillingness to assume further obligations within the frame work of the NPT as had been urged upon them since the Treaty was first drafted (1968) and offered for signature in final form (1970).

After the initial plenary session, the conferrees set up, selected the membership and assigned tasks to the First and Second Main Committees of the Review Conference. Accepting the provisional agenda conceived by the Preparatory Committee, the General Committee of the Conference divided items for evaluating the operation and role of the NPT between the two Main Committees, as follows:

(a) to Committee I:

   A. Implementation of the provisions of the Treaty relating to the non-proliferation of nuclear weapons, disarmament and international peace and security:
      1. Articles I, II and III (paragraphs 1, 2 and 4) and Preambular paragraphs 1–5
      2. Article VI and Preambular paragraphs 8–12
      3. Article VII
   B. Other provisions of the Treaty
   C. Resolution 255 (1968) of the United Nations Security Council
   D. Acceptance of the Treaty by States
   E. Measure aimed at promoting a wider acceptance of the Treaty.

(b) to Committee II:

   A. Implementation of the provisions of the Treaty relating to peaceful applications of nuclear energy:
      1. Article III and Article IV
      2. Article V and Preambular paragraphs 6 and 7.

The general debate in plenary session took place between May 6 and May 12. Going into session from May 12 to 23, the two main committees submitted their reports at the plenary meeting which resumed on May 26.

Differences between the NPT co-sponsors and the non-nuclear weapon states were apparent in the working sessions of the committees. These divergencies were reflected in the reports submitted to the plenary group. Indeed, in the general debate that followed (May 26 to 29) it seemed to many delegates that the conference might terminate without a consensual resolution to vindicate its labors. Several attempts at a compromise declaration failed, the drafts being rejected as either too strong by the NPT co-sponsors, or too weak by the non-nuclear states.

Eventually, and at the last moment, Mrs. Inga Thorsson of Sweden, President of the Conference, took the initiative to prepare a carefully drafted resolution. This document outlined the concerns of the non-nuclear states but refrained from a strong castigation of the Treaty's co-sponsors. Specifically, it did not require of the latter a contractual fulfillment of obligations under the NPT regime as urged by the non-nuclear states. With interpretative annexure and oral statements expressing the reservations of numerous States Parties, Mrs. Thorsson's draft was formally adopted as the Final Declaration of the Conference [3].

PRE-CONFERENCE DISCUSSIONS

Between March 1970, when the Treaty was offered for signature, and the convening of the Review Conference in May 1975, the NPT regime was formally extended through two lines of activity. Additional states were invited to accede to the Treaty and nuclear safeguard arrangements were urged for adoption under the aegis of the International Atomic Energy Agency.

Ninety-five states had signed and ratified the Treaty by the time the Geneva Conference ended (May 30, 1975). Sixteen states signed but did not ratify the NPT. And Thirty-eight states had chosen to remain outside of the NPT regime altogether [4]. Most states irrespective of their attitude to signature or nonsignature accepted the *technical* arguments in favour of instituting safeguards controls. Some, however, averred on the *political* follow-through which required safeguards implementation on a discriminatory basis between nuclear and non-nuclear states (e.g., India).

Several governmental and nongovernmental organizations met in the months preceding the Conference in order to propose changes in the NPT regime. The United Nations Institute for Training and Research (UNITAR) sponsored one of these meetings. A panel of experts consulted with representatives of thirty-five national delegations to the UN, at UNITAR headquarters (New York) on April 29, 1975. Eleven questions were framed to define the major tasks for the NPT Review [5].

1. What kind of security guarantees for non-nuclear weapon states would be effective? To what extent are they feasible? Should the nuclear-weapon states undertake not to use or threaten to use nuclear weapons against non-nuclear Parties to the Treaty?
2. To what extent has Article III of the NPT, dealing with IAEA safeguards, been implemented? What more should be done by supplier and receiver countries?
3. Should the peaceful activities of nuclear-weapon states also be subject to IAEA safeguards?
4. What conditions should be attached to the supply of nuclear materials to non-nuclear states that are not party to the NPT?

5. How can the problem of non-governmental proliferation be dealt with?
6. How can the peaceful uses of nuclear energy be promoted in the developing states as required by Article IV of the NPT?
7. What are the prospects for peaceful nuclear explosions? Should an international regime for PNEs be worked out now? If so, how and when should negotiations be undertaken?
8. How can the obligations undertaken in the Preamble and Article VI concerning the ending of nuclear-weapon tests, the cessation of the nuclear arms race and disarmament be made more specific and be implemented? What are the expectations for progress for a Comprehensive Test Ban (CTB) and in the Strategic Arms Limitation Talks (SALT)?
9. Should there be periodic review conferences? If so, how frequently?
10. Should some consultative machinery be created to supervise or report on the implementation of the provisions of the NPT?
11. What other measures can be taken to strengthen the effectiveness of the NPT without formal amendment to the NPT?

A second meeting was organized at Divonne, France, September 9–11, 1974, jointly by the United States Arms Control Association and the Carnegie Endowment for International Peace. The participants—experts, government officials acting in an individual capacity, eminent citizens from seventeen countries, and representatives of several well-known international non-governmental organizations—concluded that the NPT, while imperfect, remained a useful and workable instrument. Its faults resulted from failures in implementation rather than in the provisions themselves. The policy recommendations that followed, were [6]:

1. The nuclear weapon parties should take seriously their obligations to work toward disarmament, in particular by reducing numbers of nuclear delivery vehicles, limiting further missile flight testing, and negotiating a comprehensive (not a threshold) nuclear test ban treaty. The Threshold Treaty was viewed as a "disheartening step backward."
2. Better security assurances should be provided for non-nuclear powers, nuclear-weapon states should pledge themselves not to use or threaten to use nuclear weapons against non-nuclear states party to the Treaty; the establishment of nuclear-free zones should be encouraged.
3. The benefits of peaceful nuclear explosions (PNEs) are questionable. A study should be commissioned by the Secretary General of the UN of all the implications of such programs. Meanwhile, a moratorium on further PNE tests should be imposed.
4. Henceforth, provisions of nuclear supplies and technology should be made only to those recipients, whether or not parties to the NPT, which agree to place all their nuclear facilities under IAEA safeguards.
5. The system of safeguards should be strengthened and made universal, with

increased attention given to physical security applied to all materials. The problem of diversion of nuclear materials by terrorists, organized criminals and other non-governmental groups should be given immediate attention.

Certain major themes predominated in the pre-Conference discussions:

(a) security guarantees for states abjuring from nuclear weapons production;
(b) a comprehensive nuclear test ban?
(c) barriers against a 'peaceful-nuclear-explosions' path to national nuclear weapons programs; and
(d) a mandatory and perhaps time-bound acceptance by the superpowers for curbing their own nuclear-weapons-systems arms race.

While some of the debate tended to be rhetorical, a significant portion was concerned with pragmatic concern. Attempts were made to identify the obstacles confronting a serious nuclear-weapons limitation program. The substance of the unofficial nuclear debate could be viewed at one level as indicative of informed 'world opinion' on the dangers resulting for the international system from the (potential) spread of nuclear weapons. At another level, the preparatory meetings served as a socialization pressure on the national delegations attending the Geneva Review to respond to internationalist needs. At a third level, the debate that preceded and surrounded the Review provided a basis for evaluating the achievements of the various Conference participants.

REVIEWING THE NPT: CONFERENCE POLITICS

The real crescendo of the discussions originated within the western nuclear states and the non-nuclear states [7]. The non-nuclear states' assessments of the NPT regime (which took into account the official concerns of the nuclear powers) could be summarized as follows:

1. Because ninety-five states had joined the NPT regime, a stable basis existed for extending its application among the hold-outs.
2. The objections expressed by the hold-out states were not entirely without foundation, especially with regard to their 'imbalance of obligations' charge against the nuclear-weapon states.
3. The Indian nuclear explosion of 1974 was both fact and example of how other hold-out states could proceed to acquire nuclear weapons technology without sanctions or the breach of international law.
4. In spite of the possibility that many states might have national reasons for going nuclear and may eventually not be prevented from doing so, the present onus for some concrete action to reduce the probability rested on the superpowers.

5. There were scientific, technical, economic and indeed strategic grounds on which the superpowers could conceive at least minimalist changes in their weapons systems procurement and postures without endangering their intra-security objectives.
6. The system of safeguard procedures developed and administered through the International Atomic Energy Agency had worked well within limited financial means, needed to be strengthened, and could become the basis for a world nuclear policy system of acceptable controls, technical advise and expertise, and a positive means for extending the beneficial aspects of nuclear technology to all states.

As with other international conferences that have taken place in recent years (e.g., those dealing with the world's environment, population, food needs or energy supplies) divisibilities between the (nuclear) 'haves' and 'have-nots' became apparent in the initial stages of the NPT Review. As in the other international conferences, a number of the richer non-nuclear states took the side of the nuclear have-nots. By the end of the first week of the general debate, the Conference President identified the set of concerns which a majority of the states attending the Review seemed to share as:

"the general view among a majority of the non-nuclear-weapon States ... is very clearly that the nuclear-weapon States have not achieved results ... in efforts toward genuine nuclear disarmament. It seems ... that an enlightened world opinion, reflected in this case, in statements by non-nuclear-weapon States, rather impatiently awaits concrete and binding results of on-going bilateral negotiations, aiming at ending the quantitative and qualitative arms race, and reducing substantially the levels of nuclear armaments. Many have referred to the need for a time-table for results to be achieved through these negotiations. The agreement on a comprehensive test ban is clearly recognized as a most decisive element in these efforts. A least common denominator is apparent in the statements: Article VI must be implemented, in letter and spirit. I believe that this implies the existence of a strong moral pressure on the nuclear-weapon States to prove to the world, not only their genuine will, but also their capacity for disarmament" [8].

The rest of the Conference President's evaluation of the general debate could be summarized as evidencing the need for [9]:

1. A comprehensive test ban.
2. A pledge which would bind the nuclear-weapon states against the use or threat of nuclear weapons in respect of non-nuclear parties to the NPT.
3. The application of similar safeguard procedures to nuclear and non-nuclear states, whether parties to the NPT or not.
4. The provision of greater nuclear technology assistance to the developing countries in the peaceful uses of nuclear energy.

5. The establishment of an international peaceful explosions regime such that individual states, including the nuclear powers, could be dissuaded from embarking on their own peaceful explosions programs.
6. The acceptance of an international agreement to provide for the physical security of nuclear reactors, and the storage, handling and transport of fissile materials.
7. An agreement to maintain a continuous follow-up of the first NPT review and its implementation.

The coalitions of pressure on the Treaty's co-sponsors were varied and not subject to an easy depiction by the geographical, ethnic, level-of-income or ideological persuasion of the states participating in the Review. Among the non-nuclear-weapon states, prominent roles were played by Iran, Mexico, Nigeria, the Philippines, Rumania, and Yugoslavia, particularly in the Committee sessions. In company with other states, the former's representatives early on presented two draft protocols to the NPT. These sought to link an automatic suspension of underground nuclear weapons tests with a fifty percent reduction of the ceilings on superpowers' strategic delivery vehicles as agreed upon at Vladivostok, to the stage at which the number of parties signing the NPT reached one hundred. The Iranians objected to an apparent effort by the Treaty's co-sponsors aimed at blurring the distinction between nuclear and conventional arms restraint obligations in implementation of Article VI of the NPT. A proposal for the establishment of a fixed time-table to achieve specific measures of disarmament in specific areas of weapons systems, was put forward by a combination of Sweden, Iran and Australia.

Thereafter, Rumania proposed that the nuclear-weapon states undertake not to threaten the use of nuclear weapons against non-nuclear weapon states, nuclear-free zones be established, and the arrangement subsequently respected by the nuclear powers. Mexico, Nigeria, the Philippines and South Korea then tabled a resolution, requesting the creation of a special nuclear fund from which assistance could be made available to non-nuclear states in the peaceful applications of nuclear technology. Another suggestion, made by Mexico and seven other states, related to the provision of peaceful nuclear-explosives experiments by the nuclear-weapon states for others. None of these proposals were acceptable to the Treaty's co-sponsors.

The Rumanians, members of the Warsaw Pact, were unusual among the Soviet bloc representatives in their numerous agreements with resolutions and protocols tabled by nonwestern non-nuclear-weapon states. Stating their reservations to the Final Declaration of the Conference in an interpretative annexure, the Rumanians felt that the document evaluated "the past in an over-optimistic manner," that the text was "exceedingly unbalanced," that the developing countries had not received any meaningful assistance in the peaceful uses of nuclear technology, and that "the vertical proliferation of nuclear weapons and the nuclear arms race have continued and even accelerated" [10].

The Yugoslavs exhibited a more critical attitude to the inability of the nuclear-weapon states in carrying out their obligations within the NPT prior to the Conference. They also assessed the superpowers' Conference roles as the maintenance of the same effective postures. The Conference's accomplishments were evaluated skeptically, and the comments were more pungent than those of any other state—not excluding the nonwestern states. It was charged that the nuclear-weapon states had (a) not discontinued their nuclear arms race, (b) not stopped nuclear weapons tests, and (c) continued their own weapons proliferation. Further, that the superpowers" . . . and the States sharing their views have made an effort to preserve the NPT as an instrument by which they will retain all the advantages which the Treaty offers them." The Yugoslav assessment was that the Conference had "failed to reach a consensus both in the informal working groups and in the Committees on any substantive issue;" that the responsibility for the same rested "primarily with the nuclear-weapon States—the Depositaries;" and that their Government "would re-examine its attitude toward the Treaty and . . . draw corresponding conclusions" [11].

The major effort of the non-nuclear states was directed toward restitution in terms of Article VI of the NPT, i.e., the provisional undertaking by the superpowers to curb the vertical proliferation of their nuclear weapons. The Treaty had previously urged the superpowers to decelerate their nuclear arms race. Now the non-nuclear states asked that the superpowers assent, through some international agreement, to formalizing the demand. The measure itself could be modest, but it should nonetheless begin the process of translating superpower intent and non-nuclear states' faith into practice. However, while the latter tended to interpret their suggestion as minimalist, the former evaluated it as an interference which was both artificial and unreal in the context of world realities.

From among the NPT's sponsors, the significant roles were obviously those of the United States and the Soviet Union. The methods adopted by them for interaction with the Treaty's other signatories were varied. Due probably to systemic reasons of lesser control over its allies the US, in denial or affirmation of other states' proposals, had to state its views through the agency of its own representatives. The Soviet Union, on the other hand, was able to maintain a distant profile and had its positions expressed by representatives from Bulgaria, the German Democratic Republic, Hungary, Poland, Czechoslovakia and Outer Mongolia.

Differences in style of participation between the American and Soviet representatives at the Conference overlay a mutuality of interests which was apparent at the NPT Review. While there was no concatenation of their approaches during the Conference, the superpowers (along with Britain) had held a private meeting in London in April 1975. It is possible, therefore, that their disagreements, if any, had been ironed out prior to the Geneva meeting. In any case, there were no clashes between US and Soviet responses to other signatories' demands during the Review—in the plenaries, during Committee sessions, or within the deliberations of the informal working groups.

Epilogue: The NPT Review Conference 309

The US and the USSR had expressed the official limits of their flexibilities and the frameworks of their approaches prior to convening in Geneva [12]. In informal soundings from members of the two superpowers' at the Conference, and from those of others who interacted with the latter in the restricted Committee and Working Groups, a profile emerged of the US and Soviet assessments of nuclear proliferation problems. They could be summed up as follows:

1. Vertical proliferation was not as salient or as overwhelming an issue as it was made out to be. Both the superpowers had an unavoidable need to maintain their strategic flexibilities and weapons parities.
2. It was a major achievement that the Strategic Arms Limitation Talks between the US and USSR had been put on a more or less permanent footing.
3. Not enough was being said about the dangers resulting from a horizontal increase in nuclear-weapon states. The concerns were real and substantive in this category.
4. There was a real problem in that several nations were on the threshold of exploding nuclear devices. But the potential for nuclear weapons spread could not be solved by any NPT in extremely conflicted situations, e.g., if the Arabs and the Israelis sought nuclear weapons, or the Pakistanis chose to follow the Indians.
5. Article VI of the NPT was *an* issue, not *the* issue within the framework of nuclear non-proliferation concerns.
6. The superpowers had been subjected to criticism in the deliberations of the Conference, while the nonaligned states had remained safe. For instance, not a single state had thought it worthwhile to mention the name or demonstrate against the action of India in becoming the world's sixth nuclear power.
7. Nuclear deterrence between the US and the USSR had worked in the European strategic context. But the extension of similar doctrines between states with low level nuclear capabilities would only create problems in the international system. Neither viability nor credibility could really attend the latter states' acquisition of nuclear weapons.
8. The Treaty co-sponsors' efforts in providing nuclear technology to others and in building up the system of safeguards procedures through and outside the IAEA, were being underrated. Much had been done, and much would continue to be done.
9. Canada's transfer-of-technology to India had been a very sobering experience. Substantial lessons existed therein for all nuclear-technology exporting countries.

In formal annexures which were attached to the Conference's Final Declaration, the Soviet and American positions did show minor divergences of opinion. Hence, their responses were addressed to such aspects of the general debate as were separately important to either superpower. Thus, the Soviet Union stated without qualification that it was "... in favor of the cessation of all (nuclear weapon) testing, including underground testing, by all States:" whereas the American position was that "... any treaty or agreement on nuclear-weapons testing must contain provisions for adequate verification and must solve the problem of peaceful nuclear explosions." Further, the Soviet Union, "which is an advocate of nuclear disarmament," could not accede to measures in that field as prejudiced "the security of the parties concerned." Expressing the view that "the basic problems of disarmament—and especially of nuclear disarmament—can only be solved with the participation of all the nuclear powers," the Soviet interpretative statement went on to observe that "Security Council resolution 255 (1968) and the declaration made by the Soviet Union, the United States of America and the United Kingdom in relation thereto constitute an effective instrument for guaranteeing the security of Parties to the Treaty not possessing nuclear weapons."

In contrast, the American annexure to the Final Declaration did not touch upon the issues elaborated by the Soviets. Its emphasis appeared to be in the area of nuclear free zones which, many delegates had asked, be respected through superpower pledges. The American response was that "The creation of the (nuclear free) zone should not disturb necessary security arrangements; and provision must be made for adequate verification." It was felt, moreover, that it would be unrealistic "to expect nuclear-weapon States to make implied commitments... before the scope and content of any nuclear-free zone arrangement are worked out [13].

There seemed to be a consensus among the nuclear-weapon states that political commitments on their part within the clauses of the NPT were less important than agreement on technical curbs to prevent the spread of weapons technology. Thus, they tended to avoid discussions relating to issues in the former category. Effort was concentrated, instead, in areas of improving safeguards applications, coordinating nuclear technology suppliers' export policies, and persuading non-nuclear states to accept multinational nuclear fuel cycle supply centers. Not opposed to such proposals, the non-nuclear states objected to greater stringencies being placed exclusively upon their use of nuclear materials. They asked that the nuclear curbs also be made applicable in some formal measure on the nuclear-weapons powers. Reverse suggestions of the preceding type did not find favor with the Treaty's co-sponsors.

In retrospect, it seems that the US and the Soviet Union were cast in unenviable roles at the Conference. Assuming generous intent on their behalf, a formal venture to convince other states against the acquisition of nuclear weapons could not be undertaken easily. News media information and knowledge about the 'state of the art' in the superpowers arms race were too expressive and open. The basic thrust of the NPT—that the first venture to curb others' nuclear weapons proclivities be approved while the second reality of the superpowers nuclear arms race was sanitized from its application—had been rejected by many states from the very beginning of the NPT regime. By 1975 the vast gap in capabilities that would exist between the superpowers and new entrants to the nuclear club was paralleled by the gap that had grown between superpower intent and practice, and potential nuclear-weapon states' perceptions of the same. Rigidities existed on both sides. Choosing between the two viewpoints remains a function of individual assertion or preference.

ASSESSING THE NPT REVIEW CONFERENCE

The results of the Conference may be viewed as mixed. Its deliberations could not have been mandatory. Some meaning may be attached to the fact that in spite of reservations many of the convening states expressed the desire to hold another review conference in the future. It is equally meaningful that most states acclaimed the system of safeguards administered by the IAEA and asked for its extension and stabilization on an on-going basis. In structural terms, the Review Conference may have performed tasks that it was not required to accomplish, e.g., in bringing into greater focus areas of disagreement between nuclear-non-nuclear states where negotiations can in the future be pursued outside the NPT context. In the aftermath of the Conference, some activity of the preceding nature appears to have taken place, especially in the area of more stringent nuclear technology export guidelines. On the other hand, it may have convinced some states, e.g., Rumania and Yugoslavia, that the Treaty regime had run its course, and nothing further was possible for them within its terms.

Whichever conviction appears more justified, it is equally possible that the Geneva events remain too close in time for definitive judgments to be made about its failure.

BEYOND THE NPT REVIEW

There is a view that the NPT Conference amounted to little more than a cosmetic exercise staged by the Treaty's co-sponsors. Fueling such negative opinions are the recent 'fall-out" of claims to nuclear weapons capability made on behalf of various states. Spokesman for South Korea, Israel, Taiwan and South Africa have been attributed ingenuous statements about ability—but lack of present intent—to proceed with the acquisition of nuclear weapons. Japan has reaffirmed its earlier stance of not ratifying the NPT. Brazil has reached agreement with West German firms for the inception of a complete nuclear fuel industry within its territory, including the provision of uranium enrichment plants and techniques [14]. Meanwhile, all the major nuclear technology exporting countries are engaged in a series of secret-agenda meetings, unusual in that the Soviet Union and France are also participating. Little is known of the range of issues being discussed, but it is probable that they range from benign proposals for multinational nuclear fuel-cycle supply centers, to restrictive and perhaps penal nuclear export regulations [15]. Many nuclear technology recipient states may be apprehensive that nuclear export controls might lead to the formation of a worldwide nuclear exports cartel.

While conclusive evidence is not available, there are unconfirmed reports that among the strategic experts of the Treaty's co-sponsors the NPT is now evaluated as having run its course as a preventive measure. It is alleged that the two superpowers have accepted that a certain level of nuclear proliferation is likely to occur and cannot be stopped. Accordingly, their advisory personnel have been directed to consider the changes in states' interaction once such a situation comes about [16].

NOTES TO EPILOGUE

1. Formal title of the Conference: *Review Conference of the Parties to the Treaty on the Non-Proliferation of Nuclear Weapons.*

2. A list of all delegations to the Conference is contained in the Conference document NPT/Conf/Inf.5 (Annex VI of *Review Conference of the Parties to the Treaty on the Non-Proliferation of Nuclear Weapons. Final Document. Part I. Geneva, 1975.* NPT/Conf/35/1 (hereafter referred as "NPT Final Document").

3. Annex 1 of the "NPT Final Document." (See Appendix of this book.) The reservations and 'interpretative statements' were so profuse that in effect they negated the Final Declaration. (See, William Epstein, *Retrospective On The NPT Review Conference: Proposals For The Future.* The Stanley Foundation, Muscatine, Iowa. Occasional Paper No. 9, 1975, pp. 14–15.)

4. For the current status of the Non-Proliferation Treaty as regards categories of accession or non-accession to its regime, see the Appendix of this book.

5. From *Non-Proliferation Today. Issues Before the Non-Proliferation Treaty Review Conference,* No. 1, May 7, 1975. (Issued by the Carnegie Endowment for International Peace) p. 4. (The issues were not rank-ordered according to importance.)

6. Adapted from *Non-Proliferation Today.* No. 4, May 21, 1975 (op. cit.), p. 2.

7. Thirty-seven non-governmental organizations (NGO's) were accredited observers at the NPT Review in Geneva. Many of them are located in the affluent western, or nuclear-weapon, states (e.g., the United States, Britain, the Scandinavian countries, and Canada) and have been prominent and active on a worldwide basis over the years. During the conference proceedings, NGO representatives lobbied with their national delegations for affirmative action to meet the objections of the hold-out countries and release a path for extending the NPT regime. Their efforts were not particularly successful.

8. *Statement by Mrs. Inga Thorsson, President of the Review Conference of the Parties to the Treaty on the Non-Proliferation of Nuclear Weapons, at the Conclusion of the General Debate, 12 May 1975* (NPT/Conf/26. 26 May, 1975), pp. 1–2.

9. *Ibid. (passim).*

10. Rumanian Interpretative Statement in *"NPT Final Document,"* Part I. NPT/Conf/35/I. Annex II, pp. 23–24.

11. Yugoslav Interpretative Statement in *"NPT Final Document,"* Part I. NPT/Conf/35/I.

12. For the official U.S. position, see (1) U.S. Arms Control and Disarmament Agency, *Non-Proliferation Treaty Review Conference* (Washington, D.C., U.S. Arms Control and Disarmament Agency, Publication 79, April 1975); (2) U.S. Arms Control and Disarmament Agency, *The Non-Proliferation Treaty and Our Worldwide Security Structure.* Remarks by Fred C. Ikle, Director, U.S. Arms Control and Disarmament Agency at a conference on the Non-Proliferation Treaty sponsored by the Arms Control Association, Washington, D.C. (Washington D.C., U.S. Arms Control and Disarmament Agency, April 9, 1975); and (3) U.S. Arms Control and Disarmament Agency, *The Second Nuclear Era.* Remarks ddlivered by Dr. Fred C. Ikle, Director, U.S. Arms Control and Disarmament Agency before the 15th Annual Foreign Affairs Conference, U.S. Naval Academy, Annapolis, April 23, 1975. (Washington, D.C., U.S. Arms Control and Disarmament Agency, April 1975.) (Dr. Fred C. Ikle was the leader of the U.S. Delegation to the NPT Review at Geneva.)

Soviet official positions on such issues are less easily available. However, for an analysis which could be viewed in that light, see: A. Kalyadin, "The Struggle for Disarmament: New Perspectives," *Instant Research on Peace and Violence* (special issue, "The Non-Proliferation of Nuclear Weapons), 1/1975. Tampere Peace Research Institute, pp. 24–34.

13. Citations from U.S. and USSR. *Interpretative Statements in "NPT Final Document,"* Part I. NPT/Conf/35/I. Annex II, pp. 30–31; and 28–30, respectively.

14. More recent disclosures affirm that West Germany has been secretly engaged in supplying sensitive nuclear enrichment technology to South Africa. See the *New York Times,* October 12, 1975.

15. See, "U.S. Seeks Atom-Export Pact in 1975," the *New York Times,* September 23, 1975; and "The Export of Nuclear Technology," *Special Report,* No. 9. Bureau of Public Affairs. U.S. Department of Statement, Washington, D.C., October 1974. (This document contains excerpts of speeches made since the NPT Conference, by U.S. Secretary of State Kissinger; U.S. Arms Control and Disarmament Agency Director Fred C. Ikle; U.S. Under Secretary of State Joseph Sisco; and (former) Chairman of the (then) U.S. Atomic Energy Commission, Dr. Dixy Lee Ray.)

16. Possibly apocryphal, the post-Conference evaluation of the NPT Review by its President, Swedish Ambassador Inga Thorsson, was critical of the roles adopted in Geneva by the Treaty's co-sponsors. See Senator Edward M. Kennedy, "The Nuclear Non-Proliferation Treaty," *Congressional Record,* July 30, 1975, S14462–14464.

Appendixes

Appendix A

Treaty on the Non-Proliferation of Nuclear Weapons

The States concluding this Treaty, hereinafter referred to as the "Parties to the Treaty",

Considering the devastation that would be visited upon all mankind by a nuclear war and the consequent need to make every effort to avert the danger of such a war and to take measures to safeguard the security of peoples,

Believing that the proliferation of nuclear weapons would seriously enhance the danger of nuclear war,

In conformity with resolutions of the United Nations General Assembly calling for the conclusion of an agreement on the prevention of wider dissemination of nuclear weapons,

Undertaking to cooperate in facilitating the application of International Atomic Energy Agency safeguards on peaceful nuclear activities,

Expressing their support for research, development and other efforts to further the application, within the framework of the International Atomic Energy Agency safeguards system, of the principle of safeguarding effectively the flow of source and special fissionable materials by use of instruments and other techniques at certain strategic points,

Affirming the principle that the benefits of peaceful applications of nuclear technology, including any technological by-products which may be derived by nuclear-weapon States from the development of nuclear explosive devices, should be available for peaceful purposes to all Parties to the Treaty, whether nuclear-weapon or non-nuclear-weapon States,

Convinced that, in furtherance of this principle, all Parties to the Treaty are entitled to participate in the fullest possible exchange of scientific information for, and to contribute alone or in cooperation with other States to, the further development of the applications of atomic energy for peaceful purposes,

Declaring their intention to achieve at the earliest possible date the cessation of the nuclear arms race and to undertake effective measures in the direction of nuclear disarmament,

Urging the cooperation of all States in the attainment of this objective,

Recalling the determination expressed by the Parties to the 1963 Treaty banning nuclear weapon tests in the atmosphere, in outer space and under water in its Preamble to seek to achieve the discontinuance of all test explosions of nuclear weapons for all time and to continue negotiations to this end,

Desiring to further the easing of international tension and the strengthening of trust between States in order to facilitate the cessation of the manufacture of nuclear weapons, the liquidation of all their existing stockpiles, and the elimination from national arsenals of nuclear weapons and the means of their delivery pursuant to a treaty on general and complete disarmament under strict and effective international control,

Recalling that, in accordance with the Charter of the United Nations, States must refrain in their international relations from the threat or use of force against the territorial integrity or political independence of any State, or in any other manner inconsistent with the Purposes of the United Nations, and that the establishment and maintenance of international peace and security are to be promoted with the least diversion for armaments of the world's human and economic resources,

Have agreed as follows:

ARTICLE I

Each nuclear-weapon State Party to the Treaty undertakes not to transfer to any recipient whatsoever nuclear weapons or other nuclear explosive devices or control over such weapons or explosive devices directly, or indirectly; and not in any way to assist, encourage, or induce any non-nuclear-weapon State to manufacture or otherwise acquire nuclear weapons or other nuclear explosive devices, or control over such weapons or explosive devices.

ARTICLE II

Each non-nuclear-weapon State Party to the Treaty undertakes not to receive the transfer from any transferor whatsoever of nuclear weapons or other nuclear explosive devices or of control over such weapons or explosive devices directly, or indirectly; not to manufacture or otherwise acquire nuclear weapons or other nuclear explosive devices; and not to seek or receive any assistance in the manufacture of nuclear weapons or other nuclear explosive devices.

ARTICLE III

1. Each non-nuclear-weapon State Party to the Treaty undertakes to accept safeguards, as set forth in an agreement to be negotiated and concluded with the International Atomic Energy Agency in accordance with the Statute of the

International Atomic Energy Agency and the Agency's safeguards system, for the exclusive purpose of verification of the fulfillment of its obligations assumed under this Treaty with a view to preventing diversion of nuclear energy from peaceful uses to nuclear weapons or other nuclear explosive devices. Procedures for the safeguards required by this article shall be followed with respect to source or special fissionable material whether it is being produced, processed or used in any principal nuclear facility or is outside any such facility. The safeguards required by this article shall be applied on all source or special fissionable material in all peaceful nuclear activities within the territory of such State, under its jurisdiction, or carried out under its control anywhere.

2. Each State Party to the Treaty undertakes not to provide: (a) source or special fissionable material, or (b) equipment or material especially designed or prepared for the processing, use or production of special fissionable material, to any non-nuclear-weapon State for peaceful purposes, unless the source or special fissionable material shall be subject to the safeguards required by this article.

3. The safeguards required by this article shall be implemented in a manner designed to comply with article IV of this Treaty, and to avoid hampering the economic or technological development of the Parties or international cooperation in the field of peaceful nuclear activities, including the international exchange of nuclear material and equipment for the processing, use or production of nuclear material for peaceful purposes in accordance with the provisions of this article and the principle of safeguarding set forth in the Preamble of the Treaty.

4. Non-nuclear-weapon States Party to the Treaty shall conclude agreements with the International Atomic Energy Agency to meet the requirements of this article either individually or together with other States in accordance with the Statute of the International Atomic Energy Agency. Negotiation of such agreements shall commence within 180 days from the original entry into force of this Treaty. For States depositing their instruments of ratification or accession after the 180-day period, negotiation of such agreements shall commence not later than the date of such deposit. Such agreements shall enter into force not later than eighteen months after the date of initiation of negotiations.

ARTICLE IV

1. Nothing in this Treaty shall be interpreted as affecting the inalienable right of all the Parties to the Treaty to develop research, production and use of nuclear energy for peaceful purposes without discrimination and in conformity with articles I and II of this Treaty.

2. All the Parties to the Treaty undertake to facilitate, and have the right to participate in, the fullest possible exchange of equipment, materials and scientific and technological information for the peaceful uses of nuclear energy. Parties to the Treaty in a position to do so shall also cooperate in contributing

alone or together with other States or international organizations to the further development of the applications of nuclear energy for peaceful purposes, especially in the territories of non-nuclear-weapon States Party to the Treaty, with due consideration for the needs of the developing areas of the world.

ARTICLE V

Each Party to the Treaty undertakes to take appropriate measures to ensure that, in accordance with this Treaty, under appropriate international observation and through appropriate international procedures, potential benefits from any peaceful applications of nuclear explosions will be made available to non-nuclear-weapon States Party to the Treaty on a nondiscriminatory basis and that the charge to such Parties for the explosive devices used will be as low as possible and exclude any charge for research and development. Non-nuclear-weapon States Party to the Treaty shall be able to obtain such benefits, pursuant to a special international agreement or agreements, through an appropriate international body with adequate representation of non-nuclear-weapon States. Negotiations on this subject shall commence as soon as possible after the Treaty enters into force. Non-nuclear-weapon States Party to the Treaty so desiring may also obtain such benefits pursuant to bilateral agreements.

ARTICLE VI

Each of the Parties to the Treaty undertakes to pursue negotiations in good faith on effective measures relating to cessation of the nuclear arms race at an early date and to nuclear disarmament, and on a treaty on general and complete disarmament under strict and effective international control.

ARTICLE VII

Nothing in this Treaty affects the right of any groups of States to conclude regional treaties in order to assure the total absence of nuclear weapons in their respective territories.

ARTICLE VIII

1. Any Party to the Treaty may propose amendments to this Treaty. The text of any proposed amendment shall be submitted to the Depositary Governments which shall circulate it to all Parties to the Treaty. Thereupon, if requested to do so by one-third or more of the Parties to the Treaty, the Depositary Governments shall convene a conference, to which they shall invite all the Parties to the Treaty, to consider such an amendment.

2. Any amendment to this Treaty must be approved by a majority of the

votes of all the Parties to the Treaty, including the votes of all nuclear-weapon States Party to the Treaty and all other Parties which, on the date the amendment is circulated, are members of the Board of Governors of the International Atomic Energy Agency. The amendment shall enter into force for each Party that deposits its instrument of ratification of the amendment upon the deposit of such instruments of ratification by a majority of all the Parties, including the instruments of ratification of all nuclear-weapon States Party to the Treaty Treaty and all other Parties which, on the date the amendment is circulated, are members of the Board of Governors of the International Atomic Energy Agency. Thereafter, it shall enter into force for any other Party upon the deposit of its instrument of ratification of the amendment.

3. Five years after the entry into force of this Treaty, a conference of Parties to the Treaty shall be held in Geneva, Switzerland, in order to review the operation of this Treaty with a view of assuring that the purposes of the Preamble and the provisions of the Treaty are being realized. At intervals of five years thereafter, a majority of the Parties to the Treaty may obtain, by submitting a proposal to this effect to the Depositary Governments, the convening of further conferences with the same objective of reviewing the operation of the Treaty.

ARTICLE IX

1. This Treaty shall be open to all States for signature. Any State which does not sign the Treaty before its entry into force in accordance with paragraph 3 of this article may accede to it at any time.

2. This Treaty shall be subject to ratification by signatory States. Instruments of ratification and instruments of accession shall be deposited with the Governments of the United States of America, the United Kingdom of Great Britain and Northern Ireland and the Union of Soviet Socialist Republics, which are hereby designated the Depositary Governments.

3. This Treaty shall enter into force after its ratification by the States, the Governments of which are designated Depositaries of the Treaty, and forty other States signatory to this Treaty and the deposit of their instruments of ratification. For the purposes of this Treaty, a nuclear-weapon State is one which has manufactured and exploded a nuclear weapon or other nuclear explosive device prior to January 1, 1967.

4. For States whose instruments of ratification or accession are deposited subsequent to the entry into force of this Treaty, it shall enter into force on the date of the deposit of their instruments of ratification or accession.

5. The Depositary Governments shall promptly inform all signatory and acceding States of the date of each signature, the date of deposit of each instrument of ratification or of accession, the date of the entry into force of this Treaty, and the date of receipt of any requests for convening a conference or other notices.

6. This Treaty shall be registered by the Depositary Governments pursuant to article 102 of the Charter of the United Nations.

ARTICLE X

1. Each Party shall in exercising its national sovereignty have the right to withdraw from the Treaty if it decides that extraordinary events, related to the subject matter of this Treaty, have jeopardized the supreme interests of its country. It shall give notice of such withdrawal to all other Parties to the Treaty and to the United Nations Security Council three months in advance. Such notice shall include a statement of the extraordinary events it regards as having jeopardized its supreme interests.
2. Twenty-five years after the entry into force of the Treaty, a conference shall be convened to decide whether the Treaty shall continue in force indefinitely, or shall be extended for an additional fixed period or periods. This decision shll be taken by a majority of the Parties to the Treaty.

ARTICLE XI

This Treaty, the English, Russian, French, Spanish and Chinese texts of which are equally authentic, shall be deposited in the archives of the Depositary Governments. Duly certified copies of this Treaty shall be transmitted by the Depositary Governments to the Governments of the signatory and acceding States.

In witness whereof the undersigned, duly authorized, have signed this Treaty.

Done in triplicate, at the cities of Washington, London and Moscow, this first day of July one thousand nine hundred sixty-eight.

Appendix B

Final Declaration of the Review Conference of the Parties to the Treaty on the Non-Proliferation of Nuclear Weapons

PREAMBLE

The States Party to the Treaty on the Non-Proliferation of Nuclear Weapons which met in Geneva in May 1975, in accordance with the Treaty, to review the operation of the Treaty with a view to assuring that the purposes of the Preamble and the provisions of the Treaty are being realized,

Recognizing the continuing importance of the objectives of the Treaty,

Affirming the belief that universal adherence to the Treaty would greatly strengthen international peace and enhance the security of all States,

Firmly convinced that, in order to achieve this aim, it is essential to maintain, in the implementation of the Treaty, an acceptable balance of mutual responsibilities and obligations of all States Party to the Treaty, nuclear-weapon and non-nuclear-weapon States,

Recognizing that the danger of nuclear warfare remains a grave threat to the survival of mankind,

Convinced that the prevention of any further proliferation of nuclear weapons or other nuclear explosive devices remains a vital element in efforts to avert nuclear warfare, and that the promotion of this objective will be furthered by more rapid progress towards the cessation of the nuclear arms race and the limitation and reduction of existing nuclear weapons, with a view to the eventual elimination from national arsenals of nuclear weapons, pursuant to a Treaty on general and complete disarmament under strict and effective international control,

Recalling the determination expressed by the Parties to seek to achieve the discontinuance of all test explosions of nuclear weapons for all time,

Considering that the trend towards détente in relations between States provides a favourable climate within which more significant progress should be possible towards the cessation of the nuclear arms race,

Noting the important role which nuclear energy can, particularly in changing economic circumstances, play in power production and in contributing to the progressive elimination of the economic and technological gap between developing and developed States,

Recognizing that the accelerated spread and development of peaceful applications of nuclear energy will, in the absence of effective safeguards, contribute to further proliferation of nuclear explosive capability,

Recognizing the continuing necessity of full co-operation in the application and improvement of International Atomic Energy Agency (IAEA) safeguards on peaceful nuclear activities,

Recalling that all Parties to the Treaty are entitled to participate in the fullest possible exchange of scientific information for, and to contribute alone or in co-operation with other States to, the further development of the applications of atomic energy for peaceful purposes,

Reaffirming the principle that the benefits of peaceful applications of nuclear technology, including any technological by-products which may be derived by nuclear-weapon States from the development of nuclear explosive devices, should be available for peaceful purposes to all Parties to the Treaty, and

Recognizing that all States Parties have a duty to strive for the adoption of tangible and effective measures to attain the objectives of the Treaty,

Declares as follows:

PURPOSES

The States Party to the Treaty reaffirm their strong common interest in averting the further proliferation of nuclear weapons. They reaffirm their strong support for the Treaty, their continued dedication to its principles and objectives, and their commitment to implement fully and more effectively its provisions.

They reaffirm the vital role of the Treaty in international efforts to avert further proliferation of nuclear weapons,

to achieve the cessation of the nuclear arms race and to undertake effective measures in the direction of nuclear disarmament, and

to promote co-operation in the peaceful uses of nuclear energy under adequate safeguards.

REVIEW OF ARTICLES I AND II

The review undertaken by the Conference confirms that the obligations undertaken under Articles I and II of the Treaty have been faithfully observed by all Parties. The Conference is convinced that the continued strict observance of these Articles remains central to the shared objective of averting the further proliferation of nuclear weapons.

REVIEW OF ARTICLE III

The Conference notes that the verification activities of the IAEA under Article III of the Treaty respect the sovereign rights of States and do not hamper the economic, scientific or technological development of the Parties to the Treaty or international co-operation in peaceful nuclear activities. It urges that this situation be maintained. The Conference attaches considerable importance to the continued application of safeguards under Article III, 1, on a non-discriminatory basis, for the equal benefit of all States Party to the Treaty.

The Conference notes the importance of systems of accounting for and control of nuclear material, from the standpoints both of the responsibilities of States Party to the Treaty and of co-operation with the IAEA in order to facilitate the implementation of the safeguards provided for in Article III, 1. The Conference expresses the hope that all States having peaceful nuclear activities will establish and maintain effective accounting and control systems and welcomes the readiness of the IAEA to assist States in so doing.

The Conference expresses its strong support for effective IAEA safeguards. In this context it recommends that intensified efforts be made towards the standardization and the universality of application of IAEA safeguards, while ensuring that safeguards agreements with non-nuclear-weapon States not Party to the Treaty are of adequate duration, preclude diversion to any nuclear explosive devices and contain appropriate provisions for the continuance of the application of safeguards upon re-export.

The Conference recommends that more attention and fuller support be given to the improvement of safeguards techniques, instrumentation, data-handling and implementation in order, among other things, to ensure optimum cost-effectiveness. It notes with satisfaction the establishment by the Director

General of the IAEA of a standing advisory group on safeguards implementation.

The Conference emphasises the necessity for the States Party to the Treaty that have not yet done so to conclude as soon as possible safeguards agreements with the IAEA.

With regard to the implementation of Article III, 2 of the Treaty, the Conference notes that a number of States suppliers of nuclear material or equipment have adopted certain minimum, standard requirements for IAEA safeguards in connexion with their exports of certain such items to non-nuclear-weapon States not Party to the Treaty (IAEA document INFCIRC/209 and Addenda). The Conference attaches particular importance to the condition, established by those States, of an undertaking of non-diversion to nuclear weapons or other nuclear explosive devices, as included in the said requirements.

The Conference urges that:

(a) in all achievable ways, common export requirements relating to safeguards be strengthened, in particular by extending the application of safeguards to all peaceful nuclear activities in importing States not Party to the Treaty;
(b) such common requirements be accorded the widest possible measure of acceptance among all suppliers and recipients;
(c) all Parties to the Treaty should actively pursue their efforts to these ends.

The Conference takes note of:

(a) the considered view of many Parties to the Treaty that the safeguards required under Article III, 2, should extend to all peaceful nuclear activities in importing States;
(b) (i) the suggestion that it is desirable to arrange for common safeguards requirements in respect of nuclear material processed, used or produced by the use of scientific and technological information transferred in tangible form to non-nuclear-weapon States not Party to the Treaty;
(ii) the hope that this aspect of safeguards could be further examined.

The Conference recommends that, during the review of the arrangements relating to the financing of safeguards in the IAEA which is to be undertaken by its Board of Governors at an appropriate time after 1975, the less favourable financial situation of the developing countries be fully taken into account. It recommends further that, on that occasion, the Parties to the Treaty concerned seek measures that would restrict within appropriate limits the respective shares of developing countries in safeguards costs.

The Conference attaches considerable importance, so far as safeguards inspectors are concerned, to adherence by the IAEA to Article VII.D of its Statute, prescribing, among other things, that "due regard shall be paid . . . to the importance of recruiting the staff on as wide a geographical basis as possible"; it also recommends that safeguards training be made available to personnel from all geographic regions.

The Conference, convinced that nuclear materials should be effectively protected at all times, urges that action be pursued to elaborate further, within the IAEA, concrete recommendations for the physical protection of nuclear material in use, storage and transit, including principles relating to the responsibility of States, with a view to ensuring a uniform, minimum level of effective protection for such material.

It calls upon all States engaging in peaceful nuclear activities (i) to enter into such international agreements and arrangements as may be necessary to ensure such protection; and (ii) in the framework of their respective physical protection systems, to give the earliest possible effective application to the IAEA's recommendations.

REVIEW OF ARTICLE IV

The Conference reaffirms, in the framework of Article IV, 1, that nothing in the Treaty shall be interpreted as affecting, and notes with satisfaction that nothing in the Treaty has been identified as affecting, the inalienable right of all the Parties to the Treaty to develop research, production and use of nuclear energy for peaceful purposes without discrimination and in conformity with Articles I and II of the Treaty.

The Conference reaffirms, in the framework of Article IV, 2, the undertaking by all Parties to the Treaty to facilitate the fullest possible exchange of equipment, materials and scientific and technological information for the peaceful uses of nuclear energy and the right of all Parties to the Treaty to participate in such exchange and welcomes the efforts made towards that end. Noting that the Treaty constitutes a favourable framework for broadening international cooperation in the peaceful uses of nuclear energy, the Conference is convinced that on this basis, and in conformity with the Treaty, further efforts should be made to ensure that the benefits of peaceful applications of nuclear technology should be available to all Parties to the Treaty.

The Conference recognizes that there continues to be a need for the fullest possible exchange of nuclear materials, equipment and technology, including up-to-date developments, consistent with the objectives and safeguards requirements of the Treaty. The Conference reaffirms the undertaking of the Parties to the Treaty in a position to do so to co-operate in contributing, alone or together with other States or international organizations, to the further development of the applications of nuclear energy for peaceful purposes, especially in the territories of non-nuclear-weapon States Party to the Treaty, with due consideration for the needs of the developing areas of the world. Recognizing, in the context of Article IV, 2, those growing needs of developing States the Conference considers it necessary to continue and increase assistance to them in this field bilaterally and through such multilateral channels as the IAEA and the United Nations Development Programme.

The Conference is of the view that, in order to implement as fully as possible Article IV of the Treaty, developed States Party to the Treaty should consider taking measures, making contributions and establishing programmes, as soon as possible, for the provision of special assistance in the peaceful uses of nuclear energy for developing States Party to the Treaty.

The Conference recommends that, in reaching decisions on the provision of equipment, materials, services and scientific and technological information for the peaceful uses of nuclear energy, on concessional and other appropriate financial arrangements and on the furnishing of technical assistance in the nuclear field, including co-operation related to the continuous operation of peaceful nuclear facilities, States Party to the Treaty should give weight to adherence to the Treaty by recipient States. The Conference recommends, in this connexion, that any special measures of co-operation to meet the growing needs of developing States Party to the Treaty might include increased and supplemental voluntary aid provided bilaterally or through multilateral channels such as the IAEA's facilities for administering funds-in-trust and gifts-in-kind.

The Conference further recommends that States Party to the Treaty in a position to do so, meet, to the fullest extent possible, "technically sound" requests for technical assistance, submitted to the IAEA by developing States Party to the Treaty, which the IAEA is unable to finance from its own resources, as well as such "technically sound" requests as may be made by developing States Party to the Treaty which are not Members of the IAEA.

The Conference recognizes that regional or multinational nuclear fuel cycle centres may be an advantageous way to satisfy, safely and economically, the needs of many States in the course of initiating or expanding nuclear power programmes, while at the same time facilitating physical protection and the application of IAEA safeguards and contributing to the goals of the Treaty.

The Conference welcomes the IAEA's studies in this area, and recommends that they be continued as expeditiously as possible. It considers that such studies should include, among other aspects, identification of the complex practical and organizational difficulties which will need to be dealt with in connexion with such projects.

The Conference urges all Parties to the Treaty in a position to do so to co-operate in these studies, particularly by providing to the IAEA where possible economic data concerning construction and operation of facilities such as chemical reprocessing plants, plutonium fuel fabrication plants, waste management installations, and longer-term spent fuel storage, and by assistance to the IAEA to enable it to undertake feasibility studies concerning the establishment of regional nuclear fuel cycle centres in specific geographic regions.

The Conference hopes that, if these studies lead to positive findings, and if the establishment of regional or multinational nuclear fuel cycle centres is undertaken, Parties to the Treaty in a position to do so, will co-operate in, and provide assistance for, the elaboration and realization of such projects.

REVIEW OF ARTICLE V

The Conference reaffirms the obligation of Parties to the Treaty to take appropriate measures to ensure that potential benefits from any peaceful applications of nuclear explosions are made available to non-nuclear-weapon States Party to the Treaty in full accordance with the provisions of Article V and other applicable international obligations. In this connexion, the Conference also reaffirms that such services should be provided to non-nuclear-weapon States Party to the Treaty on a non-discriminatory basis and that the charge to such Parties for the explosive devices used should be as low as possible and exclude any charge for research and development.

The Conference notes that any potential benefits could be made available to non-nuclear-weapon States not Party to the Treaty by way of nuclear explosion services provided by nuclear-weapon States, as defined by the Treaty, and conducted under the appropriate international observation and international procedures called for in Article V and in accordance with other applicable international obligations. The Conference considers it imperative that access to potential benefits of nuclear explosions for peaceful purposes not lead to any proliferation of nuclear explosive capability.

The Conference considers the IAEA to be the appropriate international body, referred to in Article V of the Treaty, through which potential benefits from peaceful applications of nuclear explosions could be made available to any non-nuclear-weapon State. Accordingly, the Conference urges the IAEA to expedite work on identifying and examining the important legal issues involved in, and to commence consideration of, the structure and content of the special international agreement or agreements contemplated in Article V of the Treaty, taking into account the views of the Conference of the Committee on Disarmament (CCD) and the United Nations General Assembly and enabling States Party to the Treaty but not Members of the IAEA which would wish to do so to participate in such work.

The Conference notes that the technology of nuclear explosions for peaceful purposes is still at the stage of development and study and that there are a number of interrelated international legal and other aspects of such explosions which still need to be investigated.

The Conference commends the work in this field that has been carried out within the IAEA and looks forward to the continuance of such work pursuant to United Nations General Assembly resolution 3261 D (XXIX). It emphasizes that the IAEA should play the central role in matters relating to the provision of services for the application of nuclear explosions for peaceful purposes. It believes that the IAEA should broaden its consideration of this subject to encompass, within its area of competence, all aspects and implications of the practical applications of nuclear explosions for peaceful purposes. To this end it urges the IAEA to set up appropriate machinery within which intergovernmental discussion can take place and through which advice can be given on the Agency's work in this field.

The Conference attaches considerable importance to the consideration by the CCD, pursuant to United Nations General Assembly resolution 3261 D (XXIX) and taking due account of the views of the IAEA, of the arms control implications of nuclear explosions for peaceful purposes.

The Conference notes that the thirtieth session of the United Nations General Assembly will receive reports pursuant to United Nations General Assembly resolution 3261 D (XXIX) and will provide an opportunity for States to discuss questions related to the application of nuclear explosions for peaceful purposes. The Conference further notes that the results of discussion in the United Nations General Assembly at its thirtieth session will be available to be taken into account by the IAEA and the CCD for their further consideration.

REVIEW OF ARTICLE VI

The Conference recalls the provisions of Article VI of the Treaty under which all Parties undertook to pursue negotiations in good faith on effective measures relating

> to the cessation of the nuclear arms race at an early date and
>
> to nuclear disarmament and
>
> to a treaty on general and complete disarmament under strict and effective international control.

While welcoming the various agreements on arms limitation and disarmament elaborated and concluded over the last few years as steps contributing to the implementation of Article VI of the Treaty, the Conference expresses its serious concern that the arms race, in particular the nuclear arms race, is continuing unabated.

The Conference therefore urges constant and resolute efforts by each of the Parties to the Treaty, in particular by the nuclear-weapon States, to achieve an early and effective implementation of Article VI of the Treaty.

The Conference affirms the determination expressed in the preamble to the 1963 Partial Test Ban Treaty and reiterated in the preamble to the Non-Proliferation Treaty to achieve the discontinuance of all test explosions of nuclear weapons for all time. The Conference expresses the view that the conclusion of a treaty banning all nuclear weapons tests is one of the most important measures to halt the nuclear arms race. It expresses the hope that the nuclear-weapon

States Party to the Treaty will take the lead in reaching an early solution of the technical and political difficulties on this issue. It appeals to these States to make every effort to reach agreement on the conclusion of an effective comprehensive test ban. To this end, the desire was expressed by a considerable number of delegations at the Conference that the nuclear-weapon States Party to the Treaty should as soon as possible enter into an agreement, open to all States and containing appropriate provisions to ensure its effectiveness, to halt all nuclear weapons tests of adhering States for a specified time, whereupon the terms of such an agreement would be reviewed in the light of the opportunity, at that time, to achieve a universal and permanent cessation of all nuclear weapons tests. The Conference calls upon the nuclear-weapon States signatories of the Treaty on the Limitation of Underground Nuclear Weapons tests, meanwhile, to limit the number of their underground nuclear weapons tests to a minimum. The Conference believes that such steps would constitute an incentive of particular value to negotiations for the conclusion of a treaty banning all nuclear weapons test explosions for all time.

The Conference appeals to the nuclear-weapon States Parties to the negotiations on the limitation of strategic arms to endeavour to conclude at the earliest possible date the new agreement that was outlined by their leaders in November 1974. The Conference looks forward to the commencement of follow-on negotiations on further limitations of, and significant reductions in, their nuclear weapons systems as soon as possible following the conclusion of such an agreement.

The Conference notes that, notwithstanding earlier progress, the CCD has recently been unable to reach agreement on new substantive measures to advance the objectives of Article VI of the Treaty. It urges, therefore, all members of the CCD Party to the Treaty, in particular the nuclear-weapon States Party, to increase their efforts to achieve effective disarmament agreements on all subjects on the agenda of the CCD.

The Conference expresses the hope that all States Party to the Treaty, through the United Nations and the CCD and other negotiations in which they participate, will work with determination towards the conclusion of arms limitation and disarmament agreements which will contribute to the goal of general and complete disarmament under strict and effective international control.

The Conference expresses the view that, disarmament being a matter of general concern, the provision of information to all governments and peoples on the situation in the field of the arms race and disarmament is of great importance for the attainment of the aims of Article VI. The Conference therefore invites the United Nations to consider ways and means of improving its existing facilities for collection, compilation and dissemination of information on disarmament issues, in order to keep all governments as well as world public opinion properly informed on progress achieved in the realization of the provisions of Article VI of the Treaty.

REVIEW OF ARTICLE VII AND THE SECURITY OF NON-NUCLEAR WEAPON STATES

Recognizing that all States have need to ensure their independence, territorial integrity and sovereignty, the Conference emphasizes the particular importance of assuring and strengthening the security of non-nuclear-weapon States Parties which have renounced the acquisition of nuclear weapons. It acknowledges that States Parties find themselves in different security situations and therefore that various appropriate means are necessary to meet the security concerns of States Parties.

The Conference underlines the importance of adherence to the Treaty by non-nuclear-weapon States as the best means of reassuring one another of their renunciation of nuclear weapons and as one of the effective means of strengthening their mutual security.

The Conference takes note of the continued determination of the Depositary States to honour their statements, which were welcomed by the United Nations Security Council in resolution 255 (1968), that, to ensure the security of the non-nuclear-weapon States Party to the Treaty, they will provide or support immediate assistance, in accordance with the Charter, to any non-nuclear-weapon State Party to the Treaty which is a victim of an act or an object of a threat of aggression in which nuclear weapons are used.

The Conference, bearing in mind Article VII of the Treaty, considers that the establishment of internationally recognized nuclear-weapon-free zones on the initiative and with the agreement of the directly concerned States of the zone, represents an effective means of curbing the spread of nuclear weapons, and could contribute significantly to the security of those States. It welcomes the steps which have been taken toward the establishment of such zones.

The Conference, recognizes that for the maximum effectiveness of any Treaty arrangements for establishing a nuclear-weapon-free zone the co-operation of the nuclear-weapon States is necessary. At the Conference it was urged by a considerable number of delegations that nuclear-weapon States should provide, in an appropriate manner, binding security assurances to those States which become fully bound by the provisions of such regional arrangements.

At the Conference it was also urged that determined efforts must be made especially by the nuclear-weapon States Party to the Treaty, to ensure the security of all non-nuclear-weapon States Parties. To this end the Conference urges all States, both nuclear-weapon States and non-nuclear-weapon States to refrain, in accordance with the Charter of the United Nations, from the threat or the use of force in relations between States, involving either nuclear or non-nuclear weapons. Additionally, it stresses the responsibility of all Parties to the Treaty and especially the nuclear-weapon States, to take effective steps to strengthen the security of non-nuclear-weapon States and to promote in all appropriate fora the consideration of all practical means to this end, taking into account the views expressed at this Conference.

REVIEW OF ARTICLE VIII

The Conference invites States Party to the Treaty which are Members of the United Nations to request the Secretary-General of the United Nations to include the following item in the provisional agenda of the thirty-first session of the General Assembly: "Implementation of the conclusions of the first Review Conference of the Parties to the Treaty on the Non-Proliferation of Nuclear Weapons."

The States Party to the Treaty participating in the Conference propose to the Depositary Governments that a second Conference to review the operation of the Treaty be convened in 1980.

The Conference accordingly invites States Party to the Treaty which are Members of the United Nations to request the Secretary-General of the United Nations to include the following item in the provisional agenda of the thirty-third session of the General Assembly: "Implementation of the conclusions of the first Review Conference of the Parties to the Treaty on the Non-Proliferation of Nuclear Weapons and establishment of a preparatory committee for the second Conference."

REVIEW OF ARTICLE IX

The five years that have passed since the entry into force of the Treaty have demonstrated its wide international acceptance. The Conference welcomes the recent progress towards achieving wider adherence. At the same time, the Conference notes with concern that the Treaty has not as yet achieved universal adherence. Therefore, the Conference expresses the hope that States that have not already joined the Treaty should do so at the earliest possible date.

Appendix C

Status of the Non-Proliferation Treaty

(Review conference participants in italics)

PARTIES (95)
(58 present, one as observer)

Afghanistan*
*Australia**
*Austria**
Bahamas
*Belgium**
*Bolivia**
Botswana
*Bulgaria***
Burundi
Cameroon
*Canada***
Central African
 Republic
Chad
Costa Rica*
*Cyprus***
*Czechoslovakia***
Dahomey
*Denmark***
Dominican
 Republic**
*Ecuador***
El Salvador*
Ethiopia

*Fiji***
*Finland***
Gabon*
Gambia
Germany (GDR)**
Germany (FRG)*
*Ghana***
*Greece** (prov.)*
Grenada
Guatemala
Haiti*
*Holy See***
*Honduras**
*Hungary***
*Iceland***
*Iran***
*Iraq** (observer)*
*Ireland***
*Italy**
Ivory Coast
Jamaica
Jordan*
Kenya
Khmer Republic

Korea (RoK)
Laos
*Lebanon***
Lesotho**
Liberia
Luxembourg*
Malagasy
 Republic**
Malaysia**
Maldive Islands
Mali
Malta
*Mauritius***
*Mexico***
*Mongolia***
Morocco**
*Nepal***
Netherlands*
*New Zealand***
Nicaragua*
Nigeria
*Norway***
Paraguay
Peru

PARTIES (95) continued

Philippines**
Poland**
Romania**
Rwanda
San Marino
Senegal
Sierra Leone
Somalia
Sudan*
Swaziland*
Sweden**

*Syrian Arab
    Republic*
Taiwan
Thailand**
Togo
Tonga*
Tunisia
*Union of
    Soviet
    Socialist
    Republics*+

United Kingdom+
United States+
Upper Volta
Uruguay*
Vietnam
  (South)**
Western
  Samoa
Yugoslavia**
Zaire**

SIGNATORIES (16)
(7 present)

Barbados*
Colombia
Egypt
Indonesia
Japan*
Kuwait
Libya

Panama
Singapore
Sri Lanka
Switzerland
*Trinidad &
    Tobago*
Turkey

Venezuela
Yemen, Arab
  Republic of
Yemen,
  Democratic
  Republic

NONSIGNATORIES (38)
(7 present as observers)

Albania
Algeria
Argentina
Bahrain
Bangladesh
Bhutan
Brazil
Burma
Chile
China+
Congo
Cuba
Equatorial
  Guinea

France+
Guinea
Gunea-Bissau
Guyana
India°
Israel
Liechtenstein
Korea (DPRK)
Malawi
Mauritania
Monaco
Nauru
Niger
Oman

Pakistan
Portugal
Qatar
Saudi Arabia
South Africa
Spain
Tanzania
Uganda
United Arab
  Emirates
Vietnam (North)
Zambia

+ Nuclear weapon state
** IAEA safeguards agreements in force as required by the NPT
* IAEA safeguards agreements signed or approved by the board of governors
° India has detonated a "peaceful nuclear device"

Bibliography

ARTICLES—TECHNICAL

Boskma, P. "Uranium Enrichment Technologies and the Demand for Enriched Uranium," in *Nuclear Proliferation Problems* edited by B. Jasani, (Cambridge, MA.: MIT Press, 1975).
Brode, H. "Review of Nuclear Weapons Effects," *Annual Review of Nuclear Sciences,* 18 (1968) pp. 153–202.
Emelyanov, V. S. "On the Peaceful Use of Nuclear Explosions," in *Nuclear Proliferation Problems* edited by B. Jasani, (Cambridge, MA.: MIT Press, 1975), pp. 215–24.
Feld, B. T. "Making the World Safe for Plutonium," *The Bulletin of the Atomic Scientists* 31:5 May, 1975.
Ford, D. F. and Kendall, H. W., "Nuclear Safety," *Environment* 14:7 (September, 1972).
Gillette Robert, "Uranium Enrichment: With Help, South Africa is Progressing," *Science,* 189:4193 (June 13, 1975), pp. 1090–1092.
Inglish, David R. and Carl L. Sandler, "Prospects and Problems: The Non-Military Uses of Nuclear Explosives," *The Bulletin of the Atomic Scientists* 23:10 (December, 1967), pp. 46–53.
Jasani, B. M. "Nuclear Fuel Reprocessing Plants," in *Nuclear Proliferation Problems,* (MIT Press, Cambridge, 1974.).
Johnson, Gerald W. "Plowshare at the Crossroads," *The Bulletin of the Atomic Scientists* 26:6 (June, 1970), pp. 83–91.
Krieger, David. "Terrorists and Nuclear Technology." *The Bulletin of the Atomic Scientists* 31:6 (June, 1975), pp. 28–34.
Pastore, Senator John O. (D–RI.), "US Export of Nuclear Technology," *Congressional Record,* April 17, 1975, S 5989–5994 (includes a number of charts on world nuclear capacity and US nuclear agreements.)
Dr. Raja Ramanna. "Development of Nuclear Energy in India," *Nuclear India* (September, 1974).

"Report to the American Physical Society by the Study Group on Light Water Reactor Safety," *Reviews of Modern Physics,* 47: Suppl. 1 (1975).

Rose, David J. "Nuclear Electric Power," *Science* 184:351 (1974).

Sethna, H. N. "PNE Technology Vitally Important for India," *Nuclear India* (October, 1974).

Speth, J. G., A. R. Tamplin and T. B. Cochran, "Plutonium Recycle, the Fateful Step," *The Bulletin of the Atomic Scientists* 30:10 (November, 1974).

TECHNOLOGY AND PROLIFERATION— BOOKS AND MONOGRAPHS

Brooks, David B. and John V. Krutilla, *Peaceful Use of Nuclear Explosives: Some Economic Aspects,* (Baltimore: The Johns Hopkins Press, 1969).

Davidon, W., M. Kalkstein, and C. Hohehemser, *The Nth Country Problem and Arms Control.* (Washington, D.C.: National Planning Association), 1960.

Ebbin, Steven and Raphael Kasper, *Citizen Groups and the Nuclear Power Controversy,* (Cambridge, MA.: MIT Press, 1974).

Ford, D. F., and H. W. Kendall, "An Assessment of The Emergency Core Cooling System Rulemaking Hearings," (Cambridge, MA.: Union of Concerned Scientists, 1973).

Ford, D. F., et al. *The Nuclear Fuel Cycle,* (Cambridge, MA.: Union of Concerned Scientists, 1973).

Foreman, Harry, M.D., *Nuclear Power and the Public,* (Garden City, N.Y.: Doubleday, 1972).

Inglis, David R. *Nuclear Energy, Its Physics and its Social Challenge.* (Reading, MA: Addison–Wesley, 1973).

Jasani, Bhipendra, ed. *Nuclear Proliferation Problems.* Stockholm International Peace Research Institute monograph (Cambridge, MA: MIT Press, 1975).

Kendall, H. and Moglewar, S., ed., *Preliminary Review of The AEC Reactor Safety Study.* (Cambridge, MA: Sierra Club–Union of Concerned Scientists, November, 1974).

McPhee, John, *The Curve of Binding Energy* (a journey into the awesome and alarming world of Theodore B. Taylor) (New York: Farrar, Straus, and Giroux, 1974).

Singh, Sampooran. *India and the Nuclear Bomb.* (New Delhi: S. Chand and Co., 1971).

Taylor, Theodore B. and Mason Willrich, *Nuclear Theft: Risks and Safeguards.* (Cambridge, MA: Ballinger, 1974).

United Nations Association, *Safeguarding the Atom,* (New York: United Nations of the United States of America, 1972).

U.S. Atomic Energy Commission, *The Nuclear Industry, 1973.* (Washington, D.C.: U.S. Government Printing Office, 1973).

Willrich, Mason, ed. *International Safeguards and the Nuclear Industry.* (Baltimore: Johns Hopkins University Press, 1973).

DOCUMENTS—TECHNICAL

Effects of The Possible Use of Nuclear Weapons and The Security and Economic Implications for States of the Acquisition and Further Development of These Weapons. (New York: United Nations, 1968).
European Nuclear Energy Agency, *World Uranium and Thorium Resources.* (Paris: Organization for Economic Cooperation and Development, August, 1965).
India, Atomic Energy Commission, *Atomic Energy and Space Research, A Profile for The Decade 1970–80.* (New Delhi: 1970).
International Atomic Energy Agency, *Experience and Trends in Nuclear Law,* [Legal Series no. 8,] (Vienna, 1972).
International Atomic Energy Agency, *Peaceful Uses of Nuclear Explosions* [Bibliographical Series no. 38] (Vienna; 1970).
U.S. Congress. Joint Committee on Atomic Energy. *Commercial Plowshare Services,* [Hearings] 90th Congress, 2nd session, (Washington, D.C.: Government Printing Office, 1968).
U.S. Congress, Senate. Government Operations Committee, *A Compendium: Peaceful Nuclear Exports and Weapons Proliferation,* 94th Congress, 1st session, (Washington, D.C.: Government Printing Office, April, 1975).
U.S. Congress, Senate. Banking, Housing and Urban Affairs Committee. *Exports of Nuclear Materials and Technology,* Hearings before the Subcommittee on International Finance. 93rd Congress, 2nd session (Washington, D.C.: Government Printing Office, 1974).
U.S. Department of Defense, Defense Intelligence Agency, *Physical Vulnerability Handbook: Nuclear Weapons,* AP–550–1–2–69–INT, (Washington, D.C., 1969).

ARTICLES—POLITICAL

"The Atom Bomb in Israel: A Symposium," *New Outlook* 4:5 (March/April 1961).
Bell, J. Bowyer. "Israel's Nuclear Option," *Middle East Journal* 26:4 (Autumn, 1972), pp. 379–388.
Ben–Tzur, Avraham, "The Arabs and The Israeli Reactor," *New Outlook* 4:5 (March/April, 1961).
Bull, Hedley, "Rethinking Non-Proliferation," *International Affairs* 51:2 (April, 1975), pp. 175–189.
Bullard, Lt. Col. Monte R., USA, "Japan's Nuclear Choice," *Asian Survey* 14:9 September, 1974), pp. 845–853.
Center for Defense Information. "30,000 U.S. Nuclear Weapons." *The Defense Monitor* 4:2, (February, 1975).
_____ . "The Militarization of OuterSpace." *The Defense Monitor* 4:5 (July, 1975).
Epstein, William, "The Proliferation of Nuclear Weapons," *Scientific American* 232:4 (April, 1975), pp. 18–33.

Evron, Yair. "The Arab position in the Nuclear Field: A Study of Policies up to 1967," *Cooperation and Conflict* X:1 (1973), pp. 19–31.

_____ . "Israel and the Atom: The Uses and Misuses of Ambiguity, 1957–1967," *Orbis*, XVII:4 (Winter, 1974).

Flaphan, Simha. "Nuclear Power in the Middle East (Part I)," *New Outlook*, (July, 1974).

Flaphan, Simha. "Nuclear Power in the Middle East (Part II), The Critical Years," *New Outlook* 17:8 (October, 1974).

Freedman Lawrence. "Israel's Nuclear Policy", *Survival* 17:3 (May–June, 1975), pp. 114–120.

_____ . "A Nuclear Middle East?" *Present Tense* 3:1 (Winter, 1975).

Gharekhan, C. R. "Strategic Arms Limitation—II," *India Quarterly* 26:4 (October–December, 1970).

Halstead, Thomas A., "The Spread of Nuclear Weapons—Is the Dam About to Burst?" *The Bulletin of the Atomic Scientists* 31:5 (May, 1975), pp. 8–11.

Hodes, Aubrey. "Implications of Israel's Nuclear Capability," *Wiener Library Bulletin* 22:4 (Autumn, 1968), pp. 2–7.

Jabber, Fuad. "Not by War Alone: Curbing the Arab–Israeli Arms Race," *Middle East Journal* 28:3 (Summer, 1974), pp. 233–247.

Jones, Rodney. "India's Nuclear Posture and American Policy in Asia," in *The Next Asia: Problem for U.S. Policy,* edited by David S. Smith. (New York: Columbia University, 1969), pp. 81–120.

Kapur, Ashok. "Peace and Power in India's Nuclear Policy," *Asian Survey,* 10:9 (September, 1970), pp. 779–88.

_____ . "Non-Proliferation: Factors That India Must Weigh," *Asian Review* 2:3 (April, 1969), pp. 215–25.

Kaushik, B. M., "India and The Bomb," *South Asian Studies* (Rajasthan, India) 5:1 (January, 1970), 79–97.

Koop, Jacob, "Plowshare and the Non-Proliferation Treaty," *Orbis* 12:3, (Fall, 1968), pp. 793–815.

Krishna, Raj. "India and the Bomb," *India Quarterly* 21:2, (April–June, 1965), pp. 119–37.

Lough, Thomas S., "Peaceful Nuclear Explosions and Disarmament," *Peace Research Review* II:3 (June, 1968).

Myrdal, Alva. "The Game of Disarmament," *Impact of Science on Society.* 22:3 (July–September, 1972).

_____ . "The High Price of Nuclear Arms Monopoly," *Foreign Policy* 18 (Spring 1975), pp. 30–43.

_____ . " 'Peaceful' Nuclear Explosives," *The Bulletin of the Atomic Scientists* 31:5 (May, 1975), pp. 29–33.

Nandy, Ashis. "The Bomb, the NPT and Indian Elites," *Economic and Political Weekly* (Bombay) Special Number (August, 1972), pp. 1533–1540.

Nimrod, Y., and Korczyn, A. "Suggested Patterns for Israeli–Egyptian Agreement to Avoid Nuclear Proliferation," *New Outlook,* January, 1967.

Noorani, A. G. "India's Quest for a Nuclear Guarantee," *Asian Survey* 7:7 (July, 1967), pp. 490–502.

"Nuclear Hoax Cannot Save U.S. Imperialism and Soviet Revisionism," *Peking Review*, September 8, 1967.
"Nuclear Weapons and India's Security," *The Institute for Defence Studies and Analyses Journal* 3:1 (July, 1970).
Okimoto, Daniel I., "Japan's Non-Nuclear Policy: The Problem of the NPT," *Asian Survey* 15:5 (May, 1975), pp. 313–327.
"Proliferation and the Indian Test," *Survival* 16:5 (September–October, 1974), pp. 210–216. (Contains two articles, "A view from India," by R.V.R. Chandrasekhara Rao and "A view from Japan," by Ryukichi Imai.)
Quester, George H., "Can Proliferation Now be Stopped?" *Foreign Affairs* 53:1 (October 1974), pp. 77–97.
Rana, A. P. "The Intellectual Dimensions of India's Nonalignment," *Journal of Asian Studies* 28:2, (February, 1969).
Redick, John. "Regional Nuclear Arms Control in Latin America", *International Organization* 29:2 (Spring, 1975), pp. 415–445.
Sorenson, Jay B., "Nuclear Deterrence and Japan's Defense," *Asian Affairs* (November/December, 1974), pp. 55–69.
Subrahmanyam, K. "The Role of Nuclear Weapons: an Indian Perspective," *Security, Order, and the Bomb*, edited by John JorgenHolst, (Oslo: Universitetsforlaget, 1972), pp. 131–41.
Sullivan, Michael J., III. "Indian Attitudes on International Atomic Energy Controls," *Pacific Affairs* 43:3, (Fall, 1970), pp. 353–69.
Swamy, Subramanian, "Objectives of India's Strategic Defence," *Indian Economic Planning, an Alternative Approach,* edited by S. Swamy (New Delhi: Vikos Publications, 1971), pp. 85–103.
Tsipis, Kosta. "Physics and Calculus of Countercity and Counterforce Nuclear Attacks," *Science* 187 (February 7, 1975), pp. 393–397.
Van Cleave, William R. and Harold W. Rood. "A Technological Comparison of Two Potential Nuclear Powers: India and Japan," *Asian Survey* 7:7, (July, 1967), pp. 482–9.
Vanin, I. "Security Guarantees for Non-Nuclear Countries," *International Affairs* (Moscow) 10 (October, 1968), pp. 35–38.
Wolfers, Arnold. "National Security as an Ambiguous Symbol," *Political Science Quarterly* 67:4 (December, 1952), pp. 481–502.
Wohlstetter, Albert. "Nuclear Sharing: NATO and the N + 1 Country," *Foreign Affairs* 39:3 April, 1961, pp. 355–87.
Zoppo, Ciro E. "The Nuclear Genie in the Middle East," *New Outlook*, (February, 1975).

BOOKS AND MONOGRAPHS—POLITICAL

Bader, William B. *The United States and the Spread of Nuclear Weapons.* (New York: Pegasus, 1968).
Barnaby, C. F. (ed.) *Preventing the Spread of Nuclear Weapons.* (London: Souvenir Press, 1969).

Beaton, Leonard. *Must the Bomb Spread?* (London: Penguin Books, 1966).
Beaton, Leonard and John Maddox, *The Spread of Nuclear Weapons.* Studies in International Security 5. (New York: Frederick A. Praeger, 1962).
Boskey, Bennett and Willrich, Mason, eds. *Nuclear Proliferation: Prospects for Control.* (N.Y.: Dunellen Co. Inc., 1970).
Buchan, Alastair, ed. *A World of Nuclear Powers?* (Englewood Cliffs, N.J.: Prentice–Hall, 1966).
Dougherty, James E. and Lehman, J. F., Jr., eds. *Arms Control for the Late Sixties.* (Princeton: D. Van Nostrand, 1967).
Endicott, John E. *Japan's Nuclear Option: Political, Technological, and Strategies Factors.* Special Studies. (New York: Praeger, 1975).
Fischer, G. *The Non-Proliferation of Nuclear Weapons.* (London: Europa Publications, 1971).
Gallois, Pierre. *The Balance of Terror: Strategy for the Nuclear Age.* (Boston: Houghton Mifflin Co., 1961).
Garcia, Robles A. *The Denuclearization of Latin America.* (Washington: Carnegie Endowment for International Peace, 1967).
Gelber, Harry, *Nuclear Weapons and Chinese Policy,* Adelphi papers 99, (London: International Institute for Strategic Studies, 1973).
Grodzins, Morton and Eugene Rabinowitch, *The Atomic Age,* (New York: Simon and Schuster, 1965).
Imai, Ryukichi, *Nuclear Safeguards,* Adelphi papers 86 (London: International Institute for Strategic Studies, 1972).
Jabber, Fuad. *Israel and Nuclear Weapons.* International Institute of Strategic Studies monograph. (London: Chatto and Windus, 1971).
Jensen, Lloyd, *Return from the Nuclear Brink: National Interest and the Nuclear Non-Proliferation Treaty* (Lexington, MA: D.C. Heath and Company, 1973).
Joshua, Wynfred and Walter F. Hahn, *Nuclear Politics: America, France, and Britain,* (Washington, D.C.: The Center for Strategic and International Studies, 1973).
Kapur, Ashok. *India's Nuclear Option.* New York: Praeger, forthcoming.
Kemp, Geoffrey, *Nuclear Forces for Medium Powers: Part I: Targets and Weapons System,* Adelphia papers no. 186 (London: International Institute for Strategic Studies, 1974);
———. *Nuclear Forces for Medium Powers: Part II and III: Strategic Requirements and Options,* Adelphi papers 187, (London: International Institute for Strategic Studies, 1974).
Kemp, Geoffrey; Robert L. Pfaltzgraff, Jr., and Uri Ra'anan, *The Superpowers in a Multinuclear World,* (Lexington, MA: D.C. Heath and Company, 1973).
Kertesz, Stephen, ed. *Nuclear Proliferation in a World of Nuclear Powers.* (Notre Dame, Ind.: University of Notre Dame Press, 1967).
Lawrence, Robert M. *Arms Control and Disarmament: Practice and Promise* (Minneapolis, Minnesota: Burgess Publishing Co., 1973).
Lawrence, Robert M. and Joel Larus, eds. *Nuclear Proliferation: Phase II.* (Lawrence, Kansas: University Press of Kansas for The National Security Education Program, 1974).

Maddox, John, *Prospects for Nuclear Proliferation.* Adelphi papers 113, (London: International Institute for Strategic Studies, Spring, 1975).

Marks, Anne W. (Ed.), *NPT: Paradoxes and Problems.* (Washington: Arms Control Association and the Carnegie Endowment for International Peace, 1975).

Meeker, Thomas A., *The Proliferation of Nuclear Weapons and the Non-Proliferation Treaty: A Selective Bibliography and Source List.* (Los Angeles: Center for the Study of Armament and Disarmament, California State University, 1973).

Melman, Seymour, ed. *Disarmament, its Politics and Economics.* (Boston: The Academy of Arts and Sciences, 1962).

Metzger, H. Peter, *The Atomic Establishment* (New York: Simon and Schuster, 1972).

Mirchandani, G. G. *India's Nuclear Dilemma.* (New Delhi: Popular Book Services, 1968).

Myrdal, Alva, *The Right to Conduct Nuclear Explosions: Political Aspects and Policy Proposals,* (Stockholm: Stockholm International Peace Research Institute, 1975).

Patil, R. L. M. *India: Nuclear Weapons and International Politics* (New Delhi: National Publishing House, 1969).

Pranger, Robert J, and Dale R. Tahtinen, *Nuclear Threat in the Middle East,* Foreign Affairs Studies 23, (Washington, D.C.: The American Enterprise Institute, 1975).

Quester, George H. *Nuclear Diplomacy.* (Cambridge, MA: Dunellen, 1970).

_____. *The Politics of Nuclear Proliferation.* (Baltimore, Johns Hopkins Press, 1973).

Redick, John. *Military Potential of Latin American Nuclear Energy Programs* (Beverly Hills, CA: Sage Publications, 1972. Professional Papers. International Studies Series.

Rosecrance, Richard N. (ed.) *The Dispersion of Nuclear Weapons: Strategy and Politics* (New York: Columbia University Press, 1964).

Stockholm International Peace Research Institute. *The Arms Trade with the Third World.* (Stockholm: Almquist and Wiksell, 1971).

_____. *The Near Nuclear Countries and the NPT,* (Stockholm: Almquist and Wiksell, 1972).

_____. *The Nuclear Age,* (Stockholm: Almquist and Wiksell; Cambridge, MA: MIT Press, 1974).

_____. *Nuclear Disarmament or Nuclear War?* (Stockholm, 1975).

_____. *Preventing Nuclear-Weapon Proliferation,* (Stockholm, January, 1975).

_____. *Safeguards Against Nuclear Proliferation,* (Cambridge, MA: MIT Press, 1975).

_____. *World Armaments and Disarmament. SIPRI Yearbook 1974.* Stockholm: Almquist and Wiksell; Cambridge, MA: MIT Press, 1974).

A Strategy for India for a Credible Posture Against a Nuclear Adversary. (New Delhi: Institute for Defence Studies and Analyses, 1968).

Tsipis, Kosta. *Offensive Missiles.* Stockholm Papers 5. (Stockholm: Stockholm International Peace Research Institute, 1974).

Van Cleave, William, Harold Rood and Judith Pettenger. *Nth Country Threat Analysis: The Asiatic—Pacific Area.* (Menlo Park: Stanford Research Institute, 1966).
Williams, Shelton L. *The U.S., India and the Bomb.* (Baltimore: The Johns Hopkins Press, 1969).
Willrich, Mason, *Global Politics of Nuclear Energy,* (New York: Praeger Publishers, 1971).
_____. *Non-Proliferation Treaty: Framework for Nuclear Arms Control,* (Charlottesville: The Michie Company 1969).
_____. *Perspective on the Non-Proliferation Treaty Review Conference.* Occasional Paper 7. (Muscatine, Iowa: The Stanley Foundation, 1975).
Wilcox, Wayne. *Forecasting Asian Strategic Environments for National Decisionmaking: A Report and a Method.* Memorandum RM–6154–PR (Santa Monica, California: The Rand Corporation, June, 1970).
Young, Elizabeth. *The Control of Proliferation: The 1968 Treaty in Hindsight and Forecast.* Adelphi paper 56, (London: Institute for Strategic Studies, 1969).

DOCUMENTS—POLITICAL

Donnelly, Warren H., *Commercial Nuclear Power in Europe:* The Interaction of American Diplomacy with a New Technology, prepared for the subcommittee on National Security Policy and Scientific Developments of the Committee on Foreign Affairs, U.S. House of Representatives. Washington, D.C.: Government Printing Office, December, 1972.
Packard, David. Testimony, June 1, 1971, before the U.S. Senate Foreign Relations Committee. *Arms Control Implications of the Current Defense Budget.* Hearings before the Subcommittee on International Arms Control, International Law and Organization, June—July, 1971.
U.S. Arms Control and Disarmament Agency, *Documents on Disarmament.* Compiled and annotated by Robert Lambert and others. Washington, D.C.: U.S. Government Printing Office, 1963–.
U.S. Arms Control and Disarmament Agency. *India and Japan: The Emerging Balance of Power in Asia and Opportunities for Arms Control, 1970–75.* ACDA/IR–170. Prepared by the Southern Asian Institute and East Asian Institute of Columbia University.
U.S. Congress. House. Foreign Affairs Committee. *U.S. Foreign Policy in the Export of Nuclear Technologies to the Middle East.* Hearings before the Subcommittee on International Organizations and Movements and the Subcommittee on the Near East and South Asia. 93rd Congress, 2nd session. Washington, D.C.: U.S. Government Printing Office, 1974.
U.S. Congress. Senate. Committee on Government Operations. "Chinese Comment on Strategic Policy and Arms Limitation," Permanent Subcommittee on Investigations. Washington, D.C.: U.S. Government Printing Office, 1974.
U.S. Congress. Senate. Committee on Foreign Relations. *Nuclear Weapons and Foreign Policy.* Hearings before the Subcommittee on U.S. Security Commitments and Agreements Abroad and the Subcommittee on International Arms Control International Law and Organizations. 93rd Congress, 2nd session. Washington, D.C.: U.S. Government Printing Office, 1974.

Notes on Contributors

Edouard Bustin is professor of Political Science at Boston University in charge of the Political Science division of that institution's African Studies Center. He was senior lecturer in Public Administration and Comparative Government at the State University of the Congo (1959–1961), visiting professor at the National University of Zaire (1965–1970), visiting lecturer at UCLAs African Studies Center (1961–1963) and a guest lecturer at several Central and West African universities. His publications in English and French have appeared in the United States, in Europe and in Africa. His most recent book is *Luanda Under Belgian Rule: The Politics of Ethnicity.*

Anne Hessing Cahn is a research fellow in the Program for Science in International Affairs at Harvard University. She received her Ph.D. from the Massachusetts Institute of Technology. Dr. Cahn serves on the board of directors of the Arms Control Association and SANE, and on the executive board of the Forum on Physics and the Society of the American Physical Society. She has written on United States military policy.

Saul B. Cohen is professor and Director of the Graduate School of Geography at Clark University. Dr. Cohen is the former Executive Secretary of the Association of American Geographers, Chairman of the NAS–NRC Committee on Geography and Chairman of the Commission on College Geography. He has held the post of Visiting Professor of Geography at the United States Naval War College and has lectured extensively at other American and foreign War and Defense Colleges and Public Career Service Institutes. Dr. Cohen's writing in Political Georgraphy have focused upon theory and methodology. His publications include: *Geography and Politics in a World Divided* (Random House, 1964; Oxford University Press, 1973); *The Oxford World Atlas,* geographic editor (Clarendon Press, 1973); and, *Experiencing the Environment,* with S. Wapner and B. Kaplan (Plenum, 1975).

Christoph Hohenemser is associate professor of Physics at Clark University and chairman of the Program on Science, Technology and Society. He received his Ph.D. degree from Washington University (St. Louis), has taught at Brandeis University and spent a year in residence at the University of Groningen, The Netherlands. Dr. Hohenemser has published numerous articles in physics including an early study of nuclear proliferation, *The Nth Country Problem and Arms Control*, National Planning Association, 1960. He is currently studying decision-making and catastrophic risks in the nuclear fuel cycle.

Robert M. Lawrence is a professor of Political Science at Colorado State University (Fort Collins). He previously taught at the University of Arizona and served as a consultant to RAND Corporation and the Stanford Research Institute. Professor Lawrence's publications include: *Arms Control and Disarmament: Practice and Promise* (Burgess, 1973); with William Van Cleave and S.E. Young, *Implications of Indian and/or Japanese Nuclear Proliferation for U.S. Defense Policy Planning* (Palo Alto, Calif: Stanford Research Institute for Advanced Research Projects Agency, 1973); with Norman Wengert, "The Energy Crisis: Reality or Myth," *The Annals* (November 1973); and with Joel Larus, *Nuclear Proliferation: Phase II* (University of Kansas Press, 1974).

Stefan H. Leader is currently senior research analyst at the Center for Defense Information in Washington, D.C. Formerly assistant professor of Politics at Ithaca College (1970–1974) and research associate with the Cornell University Peace Studies Program (1973–1974), he holds a Ph.D. degree from the State University of New York at Buffalo. Dr. Leader has written on US foreign and military policy in Asia and the Persian Gulf and on Chinese foreign policy.

Onkar Marwah teaches International Relations at Clark University and is research fellow with Harvard University's Program for Science and International Affairs. He was employed by the Indian Administrative Service in the State Government of Bihar and the Union Government of India in Rural Economic Development, Political Affairs, and All-India Services Policy Planning. He studied at the University of Calcutta, the London School of Economics, Yale University, University of California, Berkeley, and has been a research associate with Cornell University's Peace Studies Program. Mr. Marwah has published *Asian Alien Pariahs*, 1974; *Modernization in India and China*, 1974.

Steven J. Rosen is assistant professor of Politics, Brandeis University, and senior research fellow at Australian National University, Canberra. He received his Ph.D. degree at the Maxwell School, Syracuse University and taught previously at the University of Pittsburgh. Dr. Rosen's publications include *The Logic of International Relations* (with Walter Jones) (Cambridge, Mass.: Winthrop, 1974); *Testing Theories of Economic Imperialism* (with James R. Kurth); *Test-*

ing the Theory of the Military-Industrial Complex (Lexington, Mass.: D.C. Heath, Lexington Books, 1973); *Alliance in International Politics* (with Julian Friedman and Christopher Bladen) (Allyn and Bacon, 1970).

H. Jon Rosenbaum is associate professor of Political Science in the Doctoral Program in Political Science, The Graduate School and Department of Political Science and at The City College, both of the City University of New York. Before assuming his present post, he received his Ph.D. degree from the Fletcher School of Law and Diplomacy and taught at Wellesley College. Dr. Rosenbaum is the co-editor of *Contemporary Brazil: Issues in Economic Development; Vigilante Politics: Perspectives on Establishment Violence;* and *Latin America: The Search for a New International Role.* Dr. Rosenbaum is the author of numerous articles, primarily on Brazilian politics, and has been a fellow or guest scholar at the Brazilian Institute of International Relations, the Woodrow Wilson International Center for Scholars, Harvard University's Center for International Affairs, the Brazilian School of Public Administration and Columbia University's Institute of Latin American Studies. Chapter 11 of this volume was written while he was serving as a Fulbright Visiting Professor at the Instituto Mexicano "Matias Romero" de Estudios Diplomaticos of the Mexican Foreign Ministry.

Yoshiyasu Sato is vice consul, Japanese Embassy, Hong Kong. Prior to assuming that post, he held diplomatic positions in Washington, D.C.; Jakarta, Indonesia; Tokyo and London. During the academic year 1974—1975, Mr. Sato was a fellow at the Center for International Affairs, Harvard University. He was graduated from the Law Faculty, Hitotsubashi University,Tokyo.

Barry R. Schneider is currently Arms Control and Military Affairs Consultant to Members of Congress for Peace through Law. He received his Ph.D. from Columbia University in 1974 and is a former faculty member of Indiana University, Purdue University and Wabash College. Dr. Schneider is also a lecturer at the University of Maryland.

Ann T. Schulz is visiting associate professor at Clark University. She received her Ph.D. degree at Yale University and taught at the University of New Hampshire before joining the Clark faculty. Dr. Schulz has authored articles on Middle East politics and comparative legislative politics. She serves on the executive council of the Society for Iranian Studies and has been a fellow at the Center for Middle Eastern Studies, Harvard.

K. Subrahmanyam is director of the Indian Institute for Defence Studies and Analyses. He joined the Indian Administrative Service in 1951 and served in the State Government of Tamil Nadu and the Union Government of India in the fields of Rural Development, Finance, and Defence Policy. He has written *The*

Liberation War, 1972; *Bangladesh and India's Security,* 1972; and *Perspectives in Defence Planning,* 1972. He is editor of the *Institute for Defence Studies and Analyses Journal,* the *Institute for Defence Studies and Analyses News Reviews,* and the *Strategic Digest.*

Theodore B. Taylor is chairman of the board of the International Research and Technology Corporation. Dr. Taylor received his Ph.D. from Cornell University and served as a nuclear physicist for the General Dynamics Corporation. He is a former Deputy Director of the Defense Atomic Support Agency and has lectured at San Diego State College. His publications include: *Nuclear Theft: Risks and Safeguards,* with M. Willrich (Ballinger, 1974); *The Restoration of the Earth* (Harper and Row, 1973); and, *The Curve of Binding Energy* (Farrar, Saraus & Giroux, 1974).

C. H. Waisman is completing his dissertation at Harvard. His field of interest is Political Sociology. He did research at several Argentine institutions in 1966–1969, at the Peace Research Institute, Oslo, in the Summer of 1971, and was associated with the Center for International Affairs, Harvard, 1972–1974. He was a Lecturer in Political Science at Yale, 1974–1975, and teaches now at the Department of Sociology, University of California, San Diego.

Conference Participants

Dr. Steven J. Baker, Research Associate
Center for International Studies
Cornell University

Mr. Frank T. J. Bray, Fellow
Foreign Policy Research Institute
Philadelphia

Dr. Shahram Chubin
Center for International Political and
    Economic Studies
Tehran, Iran

Dr. Edward Feit
Department of Political Science
University of Massachusetts

Dr. Bernard Gordon
Department of Political Science
University of New Hampshire

Dr. Farhad Kazemi
Department of Politics
New York University

Dr. Geoffrey Kemp
Fletcher School of Law and Diplomacy
Tufts University

Mr. Milton Leitenberg
Center for International Studies
Cornell University

Dr. Michael M. McClintock
Department of Physics
Clark University

Dr. James Mittelman
Institute of African Studies
Columbia University

Dr. Enid C.B. Schoettle
Department of Political Science
University of Minnesota

Dr. Dale Tahtinen
American Enterprise Institute for
    Public Policy Research
Washington, D.C.

Mr. Leslie Wilbur, Director
Nuclear Reactor Facility
Worcester Polytechnic Institute